ENVIRONMENTAL
INTELLIGENCE
UNIT

# THE EFFECTS OF OZONE DEPLETION ON AQUATIC ECOSYSTEMS

Donat-P. Häder

Friedrich-Alexander-Universität
Institut für Botanik und Pharmazeutische Biologie
Erlangen, Germany

Academic Press

R.G. LANDES COMPANY
AUSTIN

# ENVIRONMENTAL INTELLIGENCE UNIT

## THE EFFECTS OF OZONE DEPLETION ON AQUATIC ECOSYSTEMS

### R.G. LANDES COMPANY
Austin, Texas, U.S.A.

This book is printed on acid-free paper.
Copyright 1997 © by R.G. Landes Company and Academic Press, Inc.

Please address all inquiries to the Publisher:
R.G. Landes Company
810 S. Church Street, Georgetown, Texas, U.S.A. 78626
Phone: 512/ 863 7762; FAX: 512/ 863 0081

Academic Press, Inc.
525 B Street, Suite 1900, San Diego, California, U.S.A. 92101-4495

United Kingdom Edition published by Academic Press Limited
24-28 Oval Road, London NW1 7DX, United Kingdom

Library of Congress Catalog Number: 577.6'276—dc21
International Standard Book Number (ISBN): 0-12-312945-1
Transferred to Digital Printing, 2011
Printed and bound in the United Kingdom

While the authors, editors and publisher believe that drug selection and dosage and the specifications and usage of equipment and devices, as set forth in this book, are in accord with current recommendations and practice at the time of publication, they make no warranty, expressed or implied, with respect to material described in this book. In view of the ongoing research, equipment development, changes in governmental regulations and the rapid accumulation of information relating to the biomedical sciences, the reader is urged to carefully review and evaluate the information provided herein.

### Library of Congress Cataloging-in-Publication Data

Häder, Donat-Peter.
  Ozone depletion and aquatic ecosystems/ Donat-P. Häder.
    p. cm. — (Environmental intelligence unit)
  Includes bibliographical references (p. ) and index.
  ISBN (invalid) 1-57059-427-0 (alk. paper)
    1. Ozone layer—Environmental aspects. 2. Ultraviolet radiation—Environmental aspects.
3. Ozone layer depletion. 4. Aquatic ecology. I. Title. II. Series.
QH545.094H33 1997                                                    96-50981
577.6'276—dc21                                                           CIP

# Publisher's Note

R.G. Landes Company publishes six book series: *Medical Intelligence Unit, Molecular Biology Intelligence Unit, Neuroscience Intelligence Unit, Tissue Engineering Intelligence Unit, Biotechnology Intelligence Unit* and *Environmental Intelligence Unit.* The authors of our books are acknowledged leaders in their fields and the topics are unique. Almost without exception, no other similar books exist on these topics.

Our goal is to publish books in important and rapidly changing areas of bioscience for sophisticated researchers and clinicians. To achieve this goal, we have accelerated our publishing program to conform to the fast pace in which information grows in bioscience. Most of our books are published within 90 to 120 days of receipt of the manuscript. We would like to thank our readers for their continuing interest and welcome any comments or suggestions they may have for future books.

Shyamali Ghosh
Publications Director
R.G. Landes Company

# CONTENTS

# EDITOR

**Donat-P. Häder**
Friedrich-Alexander-Universität
Institut für Botanik und Pharmazeutische Biologie
Erlangen, Germany
*Chapters 1, 3, 5, 9, 11*

# CONTRIBUTORS

Andrew R. Blaustein
Department of Zoology
Oregon State University
Corvallis, Oregon, U.S.A.
*Chapter 10*

Charles R. Booth
Biospherical Instruments Inc.
San Diego, California, U.S.A.
*Chapter 4*

Giuliano Colombetti
C.N.R. Biofisica
Pisa, Italy
*Chapter 14*

John J. Cullen
Department of Oceanography
Dalhousie University
Nova Scotia, Canada
*Chapter 6*

Maria A. Häder
Friedrich-Alexander-Universität
Institut für Botanik und
   Pharmazeutische Biologie
Erlangen, Germany
*Chapter 9*

Gerhard J. Herndl
Inst. f. Zool. der Univ. Wien
Abt. Meeresbiologie
Vienna, Austria
*Chapter 8*

Gerda Horneck
Deutsche Versuchsanstalt für
   Luft- und Raumfahrt
Institut für Flugmedizin
Köln, Germany
*Chapter 7*

Joseph M. Kiesecker
Department of Zoology
Oregon State University
Corvallis, Oregon, U.S.A.
*Chapter 10*

Roberto Marangoni
C.N.R. Biofisica
Pisa, Italy
*Chapter 14*

Beatrice Martini
C.N.R. Biofisica
Pisa, Italy
*Chapter 14*

John H. Morrow
Biospherical Instruments Inc.
San Diego, California, U.S.A.
*Chapter 4*

Patrick J. Neale
Smithsonian Environmental
    Research Center
Edgewater, Maryland, U.S.A.
*Chapter 6*

Canice Nolan
Commission of the European
    Communities
Brussels, Belgium
*Chapter 2*

Helmut Piazena
Friedrich-Alexander-Universität
Institut für Botanik und
    Pharmazeutische Biologie
Erlangen, Germany
*Chapter 5*

Regas Santas
Oikotechnics
Athens, Greece
*Chapter 13*

Johanne-Sophie Selmer
Department of General
    and Marine Microbiology
Göteborg University
Göteborg, Sweden
*Chapter 12*

Rajeshwar P. Sinha
Department of Genetics
    and Plant Breeding
Institute of Agricultural Sciences
Banaras Hindu University
Varanasi, India
*Chapter 11*

Raymond C. Smith
CSL/Center for Remote Sensing
    and Environmental Optics
University of California
Santa Barbara, California, U.S.A
*Chapter 15*

Maria Vernet
Marine Research Division
Scripps Institution of Oceanography
La Jolla, California, U.S.A.
*Chapter 15*

Sten-Åke Wängberg
Department of Plant Physiology
Göteborg University
Göteborg, Sweden
*Chapter 12*

Robert C. Worrest
Consortium for International Earth
    Science
Washington, D.C., U.S.A.
*Chapter 3*

# PREFACE

Stratospheric ozone depletion and the resulting increased solar short-wavelength ultraviolet radiation are of major concern to both scientists and politicians. After some of the adverse effects on all living matter were discovered the general public also became aware of this major environmental problem. While the risks for human health, including increased skin cancer and decreased immune system, can be minimized by preventive measures, the effects on the biota are difficult or impossible to reduce.

In 1982 many developed and developing countries signed a treaty in Montreal, which has been augmented by several amendments in Vienna, London and Copenhagen, to reduce the production and emission of anthropogenic trace gases, including the chlorofluorocarbons (CFC) which have been identified to be responsible for stratospheric ozone depletion. Simultaneously, the United Nations established a panel of scientists to study and report the effects of increased solar UV-B radiation.

While the effects of increased ultraviolet-B radiation (280-315 nm) on human health and terrestrial ecosystems have been summarized in recent books and review articles, the effects on aquatic ecosystems are not adequately covered even though their importance for food production and their impact on global climate are well recognized. Aquatic ecosystems equal terrestrial ecosystems in their productivity, and 50% of atmospheric carbon is taken up by mainly marine ecosystems. Most of the primary productivity in the oceans is due to microscopic plants, cumulatively called phytoplankton, which dwell in the top layers of the water in order to harvest solar radiation for their photosynthesis. Here they are likewise exposed to UV-B radiation from which they are not protected by an epidermal layer.

Any substantial decrease in primary productivity will result in a number of detrimental effects, both for the ecosystem itself and for the primary and secondary consumers dependent directly or indirectly on phytoplankton. In addition to reduced productivity, which will affect all subsequent members of the intricate food web, changes in species composition and reduced sink capacity for atmospheric carbon dioxide, resulting in an augmented climate change, are possible consequences.

This book covers all important aspects of UV-B effects on aquatic ecosystems, from light penetration into the water column to effects on aquatic plants and animals. All major aquatic ecosystems ranging from the Antarctic Ocean to freshwater ecosystems are discussed. In addition to the important results obtained by the efforts of the international scientific community, the sophisticated methods and tools used are also covered in detail.

# ABBREVIATIONS

| | |
|---|---|
| 7-DHC | 7-dehydrocholesterol (provitamin D3) |
| ATP | adenosine triphosphate |
| ATS | algal turf scrubber |
| BSP | bacterial secondary production |
| BWF | biological weighting function |
| CC | chlorocarbons |
| CFC | chlorinated fluorocarbons |
| Chl | chlorophyll |
| CPD | cyclobutane pyrimidine dimer |
| CZCS | Coastal Zone Color Scanner |
| DCMU | (3-(3,4-dichlorophenyl)-1,1-dimethylurea |
| DGGE | denaturing gradient gel electrophoresis |
| DHC | dehydrocholesterol |
| DMS | dimethylsulfide |
| DMSO | dimethyl sulfoxide |
| DMSP | dimethyl sulfoniopropionate |
| DNA | desoxyribonucleic acid |
| DOC | dissolved organic carbon |
| DOM | dissolved organic matter |
| DU | Dobson Unit |
| EC50 | concentration resulting in 50% of specified effect |
| ERC | exposure response curve |
| EXTRA | experimental troughs apparatus |

| | |
|---|---|
| FWHM | full width half max, commonly used to describe the bandpass of a filter at the point which is 50% of the peak height |
| GS | glutamine synthetase |
| Gt | gigatons |
| HCFC | hydrochlorofluorocarbon |
| HFC | hydrofluorocarbon |
| ISLSCP | international satellite land surface climatology project |
| LUVSS | light and ultraviolet submersible spectroradiometer system |
| MAA | mycosporine-like amino acids |
| MED | minimal erythemal dose |
| MERIS | medium resolution imaging spectrometer |
| mRNA | messenger ribonucleic acid |
| NADP | nicotine amide dinucleotide phosphate |
| NIST | National Institute of Standards and Technology |
| NR | nitrate reductase |
| NSF | National Science Foundation |
| OD | optical density |
| OB | organo-bromides |
| ODEX | optical dynamics experiment |
| PAM | pulse amplitude modulation |
| PAR | photosynthetic active radiation (400-700 nm) |
| PCA | principal component analysis |

| | |
|---|---|
| PER | photosynthetic extracellular release |
| PFB | paraflagellar body |
| PIH | photoinhibition |
| PMT | photomultiplier tube |
| POM | particulate organic matter |
| PS | polysulphone |
| PSII | photosystem II |
| PUR | photosynthetically utilizable radiation |
| RAF | radiation amplification factor |
| RB meter | Robertson-Berger meter |
| RER | repair effective radiation (390-470 nm) |
| RNA | ribonucleic acid |
| ROV | remotely operated vehicle |
| RuBisCO | ribulose 1,5-bisphosphate carboxylase |
| SDS-PAGE | sodium dodecyl sulfate polyacrylamid gel electrophoresis |
| SeaWiFS | Sea-viewing Wide Field-of View Sensor |
| SI | Système International d'Unités |
| SWF | spectral weighting function |
| SZA | solar zenith angle |
| TFA | trifluoroacetic acid |
| UV | ultraviolet |
| UV-A | ultraviolet-A radiation (CIE definition: 315- 400 nm, sometimes 320-400) |

| | |
|---|---|
| UV-B | ultraviolet-B radiation (CIE definition: 280-315 nm, sometimes 290-320) |
| UV-C | ultraviolet-C radiation (200-280 nm) |
| UVR | ultraviolet radiation |
| VIS | visible radiation |
| WSC | Weddell-Scotia Confluence |
| ZA | zenith angle |

# CHAPTER 1

# Stratospheric Ozone Depletion and Increase in Ultraviolet Radiation

## Donat-P. Häder

Since its discovery in the 1970s the Antarctic ozone hole has dramatically grown in size and depth, and at mid latitudes the ozone concentration also decreases continuously. There is no longer any reasonable doubt that trace gases of anthropogenic origin (chlorinated fluorocarbons, CFCs) are responsible for this development. The resulting increased UV-B radiation (280-315 nm) threatens humans, animals, terrestrial and aquatic ecosystems well into the next century.

Since their discovery, CFCs were regarded as inexpensive, versatile and absolutely nonhazardous chemicals. In fact, they are chemically inert and react only very sluggishly with other substances. In addition, they can be used for many purposes: as propellants in spray cans, as coolants in refrigerators and air conditioning, as propellants in the production of foams, as cleansing agents for wafers and in the production of integrated electronic circuits as well as in fire extinguishers (Fig. 1.1).

Only when Rowland and Molina published their concern that these substances are responsible for the decrease of the ozone layer in the stratosphere, atmospheric and environmental researchers started to become interested in CFCs. Together with Paul Crutzen, Rowland and Molina received the 1995 Nobel prize in Chemistry for their revolutionary results.

## HOW IS THE OZONE LAYER AFFECTED?

In the last few years ozone has made headlines in the newspapers since it may pose a health risk for humans and animals when it is produced in the lower atmosphere during the occurrence of smog episodes in the summer. However, in this book we are mainly interested in the stratospheric ozone layer which extends between 15 and 40 km above the earth's surface (Fig. 1.2). The concentration of ozone is extremely

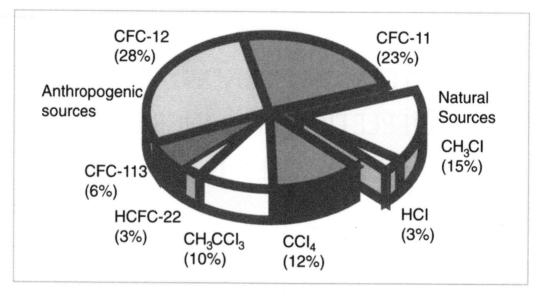

*Fig. 1.1. Substances which destroy the stratospheric ozone layer from natural (detached sectors) and anthropogenic sources.*

small: if we could concentrate the total ozone under atmospheric pressure it would form a layer only 3-5 mm thick.

Ozone undergoes a cycle of generation and breakdown: it is generated by short-wavelength ultraviolet solar radiation and is destroyed by longer wavelengths. This natural cycle in the stratosphere is disturbed by the injection of CFCs, which have a lifetime of about 100 years and can be transported from the lower atmosphere into the stratosphere. Solar UV radiation splits chlorine from the CFCs which remove an oxygen atom from an ozone ($O_3$) molecule, leaving molecular oxygen ($O_2$). Subsequently the chlorine is recycled and can catalytically destroy hundreds of thousands of ozone molecules before it is washed out of the stratosphere. It is because of this catalytic action that the effects of the CFCs are so dramatic, even though their production amounts to only about one million tons annually.[1]

The first and best known example of the effect of CFC release is the ozone hole which opens up every year since 1979 over the Antarctic during the Austral spring (starting in September) and covers an area the size of the continental U.S. Since its first appearance it has grown almost every

year in size and duration and reached a new record in 1995 with a value below 100 Dobson units (100 Dobson units correspond to 1 mm thickness of the ozone layer). Extensions of the Antarctic ozone holes reach to South America, Australia and New Zealand. There are no longer any serious doubts about the anthropogenic origin of the ozone depletion by trace gases. A close correlation between the concentration of the trace gases and ozone depletion was found with the aid of airplanes which can fly in the stratosphere (Fig. 1.3).[2] A gradual decrease in stratospheric ozone has also been observed during the last few years over the Arctic and at mid latitudes.[3]

## UV INCREASE WAS PARTIALLY MASKED BY AIR POLLUTION

The ozone layer is an effective filter of short-wavelength solar UV radiation in the range between 280 and 315 nm (called UV-B). In fact, without it life would not be possible on our planet in its current form, since this high energy radiation would kill life within a very short time. Since the discovery of the ozone hole, a steep increase in UV radiation has been found in the Antarctic. In contrast, this effect could not be measured at mid

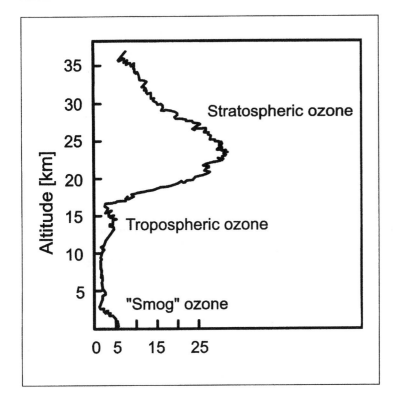

Fig. 1.2. Distribution of the ozone in the troposphere and in the stratosphere.

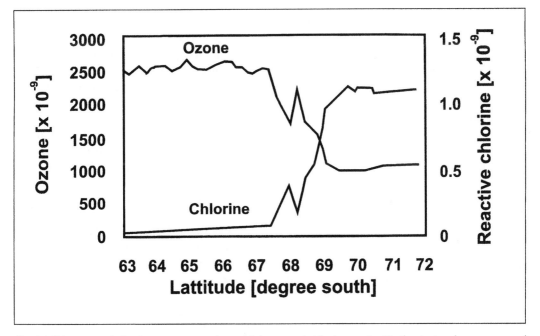

Fig. 1.3. Direct correlation between ozone layer thickness and chlorine content in the stratosphere, measured with a research airplane during a flight from South America to Antarctica.

latitudes, specifically in the industrialized areas of the Northern hemisphere. Today we know that the general air pollution in the lower atmosphere had partially compensated the increase in UV radiation. Ironically, the decrease in air pollution by the extensive installment of filters in industry and power plants has resulted in a significant increase in UV radiation.[4]

The political discussion concerning the phasing out of CFC technology has lasted for several decades. Today especially the developing countries are reluctant to phase out of usage of CFCs in order to maintain and increase their quality of life.

These problems and the long lifetime of CFCs in the stratosphere will result in a further increase in ozone depletion well beyond the year 2000. Indeed, according to recent model calculations ozone values of 1979 will not be reached before the year 2060.[5]

## NEGATIVE EFFECTS OF INCREASED SOLAR UV RADIATION ON LIFE

The short-wavelength solar UV radiation affects almost all forms of life on our planet. Dermatologists were the first to discuss an increase in skin cancer, especially in people of fair complexion. Later investigations showed that an increase in cataracts as well as a weakening of the immune system will also occur.[6] In contrast to humans, most animals seem to be protected from increased UV-B radiation by the possession of feathers, fur or a shell.

As another consequence of increased solar UV-B radiation, effects on growth and productivity of higher plants was expected. Initially, only crop plants were investigated,[7] and subsequently wild plants and whole terrestrial ecosystems were studied. The research over the last few years has indicated that about 50% of the 600 odd plants and cultivars investigated so far are sensitive while the other half is more or less resistant. Increased UV-B radiation affects photosynthesis, inhibits growth and reduces food quality.[8]

The third area of concern are aquatic ecosystems with their primary production by photosynthetically active algae in fresh water and marine habitats.[9] Despite their microscopic dimensions these organisms produce about half of the biomass on our planet. In addition, they are responsible for the uptake of huge amounts of carbon dioxide from the atmosphere and thus limit the greenhouse effect.

## REFERENCES

1. Rowland FS. Chlorofluorocarbons and the depletion of stratospheric ozone. Am Scientist 1989; 77:36-46.
2. Schnell RC, Liu SC, Oltmans SJ et al. Decrease of summer tropospheric ozone concentrations in Antarctica. Nature 1991; 351:726-729.
3. Hough AM, Derwent RG. Changes in the global concentration of tropospheric ozone due to human activities. Nature 1990; 344:645-648.
4. Kerr JB, McElroy CT. Evidence for large upward trends of ultraviolet-B radiation linked to ozone depletion. Science 1993; 262:1032-1034.
5. Madronich S, Björn LO, Iliyas M et al. Changes in biologically active ultraviolet radiation reaching the earth's surface. Environmental Effects Panel Report, United Nations, Environmental Program, 1991: 1-13.
6. Longstreth JD, de Gruijl FR, Takizawa Y et al. Human Health. UNEP Environmental Effects Panel Report, 1991:33-40.
7. Tevini M, Teramura AH. UV-B effects on terrestrial plants. Photochem Photobiol 1989; 50:479-487.
8. Caldwell MM, Teramura AH, Tevini M et al. Effects of increased solar ultraviolet radiation on terrestrial plants. UNEP Environmental Effects Panel Report, 1994: 49-64.
9. Häder D-P, Worrest RC, Kumar HD et al. Effects of increased solar ultraviolet radiation on aquatic ecosystems. Ambio 1995; 24:174-180.

# CHAPTER 2

# MAKING LINKS: FROM CAUSES TO EFFECTS TO ACTION

Canice Nolan[†]

Ozone depletion and its consequences is an issue which has received considerable attention in the scientific and lay press in recent years. Yet, I remember having conducted a straw poll in 1990 among a group of marine scientists where I asked them "is UV-B a significant factor in marine productivity and would stratospheric ozone depletion be expected to have serious consequences for the marine environment?" The consensus answers were (with appropriate caveats to cover their uncertainty) no and no. Were they to be asked again today, how different would their responses be? Although the respondents were not photobiologists (who would no doubt have given different replies) it *was* known at that time that impacts could be expected in aquatic systems and it did give me some concern. What then, does their response then tell us about scientific awareness and the communication of science to the public and to policy-makers? Were the negative replies due to a lack of interest in UV among the wider marine community or were they due to an ineffective communication of the potential UV hazards by the narrower UV effects research community?

Recognition of the hazardous nature of UV-B and of the potential for future increased UV-B fluxes in the future as a result of stratospheric ozone depletion was a key factor in spurring international action to protect the ozone layer as long ago as 1985 with the creation of the Vienna Convention. This had the aim of protecting the ozone layer and was signed by 22 countries as well as by the European Community. This was followed in 1987 by the Montreal Protocol and by its later revisions. These political actions had the objective and effect of limiting the production and use of ozone-depleting chemicals—particularly chlorofluorocarbons (CFCs).

---

[†] The views expressed in this article are those of the author and do not necessarily represent those of the European Commission.

*The Effects of Ozone Depletion on Aquatic Ecosystems*, edited by Donat-P. Häder.
© 1997 R.G. Landes Company.

The significance of the Protocol should not be underestimated. Both it and the Vienna Convention are the first global agreements and preventative actions to protect the atmosphere. Not only were they based on the results of scientific research, they impose an obligation on signatories to undertake research on the causes and consequences of ozone depletion. The Vienna Convention and the Montreal Protocol were also special in that they were negotiated and ratified in the face of considerable scientific uncertainty—we simply did not know, in 1985, the consequences of stratospheric ozone depletion. To paraphrase Edmund Burke: where there is uncertainty, there is room for mischief. How often we have seen international agreements on global change flounder or be delayed because some parties advocate inaction (pending the results of future research) wherever there is uncertainty. The Protocol and its amendments are a good example of the precautionary principle prevailing.

International action to protect the ozone layer is not without costs to society. The substances being regulated, which were useful and which provided tangible benefits, have been foregone to protect the global environment and society against... well, against what?

We knew that stratospheric ozone is depleted through a complex series of physical and chemical reactions involving CFCs and related compounds and that there is a lag-time of years-to-decades between action at ground level and consequences in the stratosphere.

We knew that stratospheric ozone attenuates solar UV radiation and could estimate that a 10% reduction in stratospheric ozone would result in a 0.5% increase in solar energy (and a 1% increase in UV energy) reaching the earth's surface and penetrating surface waters.

We knew that under controlled (but arguably unnatural or unrealistic) conditions, UV-B could interact with biological systems, including the maintenance of DNA integrity. Additionally, we knew that there was considerable variation among plant and animal species in their sensitivity to ultraviolet radiation.

The above is somewhat simplistic since in many instances, our knowledge was quite detailed. However, we also know that our qualitative knowledge was incomplete and that our quantitative knowledge was blurred by uncertainty. In many cases we were not even aware of the level of uncertainty. Furthermore, we did not know enough about global systems and processes to be sure that the agreements reached at Vienna and Montreal were adequate to protect the planet, its people, and its resources.

This was recognized at the time and the Protocol imposed an obligation on signatories to undertake research on the causes and consequences of ozone depletion. In the last decade there have been concerted and sustained research efforts at national and at international levels to improve our knowledge and to reduce our uncertainty in this area. There has been a concurrent explosion in the number of scientific publications, conferences and assessments produced on the subject. As our scientific knowledge has progressed so also has the Protocol been amended during this time. The European Commission, as a signatory, has also played its part both in the support of scientific research (through the Environment & Climate Research Programmes) and the promotion of a continued review (and amendment where necessary) of the adequacy of the current agreements to protect the ozone layer and of legislation to protect populations and ecosystems.

It is not my intention in this introductory chapter to make another overview or assessment of where we currently stand with respect to ozone and UV radiation. That has been done much more effectively in the chapters that follow, by others vastly more competent to do so. I note, however, from the work herein, that our knowledge has progressed remarkably over the last few years and I applaud the scientific community for their sustained and successful efforts.

Rather, I content myself to draw the readers' attention to two areas which I think important. The first of these, paradoxically perhaps, relates to what we still do not know and the second relates to how we communicate that which we *do* know and promote its translation into positive and protective action.

We do not know stratospheric processes in sufficient detail to confidently assert when and how to prevent ozone depletion. Limitation of the releases of CFCs and related substances into the atmosphere was a tremendous step forward in reducing the risks of the development of ozone holes in the future, but we still have a lot to learn about processes and their time scales in the upper atmosphere. We are still not sure about the effects of replacement substances, of the effects of natural processes, e.g. injection of volcanic dusts and aerosols on ozone, of our ability to predict ozone holes, of the significance of "mini-ozone holes" and of our ability to predict UV fluxes at the earth's surface (as evidenced by the ever-growing number of ground-level UV monitoring stations). We have yet to convince all the actors involved that there is a downward trend in ozone-levels and hence on upward trend in UV fluxes at the earth's surface (is time the only solution?). We have almost, but not quite solved the problem of intercomparability among UV-monitoring instruments and we still need to develop and to provide users with cheap, rugged and reliable monitors that provide adequately precise and accurate data. In this regard the aquatic effects community is currently particularly poorly served. Especially critical is the need for good spectral resolution in measurements and a sufficient understanding of atmospheric systems and processes to permit good prediction of spectral changes rather than gross flux changes. UV-monitoring stations are fairly thick on the ground in developed countries but these cover only a small fraction of the earth's surface. Can and should we interpolate and extrapolate or should we try to broaden the spread of the measurements database?

The troposphere also attenuates UV radiation and in doing so is itself effected by it. However, tropospheric air quality is changing in line with industrial and demographic trends. Our knowledge of these trends and of tropospheric processes is still inadequate to incorporate them into global or even regional long-term models to predict fluxes, their variation in space and time as well as their response to changes in forcing factors.

To date, at the biological level, effects models and dose-response models have relied on general effects or overly specific action spectra—often constructed on a broad-band basis and without full information on associated uncertainty. We cannot, for instance, predict with confidence for a given degree of stratospheric ozone depletion, what would be: (a) the increase in skin cancer incidence; (b) the changes in terrestrial ecosystems; (c) changes in oceanic productivity; and (d) changes in $CO_2$ budgets and cycling. There are of course other impacts and changes that would be expected, but if we cannot answer the priority questions then why bother with the others? After years of research support, policy-makers cannot be blamed for demanding answers to real, relevant and valid questions. Why can we not answer these questions? Once again, the answer lies not so much on our knowledge (or lack of it) of UV-impacts on specific molecules, organisms, or processes as it does with our understanding of the systems in which these are embedded and with where we draw the boundaries of these systems. We do know that ecosystems are sufficiently resilient to adapt to altered UV; we do not know if such adaptation will be to the advantage of society (I suspect not).

I have not specifically discussed aquatic systems above. Given their role as a source of nutrients and in modulating global climate their continued well-being is of course extremely important for society and the planet.

We know, qualitatively but not quantitatively, that sensitive larval stages of many species spend a vulnerable life-stage close to the surface and that both phytoplankton and organic matter cycling are affected by UV. Nonetheless, they have always been affected by UV. To the policymaker the question is to what extent they will be affected by altered UV as a result of stratospheric ozone depletion. At the global level, if we cannot estimate the former then how can we assess the latter?

At the policy-making level and the simplest case for inaction, it can be argued that skin cancer incidence is due more to individual lifestyles and behavior than it is to ozone depletion. It can also be argued that, if an Antarctic ozone hole decreases primary productivity there by 5% in any one year, then this is not truly important against a *natural* annual variation of up to 25%, and that society should not incur real costs to prevent it. These arguments have some validity but they should be refuted. Going back to the philosophy behind the Vienna Convention and the Montreal Protocol, lack of knowledge is not a substitute for lack of action.

The problem is that the consequences of our actions lie in the future and that without an adequate understanding of the present we shall not be able to construct solutions, no matter how many scenarios we simulate. We need to improve our knowledge of basic systems and processes perhaps more urgently than we need to improve our knowledge of the impact of UV on specific molecules or organisms. Practically speaking, however, these will be done in parallel. This means that the answers will not be provided by the "UV-effects" research community acting alone: they will be provided by international multi-disciplinary action involving many cultures, many fields of research, many countries, and over many years.

I move now to the second point that I think should be made—communication of scientific knowledge to user groups, i.e. other scientists, the public, and the policy-

maker/regulator. The Vienna Convention and Montreal Protocol were significant steps forward for the environment because, for the first time in the environmental field, they were negotiated and ratified, at the international level, in the face of significant scientific uncertainty—they demonstrated a common vision among policymakers almost unprecedented in environmental matters before that time.

The negotiations leading to the Convention and the Protocol were not without difficulties—a situation repeated every time that amendments are discussed. Cost-benefit calculations where the environment is concerned are fraught with difficulty even today and at that time environmental economics was still in its infancy. The cost-benefit equations also differ between temperate, (post-)industrial societies and (sub-)tropical developing societies and, of course, change with time.

Since chlorine and bromine loading of the stratosphere is expected to peak in the next decade and then return to 'normal' levels in about 50 years time, the cost-benefit arguments will not go away. Indeed, if and as the more severe of the original predicted consequences are demonstrated to be more and more unlikely, they will ever more glamorously recur.

If the wisest and most effective decisions on actions to minimize ozone depletion are to be taken, then they are best discussed beforehand on the basis of the fullest possible scientific knowledge. It is also preferable that the scientific community—the whole scientific community—speak objectively and with one voice. A broad swathe of disciplines are required to understand the problem—they are also required to reach the solution.

Wending my way back to the opening paragraph, not all disciplines appeared to be adequately informed of the problem six years ago. If scientists are not informed, can we, as scientists, be confident that policy-makers are? I am convinced that the situation has improved considerably since then as the necessity for interdisciplinary approaches to environmental problems has become accepted by all. The need to communicate is as much with us now as it ever was, however. I welcome this book not only as a contribution to the information exchange process but also as a contribution to the scientific assessment of where we are and the scientific debate about where we should be going.

# CHAPTER 3

# CONSEQUENCES OF THE EFFECTS OF INCREASED SOLAR ULTRAVIOLET RADIATION ON AQUATIC ECOSYSTEMS

Donat-P. Häder and Robert C. Worrest

## IMPORTANCE OF AQUATIC ECOSYSTEMS

The biomass production of aquatic ecosystems equals that of terrestrial ecosystems. It is assumed that aquatic ecosystems incorporate between 90 and 100 gigatons (Gt, $10^9$ tons) of atmospheric carbon annually into organic material.[1,2] Thus, it is of great importance to know what effect increased solar UV-B irradiation has on marine productivity[3,4] as well as the climatological processes linked with it. Only 0.5% of the water surface is represented by freshwater bodies. Although marine systems are by far the most important, freshwater ecosystems are excellent model systems for studying larger marine environments, and many of these systems are locally important.

Since most macroalgae are restricted to coastal areas, the largest share in biomass production can be attributed to phytoplankton. Phytoplankton constitute the basis for the intricate food web in the oceans and are thus a prerequisite for the crop of fish, crustaceans and mollusks. Furthermore, whereas they are responsible for the uptake of half of the carbon dioxide from the atmosphere, any reduction in the uptake capacity would result in an increase in the greenhouse effect, with subsequent impacts on global climate change.

## EFFECTS OF SOLAR UV-B ON AQUATIC ECOSYSTEMS

Recent investigations indicate that many aquatic ecosystems are under considerable UV-B stress even at current levels.[5-7] Most primary

*The Effects of Ozone Depletion on Aquatic Ecosystems*, edited by Donat-P. Häder.
© 1997 R.G. Landes Company.

producers depend on solar energy for their photosynthesis and are therefore restricted to the upper layers in the water column. At that depth they are simultaneously exposed to high levels of ultraviolet radiation. UV-B effects on aquatic ecosystems have been covered in several recent reviews.[8-21] Studies conducted under the Antarctic ozone hole have shown a significant impact on primary productivity. Since depletion of global stratospheric ozone occurs at all latitudes except near the equator and since the depletion is expected to continue well into the next century,[22] the effects of increased UV-B on marine primary productivity may also materialize outside polar regions.

In laboratory studies UV-B radiation affects cellular DNA, impairs photosynthesis, enzyme activity and nitrogen incorporation, bleaches cellular pigments, and inhibits motility and orientation.[23-27] To determine the effects of solar UV-B radiation on natural aquatic ecosystems a number of open questions need to be answered:

- What are the predicted changes in ozone concentration and UV-B radiation on a global basis during the decades to come?
- What is the spectral penetration of solar short-wavelength radiation as a function of depth in different water types?
- What is the vertical distribution of the aquatic organisms in the water column for major water types?
- What is the biologically and spectrally weighted sensitivity of the organisms affected by solar UV-B?
- What are the extent and limits of UV repair and adaptation as well as the effects of other environmental factors in mitigating or augmenting UV effects?

## GLOBAL DISTRIBUTION OF PHYTOPLANKTON

Phytoplankton are not uniformly distributed in the oceans of the world: the highest concentrations are found in the circumpolar regions at both Arctic and Antarctic high latitudes, while the biomass concentrations are 10 to 100 times lower in the tropics and subtropics. Only in the upwelling areas on the continental shelves are higher biomass concentrations found. This uneven distribution is probably due to a number of factors, including nutrients, light availability and water column stability. Moreover, UV-B radiation may also play a role, as irradiances are several times higher than in circumpolar areas.[8] Another indication that phytoplankton are indeed affected by solar UV-B radiation even at current levels is the fact that in temperate oceans phytoplankton blooms occur in spring and subside during summer. Sometimes there is a second bloom in autumn.

## PENETRATION OF SOLAR RADIATION INTO THE WATER COLUMN

The transparency of the water strongly depends on the amount of seston (particulate substances) and gelbstoff (yellow dissolved organic substances). Jerlov[28] suggested describing the optical properties of bodies of water by classifying them into several types: Open oceanic waters range from I (clearest waters) to III (highest attenuation coefficient) and coastal waters from 1 (highest transparency) to 9 (highest absorption). In eutrophic ponds and lakes as well as in turbid coastal waters UV-B radiation may penetrate less than 1 m to the 1% level; in contrast, in clear oceanic waters penetration to several tens of meters has been measured.[7,29]

Until recently only few underwater optical sensors capable of accurately measuring UV-B as a function of depth in aquatic systems were available. This problem was solved by the development of new equipment. Two different techniques are being employed: either the underwater irradiance distribution is measured with high spectral resolution using a double monochromator spectroradiometer, or a filter radiometer is used which allows a fast determination in several defined spectral bands as a function of depth. Since especially short wavelength radiation is subject to multiple scattering, underwater radiation is far from being direct, and instead impinges from all directions on an organism. For this reason a new sensor with a $4\pi$ geometry was developed which collects radiation from all directions and guides it to the entrance slit of a scanning double monochromator spectroradiometer (Fig. 3.1).[30] Recent comparisons by Kirk et al[31] suggest that several commercial instruments may be used to obtain quantitative underwater UV-B information. Figure 3.2 shows typical examples for the spectral dependence of underwater solar radiation in the Baltic Sea and the Mediterranean.

A useful term in the definition of the vertical distribution of phytoplankton in the water column is that of the euphotic zone, which describes the thickness of the top layer in which the incident irradiance at the surface is attenuated to 0.1%. This, generally, is the lower limit of where photosynthesis balances respiration. In physical terms the depth of the euphotic zone can vary between a few decimeters for turbid coastal and eutrophic freshwater habitats and several tens of meters for clear oceanic waters. The organisms are passively distributed by wind and waves in the mixing layer which extends down to the thermocline. However, many organisms are capable of active movements using cilia or flagella or changes in buoyancy by producing gas vacuoles or oil droplets;[32-34] daily vertical migrations of up to 15 m have been measured. Therefore, in most cases typical vertical distribution patterns are found in both freshwater and marine ecosystems.[35-37]

The movements of the organisms are guided by a number of external factors, of which light and gravity are the most

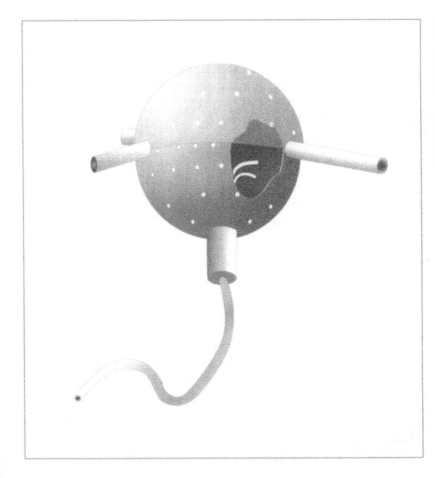

Fig. 3.1. A new sensor with a $4\pi$ geometry was developed to measure radiation under water from all directions and to guide it to the entrance slit of a scanning double monochromator spectroradiometer.

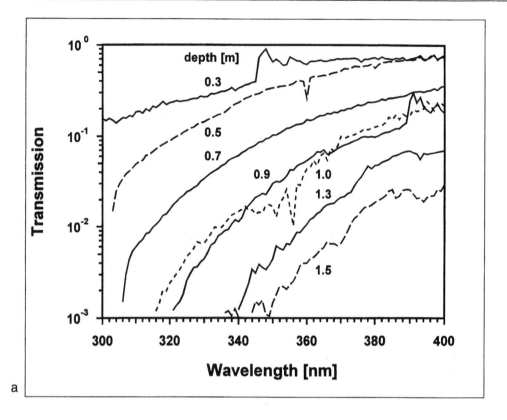

a

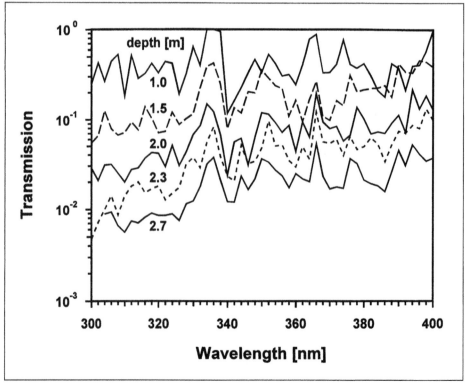

b

Fig. 3.2. Spectral irradiance of underwater solar radiation in the Baltic Sea (a) and the Mediterranean (b).[30]

important. Many species have been found to swim upwards before dawn due to negative gravitaxis and during early morning hours triggered by positive phototaxis. By these orientation mechanisms the cells gather close to the water surface (Fig. 3.3). Often the irradiance becomes excessive around noon and consequently, the cells move to deeper water layers employing negative phototaxis. Phytoplankton use various wavelength bands in the visible and UV-A range to orient themselves with respect to light while UV-B is not used for photo-orientation; thus, they are in a situation similar to humans who do not detect UV-B radiation. Therefore, during a selective increase in UV-B radiation the organisms cannot escape this stress situation by simply moving to lower levels in the water column.

Solar UV-B irradiation affects both motility and the orientation mechanisms in many phytoplankton species.[5,8] Figure 3.4 shows a typical example of the effect of solar radiation on the percentage of motile cells in a population of the marine flagellate *Cryptomonas maculata*. At the same time the swimming velocity decreases drastically in the still motile cells. The precision of orientation is also impaired by solar ultraviolet radiation: both positive and negative

phototaxis decrease, as quantified by the Rayleigh test, with increasing exposure time. The reason for this effect was investigated in one test system. The flagellate *Euglena gracilis* orients in light using an organelle located near the base of its long trailing flagellum. This so-called paraflagellar body (PFB) is composed of four major proteins in the range of 27-33 kDa[38] which carry pterins and flavins as chromophoric groups.[39] Exposure of the cells to solar or artificial ultraviolet radiation causes a drastic decrease in the concentration of the PFB proteins (Fig. 3.5).[40,41] Gravitactic orientation also decreases with exposure to solar or artificial UV-B radiation. As a result of prolonged exposure to solar radiation, those cells which are still motile, move in random directions. These effects have negative consequences for phytoplankton which depend on motility and orientation to move to and stay at a depth with favorable light conditions for growth and biomass production. Some dinoflagellates may be the exception to the rule since they seem to have developed an escape mechanism for excessive radiation. When the marine *Gymnodinium* (strain Y100) is exposed to strong solar radiation it reverses its gravitactic orientation from negative (upward movement) to positive (downward

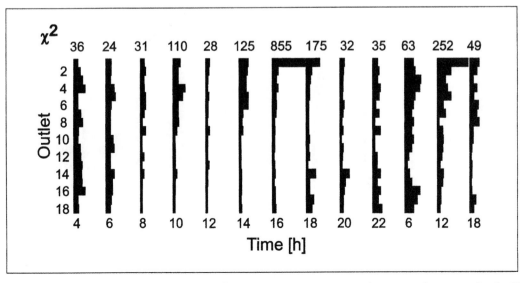

*Fig. 3.3. Vertical distributions of the dinoflagellate* Prorocentrum micans *in the water column over the day.[36]*

swimming) and thus escapes the unfiltered radiation encountered at the surface.[42]

In the past, nano- and picoplankton have been grossly neglected even though their contribution to the total biomass is estimated to be at least 40%. The reasons for this neglect may be the small size of the organisms and technical problems that occur during harvesting. Likewise, only in recent years has the role of bacterioplankton been elucidated. These organisms are responsible for degradation and cycling of organic matter in the sea. Bacterioplankton, as well as their extracellular enzyme activity in subsurface waters, are strongly affected by solar UV-B radiation.[43] It has been shown that small organisms (bacteria and microalgae) are more susceptible to UV

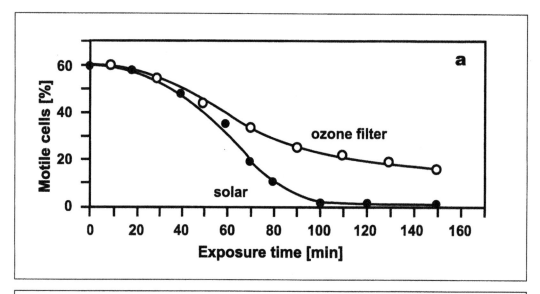

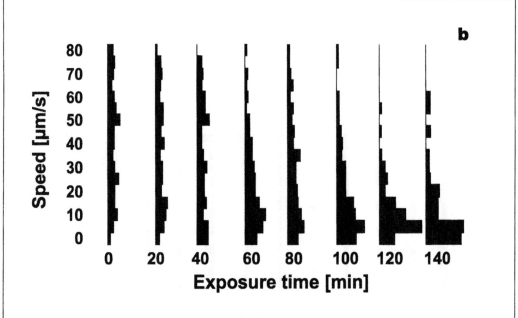

Fig. 3.4. Effect of solar radiation on the percentage of motile cells (a) under unfiltered radiation (closed symbols) and under a layer of artificially produced ozone (open symbols) and the swimming velocity distribution (b) in the marine flagellate Cryptomonas maculata.

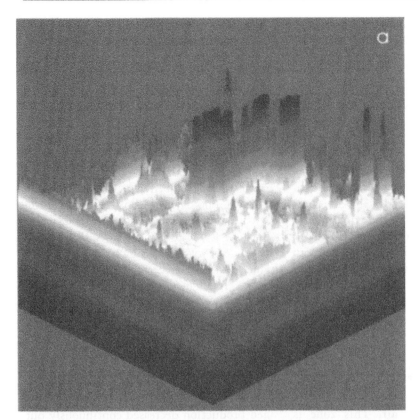

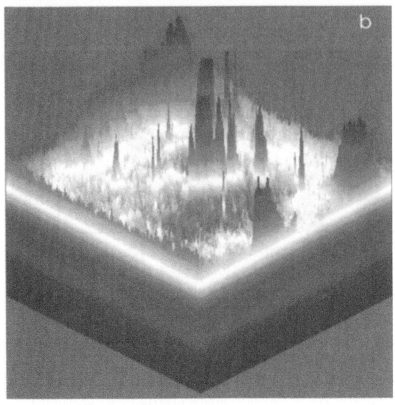

Fig. 3.5. 3D representation of 2D gels of the proteins isolated from the paraflagellar body, the putative photoreceptor of the flagellate Euglena gracilis, before (a) and after (b) UV-B irradiation.[41] See color figure in insert.

stress than larger organisms.[20] Therefore it is expected that increased UV-B radiation may alter species compositions in marine ecosystems.

## MACROALGAE AND SEAGRASSES

Macroalgae (seaweeds) and seagrasses are large biomass producers in almost all coastal areas of the oceans and contribute both to the natural food chain and human usage. Being sessile, these plants are subject to the ambient radiation at their growth site. In fact, it is believed that light has a decisive role in controlling the depth distribution.[44] Some algae are adapted to the area above the water surface which is reached only by spray water or by occasional high tides. In addition to direct solar exposure, these algae need to adapt to extreme differences in salinity and temperature. Most algae are restricted to the eulittoral (intertidal zone) which falls dry once during the tidal cycle. Other algae populate the sublittoral zone which is rarely exposed to the air. This zone is populated by the large kelp species on many rocky shores of the oceans. Substory

algae and deep water algae are even more protected from direct solar radiation.

If the UV-B/photosynthetic active radiation (PAR) ratio increases, the algae will be exposed to enhanced short-wavelength radiation to which they may not be adapted. In recent studies little effect on respiration was found, while photosynthesis was inhibited in many red, brown and green benthic algae. Recently, an instrument was developed which allows one to measure photosynthetic oxygen production within the water column under solar radiation.[45,46] Figure 3.6 shows the oxygen production of the common Mediterranean alga *Halopteris scoparia* in relation to the depth. After measurement of dark respiration the organisms were exposed at different depths from 5 m to 1 m. Even though the irradiance is highest at the surface, photosynthesis does not change much between depths; slightly higher values were even found at greater depths. When exposed at the surface, photosynthetic oxygen production decreased continuously but never became zero (Fig. 3.7). This behavior is due to the phenomenon of photo-

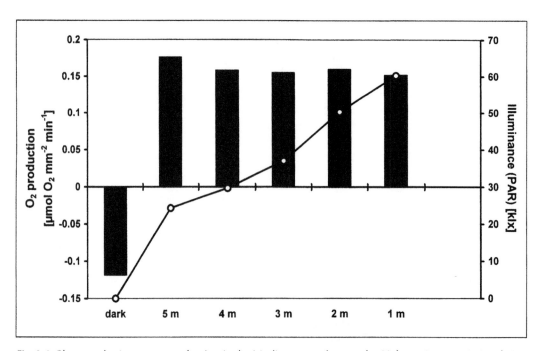

Fig. 3.6. Photosynthetic oxygen production in the Mediterranean brown alga Halopteris scoparia *in relation to the depth in the water column.*

inhibition, the exact molecular mechanism of which is still under debate. The D1 protein of photosystem II seems to play a key role in this process, in which the photosynthetic electron transport is reduced to protect the cells from excessive solar radiation.

When using PAM (pulse amplitude modulation) fluorescence measurements, deep-water benthic algae were most sensitive while intertidal algae were least sensitive.[47-50] The recent development of a portable PAM instrument allows one to measure the photosynthetic quantum efficiency in algae in their natural habitat. Earlier measurements were hampered by the necessity that the specimen be transferred to the laboratory, which induced a number of stress factors including changes in temperature, salinity and irradiation. Thalli of the common green macroalga *Valonia utricularia* were harvested by diving and were dark-adapted for 30 min, after which the optimal photosynthetic quantum yield was determined (Fig. 3.8). After this treatment the samples were exposed to solar radiation close to the surface in a container made from UV transparent Plexiglas which kept the thalli in place and allowed sea water to circulate. After 30 min of exposure the quantum yield had dropped to less than 0.2, and it took several hours under shaded conditions for the

organisms to recover. Even at the end of six hours, recovery was only partial. The rightmost bar in Figure 3.8 shows the quantum yield of a sample subjected to the same experimental procedure except for solar radiation in order to determine whether there were other stress factors in addition to excessive radiation; obviously, the conditions and handling had no effect on photosynthetic quantum yield.

The following experiment is ecologically even more relevant as it determines whether or not macroalgae are affected by solar radiation with current UV-B levels at their natural growth sites. Using the portable PAM instrument, the quantum yield was determined at 1-h intervals from dawn to dusk in *Valonia utricularia* immediately after harvesting samples from their growth sites. Early in the morning the quantum yield was found to be close to its optimal value of 0.7. As the sun rose and solar UV-B radiation increased, the yield decreased and only recovered later in the afternoon for specimens harvested from 2 m depth (Fig. 3.9).

## CYANOBACTERIA

Algae and higher plants are not capable of utilizing atmospheric nitrogen; rather they need nitrate, nitrite or other soluble nitrogen compounds. In contrast, cyanobacteria (blue-green algae) and other

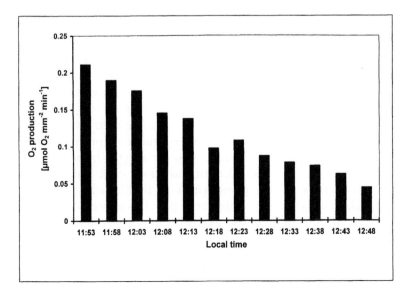

*Fig. 3.7. Photoinhibition of the photosynthetic oxygen production in the Mediterranean brown alga* Halopteris scoparia *exposed immediately below the water surface.*

*Fig. 3.8. Photoinhibition and recovery of the photosynthetic quantum yield in the green macroalga Valonia utricularia measured by PAM fluorescence.*

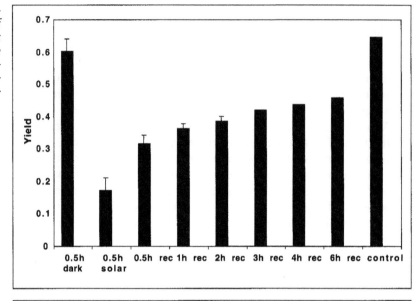

*Fig. 3.9. Photosynthetic quantum yield of the green macroalga Valonia utricularia measured by PAM fluorescence in material harvested from the growth site from dawn to dusk.*

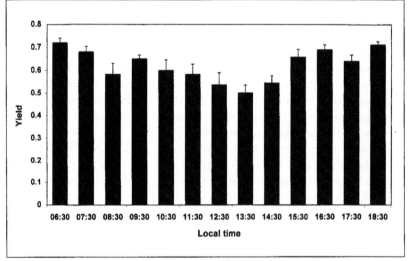

prokaryotic organisms are able to fix atmospheric nitrogen. The enzymes used for this process are highly sensitive to solar UV-B radiation.[51] The incorporation of atmospheric nitrogen by cyanobacteria is an important source of nitrogen available for phytoplankton and macroalgae. This process is important also in terrestrial habitats. It is assumed that cyanobacteria in tropical rice paddy fields make available about 100 million tons of nitrogen.[52] In comparison, about 30 million tons of artificial nitrogen fertilizer are produced per year using the Haber-Bosch process. Additionally, in farm land of temperate zones, cyanobacteria contribute significant amounts of atmospheric nitrogen.

Most cyanobacteria are very sensitive to solar UV-B radiation: their motility, orientation, photosynthetic pigments and oxygen production are affected by solar short-wavelength radiation, even during short exposure times. A sizable reduction of the cyanobacterial populations would have significant consequences for agricultural production, especially in developing countries which may face problems in financing the necessary replacement by artificial fertilizers.

Field studies in Ghana have shown that both velocity and orientation mechanisms of several cyanobacteria are affected by solar radiation, even after short exposure times. The damage was only partially repaired in subsequent dim light and only after short exposure times (Fig. 3.10). In contrast, longer exposure times even resulted in increasing damage well after the exposure.[53,54]

Work carried out in India has indicated that solar UV-B affects the nitrogenase activity and carbon dioxide uptake in rice paddy cyanobacteria.[55,56] Survival, growth, protein content and enzyme activity are affected by both solar and artificial UV-B radiation.[57,58] Cyanobacteria produce photosynthetic, accessory pigments, the phycobilins, which harvest radiation and funnel it to the reaction centers of photosystem II. These proteins are organized in supramolecular assemblies which are affected easily by UV-B radiation.

Some cyanobacteria, however, characterized by a brown color, seem to be better adapted to high solar radiation than closely related green forms.

## SCREENING PIGMENTS

DNA is one of the targets of radiation, and the most abundant damage seems to be the formation of thymine dimers. But, in addition, a multitude of other chromophores and proteins are affected. Thus, UV-B does not only damage one key target in phytoplankton but has many other deleterious effects which differ in their action spectra. The action spectra are further complicated by antagonistic and repair processes stimulated by UV-A and visible radiation. Figure 3.11 shows the action spectrum based on irradiance response curves of UV inhibition of photosynthetic oxygen production in a mass biomass producer, the cyanobacterium *Nodularia spumigena*, isolated from the Baltic Sea.

The induction of screening pigments has been found in marine and freshwater organisms.[59] The pigment scytonemin has been isolated from cyanobacteria where it is induced by UV radiation.[60] In addition,

cyanobacteria as well as eukaryotic phytoplankton use several water soluble, UV-absorbing mycosporines as screening pigments. Other phytoplankton use carotenoids to dissipate the excess radiation energy from the photosynthetic pigments, and some have been found even to tolerate the unfiltered solar radiation at the water surface in tropical oceans. As in phytoplankton, the occurrence and induction by UV of screening pigments has been recorded in tropical red algae.[61]

## CARBON DIOXIDE UPTAKE AND ITS ROLE IN GLOBAL WARMING

A global warming of mean surface temperatures by 1.5-4.5°C is predicted for a doubling of the $CO_2$ concentration by the year 2080. This is predicted to be accompanied by a 1 m rise in sea level.[62,63] The oceans play a key role with respect to global warming. Marine ecosystems are a major sink for atmospheric $CO_2$ and thus they play a prominent role in future trends of carbon dioxide concentrations in the atmosphere.[64,65]

The uptake and release of atmospheric carbon were balanced and the concentration of $CO_2$ in the atmosphere was constant for extended periods of time before the system was disturbed by human activities (Fig. 3.12). Both aquatic and terrestrial ecosystems incorporate about 100 Gt of carbon in the form of carbon dioxide per year. Fossil fuel burning and deforestation contribute about 7 Gt of carbon per year. However, long-term measurements on the Mauna Loa (Hawaii) indicate only an annual increase of 3 Gt in the atmosphere. Therefore it was suggested that the remaining 4 Gt are removed from the cycle by the biological pump in the ocean, a net uptake of $CO_2$ by the terrestrial biosphere, or a combination of both.[66] The relative importance of the two uptake modes is still under debate.[67,68]

## CONSUMERS

The primary producers, phytoplankton and macroalgae, are the basis for the

Fig. 3.10. Inhibition of motility (a) and its recovery (b) in the cyanobacterium Phormidium uncinatum (strain Tübingen) after exposure to solar radiation of different duration in Ghana.[54]

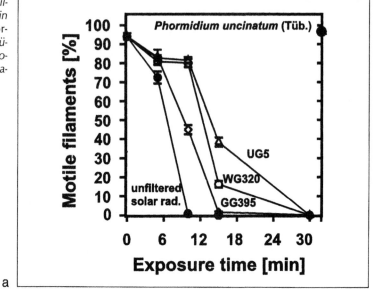

a

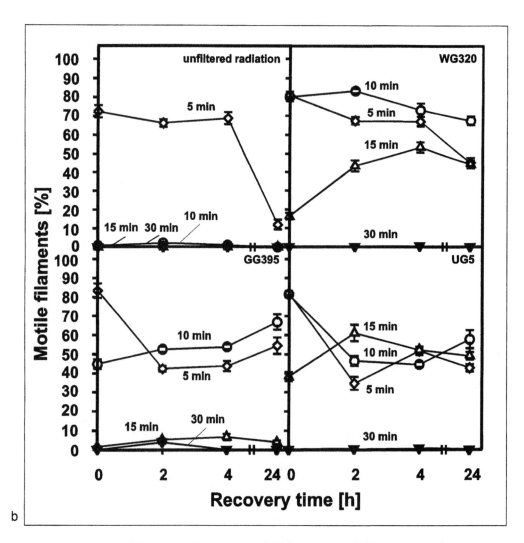

b

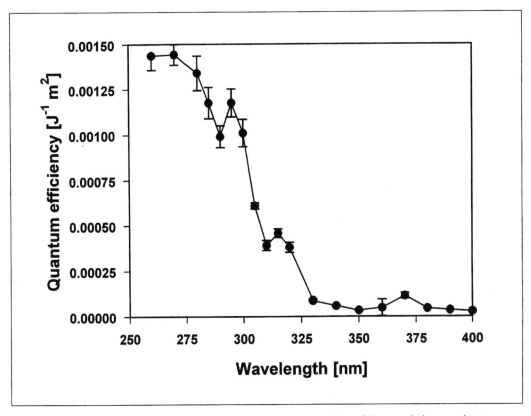

Fig. 3.11. Action spectrum based on irradiance response curves of UV inhibition of photosynthetic oxygen production in the cyanobacterium Nodularia spumigena.[66]

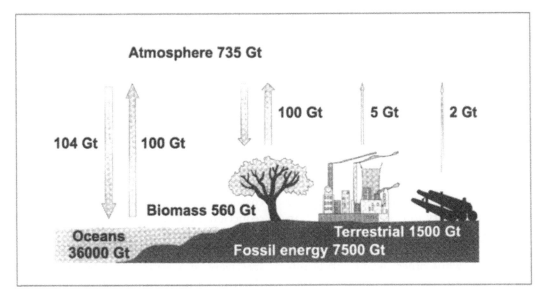

Fig. 3.12. Global carbon fluxes (in Gt) between the atmosphere, biosphere and aquasphere as well as the contributions by human activities and the major deposits of carbon.[66]

intricate marine food webs. Any losses in biomass production will therefore affect the next higher trophic levels and ultimately cause losses in fisheries yield.[69,70] In addition, solar UV-B radiation has been found to cause direct damage to fish, shrimp, crab and other animals. Early developmental stages are especially prone to UV-B effects. The most detrimental effects are decreased reproductive capacity and impaired larval development (see also chapter 10 by Blaustein and Kiesecker).[71] As in phytoplankton and macroalgae, solar UV-B radiation is a limiting factor even at current levels, and small increases in UV-B exposure could result in significant damage to the consumer populations.[72] Freshwater crustaceans are also affected by solar UV-B. However, it is interesting to note that *Daphnia* species from an alpine lake are more intensely colored and tolerate higher UV-B doses than in lowland lakes where UV-B doses are considerably lower.[73]

Zooplankton communities in clear north temperate lakes may be prevented by UV-B radiation from exploiting warmer surface waters during periods of summer stratification.[74] As a consequence, UV-B may alter their ecological interactions with food resources, predators and other environmental variables. An unexpected effect was observed in a recent ecosystem study: Algal growth in an artificial stream was higher under UV-B than in the control due to the fact that the grazers, larval chironomids, were more sensitive to UV-B than the algae.[75] The result confirms the opinion that predictions of ecosystem responses should not be based on single-species assessments.

Various marine organisms differ greatly in their sensitivity to UV-B radiation.[76] In one crustacean about 50% mortality was found at current UV-B irradiances at the sea surface, while other shrimp larvae tolerate even irradiances predicted for a 16% ozone depletion.[77] The threshold sensitivity of the crustacean *Thysanoessa raschii* clearly exceeds radiation levels encountered in spring, while in summer a 50% mortality would occur within 5 days assuming

a 16% ozone depletion. Current UV-B kills most individuals of the common copepod *Acartia clausii* in culture, and similar inhibitions have been found in shrimp-like crustaceans and crab larvae in the Pacific Northwest. The available results indicate that at high latitudes (over 40°) the recently measured increases in UV-B during late spring may affect the critical developmental phases of some species. Even sustained small increases or temporary fluctuations in UV-B may affect especially sensitive species.

Cleavage in sea urchin eggs is also impaired by ultraviolet radiation,[78] and coral reef organisms, such as sponges, bryozoans and tunicates are severely affected. Some organisms seem to utilize melanins as screening pigments; some colored corals withstand high levels of radiation by production of the pigment S-320. UV-B sensitivity in corals differs depending on the depth at which they grow[73] and the amount of the UV screening substances increases towards the surface.[79,80] Colonies of *Motastrea annularis* showed a gradual reduction in mycosporine amino acids with depth, and transplant experiments indicated that the organisms could not adapt to higher UV levels within a period of 21 days. UV-B radiation seems to affect invertebrate organisms such as sea anemones and octocorals by oxidative stress.[81] The recently observed coral bleaching in the Bahamas is also associated with UV radiation under calm, clear water column conditions.[82]

Growth and survival of larval fish is affected by solar UV-B radiation.[76] Predictions for the North American Pacific coastal shelf indicate that a 16% ozone reduction in June would result in larval mortality of 50%, 82% and 100% for anchovy larvae of ages 2, 4 and 12 days, respectively, at a depth of half a meter. Anchovy larvae grow in many oceans between June and August with a peak in July coincident with high radiation levels. Virtually all larvae populate the upper 0.5 m; thus a significant ozone reduction level could result in enhanced larval mortality.

Also in rainbow trout and two threatened salmonids UV-B radiation is an important environmental stress[83] with some individuals showing skin injury as well as apparent suppression of their immune system.

## SUBSTITUTES OF CFCs AND THEIR DEGRADATION PRODUCTS

Recently, partially halogenated substances have been introduced as substitutes for CFCs in several applications. Some HFCs and HCFCs, notably HFC134a, HCFC123 and HCFC124, generate trifluoroacetic acid (TFA, $CF_3COOH$) during degradation as their main product. TFA is a very inert chemical and is neither photolyzed nor does it undergo any other physicochemical degradation. It is also apparently not metabolized by plants. However, several microorganisms such as methanogens, sulfate reducers, and aerobic soil bacteria have been found to degrade TFA.[84] Therefore TFA may accumulate in air, water and soil over long periods. TFA is mildly toxic to the few investigated marine and freshwater phytoplankton (EC50 1200-2400 mg $l^{-1}$). However, the freshwater green alga, *Selenastrum capricornutum*, showed an EC50 of 4.8 mg $l^{-1}$.[85] It is still uncertain whether or not TFA is concentrated in the food web. Current levels of TFA in an industrialized area of Southern Germany are 0.01-0.05 ng $m^{-3}$ in air, and 40 times these values are predicted for the year 2010. But even with the event of continued production well into the next century, TFA is unlikely to reach toxic levels for phytoplankton in the oceans. However, in environmental niches such as vernal pools, TFA could accumulate to critical levels.[86]

## CONSEQUENCES

Investigations during the past one or two decades have provided ample evidence that solar UV-B radiation even at current levels constitutes a stress for many marine and freshwater ecosystems. However, quantitative estimates of possible damage due to expected increases in solar UV-B radiation are currently not possible. Further, concentrated research is necessary to evaluate the effects in important aquatic ecosystems on a global scale. Recent assessments have indicated that remote sensing by satellite is not a sufficient tool to measure the biomass and productivity, since the information basically reflects the surface concentration of phytoplankton. Ground-truth measurements are necessary to evaluate the vertical distribution and the species composition of the investigated populations. Increased work on important biomass producers such as diatoms, dinoflagellates and cyanobacteria, as well as nano- and picoplankton, is necessary. Another important field is the investigation of indirect effects of UV-B irradiation on humic acids and other organic material in the water column, which changes its biodegradability.

A major loss in primary biomass productivity has significant consequences for the intricate food web in aquatic ecosystems and will affect food productivity and quality. A 16% ozone depletion has been estimated to result in a 5% loss in phytoplankton productivity. This would cause a reduction in fishery and aquaculture yields of about 7 million tons of fish per year.[69] Biological effects of small changes in UV-B may be difficult to determine because the biological uncertainties and variations are large and furthermore, the baseline productivity for pre-ozone-loss eras is not well established. Predictions of future trends are further complicated by unpredicted feedback loops involving temperature, salinity, $CO_2$ concentration and different irradiation patterns caused by changing cloud cover.

The second major consequence of decreased phytoplankton productivity may be a reduced sink capacity for atmospheric carbon dioxide, would result in a faster development of the greenhouse effect and global climate change.

Cyanobacteria and other prokaryotic microorganisms are capable of fixing atmospheric nitrogen, while eukaryotic organisms such as higher plants can only utilize ammonia, nitrate or nitrite. Cyanobacteria contribute large amounts of nitrogen to farmland and especially tropical rice paddy fields. Since cyanobacteria have been found to be highly sensitive to solar UV-B radiation, an increase in solar UV-B radiation may cause decreases in the cyanobacterial populations, resulting in decreased nitrogen assimilation. This effect may lead to a nitrogen deficiency for higher plant ecosystems. Consequently, losses in nitrogen fixation due to increases in UV-B radiation may need to be compensated for by artificial nitrogen fertilization, stretching the financial capabilities especially of developing countries.

Macroalgae and phytoplankton are known to release organic sulfur compounds such as dimethylsulfide (DMS) which enter the atmosphere and serve as cloud condensation nuclei. Changes in DMS production may affect the atmospheric radiation balance.

Compared to past natural changes of the earth, the predicted changes in the ozone layer occur over extremely short times—too short for genetic adaptation to higher UV-B levels. Different species differ in their sensitivity toward solar short-wavelength radiation; therefore, shifts in species composition of aquatic ecosystems may be a consequence. Generally speaking, UV seems to affect smaller phytoplankton more than larger organisms. As primary feeders prey by size and not by species preference, this effect may also alter the subsequent links in the food web and the species composition of the ecosystem.

## Acknowledgments

This work was supported by financial support from the European Community (EV5V-CT91-0026; EV5V-CT94-0425; DG XII, Environmental Programme) to D-P Häder. The authors gratefully acknowledge the skillful technical assistance of J. Schäfer and H. Wagner.

## References

1. Houghton RA, Woodwell GM. Global climatic change. Sci Amer 1989 (Apr); 260:18-26.

2. Siegenthaler U, Sarmiento JL. Atmospheric carbon dioxide and the ocean. Nature 1993; 365:119-125.

3. Smith RC. Ozone, middle ultraviolet radiation and the aquatic environment. Photochem Photobiol 1989; 50:459-468.

4. Prézelin BB, Boucher NP, Smith RC. Marine primary production under the influence of the Antarctic ozone hole. Ultraviol Radiat Biolog Res Antarct 1993; 1-62.

5. Häder D-P. Effects of enhanced solar ultraviolet radiation on aquatic ecosystems. In: Tevini M, ed. UV-B Radiation and Ozone Depletion. Effects on Humans, Animals, Plants, Microorganisms, and Materials. Boca Raton, Ann Arbor, London, Tokyo: Lewis Publ, 1993:155-192.

6. Cullen JJ, Lesser MP. Inhibition of photosynthesis by ultraviolet radiation as a function of dose and dosage rate: results for a marine diatom. Marine Biol 1991; 111: 183-190.

7. Smith RC, Prézelin BB, Baker KS et al. Ozone depletion: Ultraviolet radiation and phytoplankton biology in Antarctic waters. Science 1992; 255:952-959.

8. Häder D-P. Risks of enhanced solar ultraviolet radiation for aquatic ecosystems. Progress in Phycological Research 1993; 9:1-45.

9. Acevedo J, Nolan C. Environmental UV Radiation. Brussels: Commission of the European Communities, Directorate-General XII for Science, Research and Development, 1993.

10. Weiler CS, Penhale PA, eds. Ultraviolet Radiation in Antarctica: Measurements and Biological Effects. Washington, D.C. American Geophysical Union, 1994.

11. Cullen JJ, Neale PJ. Ultraviolet radiation, ozone depletion, and marine photosynthesis. Photosynthesis Res 1994; 39:303-320.

12. U.S. Dept of Energy, Environmental Sciences Div, Washington, D.C. Overview of the DOE Atmospheric Chemistry Program's Ozone Project, DE93-007416 DOE/ER-0575T. Jan 1993.

13. SCOPE. Effects of increased ultraviolet radiation on global ecosystems, Proceedings of a workshop arranged by the Scientific Committee on Problems of the Environment (SCOPE), Tramariglio, Sardinia, October 1992.

14. SCOPE. Effects of increased ultraviolet radiation on biological systems, Scientific Committee on Problems of the Environment (SCOPE), Paris. 1992.

15. Holm-Hansen O, Helbling EW, Lubin D. Ultraviolet radiation in Antarctica: inhibition of primary production. Photochem Photobiol 1993; 58:567-570.

16. Holm-Hansen O, Lubin D, Helbling EW. UVR and its effects on organisms in aquatic environments. In: Young AR, Björn O, Moan J, Nultsch W, eds. Environmental UV Photobiology. New York: Plenum, 1993:379-425.

17. Tevini M, ed. UV-B Radiation and Ozone Depletion. Effects on Humans, Animals, Plants, Microorganisms, and Materials. Boca Raton, Ann Arbor, London, Tokyo: Lewis Publ, 1993.

18. Biggs RH, Joyner MEB, eds. Stratospheric Ozone Depletion/UV-B Radiation in the Biosphere. Springer, Berlin, Heidelberg: NATO ASI Series, 1994.

19. Williamson CE, Zagarese HE, eds. Impacts of UV-B Radiation on Freshwater Organisms. Arch Hydrobiol 1994; Beiheft 43 (special issue).

20. Karentz D, Bothwell ML, Coffin RB et al. Impact of UV-B radiation on pelagic freshwater ecosystems: report of working group on bacteria and phytoplankton. Arch Hydrobiol 1994; Beiheft 43:31-69.

21. Smith RC, Cullen JJ. Implications of increased solar UVB for aquatic ecosystems. U.S. National Report to the IUGG (1991-1994), American Geophysical Union. 1995.

22. Stolarski R, Bojkov R, Bishop L et al. Measured trends in stratospheric ozone. Science 1992; 256:342-349.

23. Döhler G, Hagmeier E, Grigoleit E et al. Impact of solar UV radiation on uptake of $^{15}$N-ammonia and $^{15}$N-nitrate by marine diatoms and natural phytoplankton. Biol Phys Pfl 1991; 187:293-303.

24. Häder D-P, Worrest RC, Kumar HD. Aquatic ecosystems. UNEP Environmental Effects Panel Report. 1989:39-48.

25. Häder D-P, Worrest RC, Kumar HD. Aquatic ecosystems. UNEP Environmental Effects Panel Report. 1991:33-40.

26. Worrest RC, Häder D-P. Effects of stratospheric ozone depletion on marine organisms. Environmental Conservation 1989; 16:261-263.

27. Häder D-P, Worrest RC. Effects of enhanced solar ultraviolet radiation on aquatic ecosystems. Photochem Photobiol 1991; 53:717-725.

28. Jerlov NG. Light—general introduction. In: Kinne O, ed. Marine Ecology 1970 Vol 1, Wiley:London, New York; 1:95-102.

29. Smith RC, Baker KS. Penetration of UV-B and biologically effective dose-rates in natural waters. Photochem Photobiol 1979; 29:311-323.

30. Piazena H, Häder D-P. Penetration of solar UV irradiation in coastal lagoons of the Southern Baltic Sea and its effect on phytoplankton communities. Photochem Photobiol 1994; 60:463-469.

31. Kirk JTO, Hargreaves BR, Morris DP et al. Measurement of UV-B radiation in freshwater ecosystems. Arch Hydrobiol 1994; Beiheft 43:77-99.

32. Walsby AE. Mechanisms of buoyancy regulation by planktonic cyanobacteria with gas vesicles. In: Fay P, Van Baalen C, eds. The Cyanobacteria. Amsterdam: Elsevier Science Publishers, 1987:385-392.

33. Walsby AE, Kinsman R, George KI. The measurement of gas volume and buoyant density in planktonic bacteria. J Microbiol Meth 1992; 15:293-309.

34. Gosink JJ, Irgens RL, Staley JT. Vertical distribution of bacteria in Arctic sea ice. FEMS Microbiol Ecol 1993; 102:85-90.

35. Lindholm T. Ecological role of depth maxima of phytoplankton. Arch Hydrobiol Beih Ergebn Limnol 1992; 35:33-45.

36. Eggersdorfer B, Häder D-P. Phototaxis, gravitaxis and vertical migrations in the marine dinoflagellate, *Prorocentrum micans*. Eur J Biophys 1991; 85:319-326.

37. Eggersdorfer B, Häder D-P. Phototaxis, gravitaxis and vertical migrations in the marine dinoflagellates, *Peridinium faeroense* and *Amphidinium caterii*. Acta Protozool 1991; 30:63-71.

38. Brodhun B, Häder D-P. Photoreceptor proteins and pigments in the paraflagellar body of the flagellate, *Euglena gracilis*. Photochem Photobiol 1990; 52:865-871.

39. Galland P, Keiner P, Dörnemann D et al. Pterin- and flavin-like fluorescence associ-

ated with isolated flagella of *Euglena gracilis*. Photochem Photobiol 1990; 51:675-680.

40. Häder D-P, Brodhun B. Effects of ultraviolet radiation on the photoreceptor proteins and pigments in the paraflagellar body of the flagellate, *Euglena gracilis*. J Plant Phys 1991; 137:641-646.

41. Brodhun B, Häder D-P. UV induced damage of proteins in the paraflagellar body of *Euglena gracilis*. Photochem Photobiol 1993; 58:270-274.

42. Tirlapur U, Scheuerlein R, Häder D-P. Motility and orientation of a dinoflagellate, *Gymnodinium*, impaired by solar and ultraviolet radiation. FEMS Microbiol Ecol 1993; 102:167-174.

43. Herndl GJ, Müller-Niklas G, Frick J. Major role of ultraviolet-B in controlling bacterioplankton growth in the surface layer of the ocean. Nature 1993; 361:717-719.

44. Lüning K. Meeresbotanik. G Thieme. New York: Stuttgart, 1985.

45. Häder D-P, Schäfer J. *In-situ* measurement of photosynthetic oxygen production in the water column. Environm Monitor Assessm 1994; 32:259-268.

46. Häder D-P, Schäfer J. Photosynthetic oxygen production in macroalgae and phytoplankton under solar irradiation. J Plant Physiol 1994; 144:293-299.

47. Herrmann H, Ghetti F, Scheuerlein R et al. Photosynthetic oxygen and fluorescence measurements in *Ulva laetevirens* affected by solar irradiation. J Plant Physiol 1995; 145:221-227.

48. Häder D-P, Herrmann H, Santas R: Effects of solar radiation and solar radiation deprived of UV-B and total UV on photosynthetic oxygen production and pulse amplitude modulated fluorescence in the brown alga *Padiana pavonia*. FEMS Microbiol Ecol 1996; 19:53-61.

49. Larkum AWD, Wood WF. The effect of UV-B radiation on photosynthesis and respiration of phytoplankton, benthic macroalgae and seagrasses. Photosynth Res 1993; 36:17-23.

50. Maegawa M, Kunieda M, Kida W. The influence of ultraviolet radiation on the photosynthetic activity of several red algae from different depths. Jpn J Phycol 1993; 41:207-214.

51. Döhler G. Effect of UV-B (290-320 nm) radiation on uptake of $^{15}N$-nitrate by marine diatoms. In: Ullrich WR, Rigano C, Fuggi A, Apariciokof JP, eds. Inorganic Nitrogen in Plants and Microorganisms. Uptake and Metabolism. Berlin, Heidelberg, New York: Springer Verlag, 1990:359-354.

52. Kumar A, Kumar HD. Nitrogen fixation by blue-green algae. In: Seu SP, ed. Plant Physiology Research. New Dehli: Society for Plant Physiology and Biochemistry, 1st International Congress of Plant Physiology. 1988:15-22.

53. Donkor VA, Amewowor DHAK, Häder D-P. Effects of tropical solar radiation on the motility of filamentous cyanobacteria. FEMS Microbiol Ecol 1993; 12:143-148.

54. Donkor VA, Amewowor DHAK, Häder D-P. Effects of tropical solar radiation on the velocity and photophobic behavior of filamentous gliding cyanobacteria. Acta Protozool 1993; 32:67-72.

55. Tyagi R, Srinivas G, Vyas D et al. Differential effect of ultraviolet-B radiation on certain metabolic processes in a chromatically adapting *Nostoc*. Photochem Photobiol 1992; 55:401-407.

56. Tyagi R, Kumar HD, Vyas D et al. Effects of ultraviolet-B radiation on growth, pigmentation, $NaH^{14}CO_3$ uptake and nitrogen metabolism in *Nostoc muscorum*. In: Abrol YP, Wattal PN, Gnanam A, Govindjee, Ort DR, Teramura AH, eds. Impact of Global Climatic Changes on Photosynthesis and Plant Productivity. Proceedings of the Indo-US Workshop held on January 8-12, 1991 at New Delhi, India. New Delhi, Bombay, Calcutta: Oxford and IBH Publishing Co, 1991:109-124.

57. Sinha RP, Lebert M, Kumar A et al. Spectroscopic and biochemical analyses of UV effects of phycobiliproteins of *Anabaena* sp. and *Nostoc carmium*. Bot Acta 1995; 108:87-92.

58. Sinha RP, Kumar HD, Kumar A et al. Effects of UV-B irradiation on growth, survival, pigmentation and nitrogen metabolism enzymes in cyanobacteria. Acta Protozool 1995; 34:187-192.

59. Karentz D, Cleaver JE, Mitchell DL. Cell survival characteristics and molecular re-

sponses of Antarctic phytoplankton to ultraviolet-B radiation. Phycol 1991; 27:326-341.

60. Garcia-Pichel F, Castenholz RW. Characterization and biological implications of scytonemin, a cyanobacterial sheath pigment. J Phycol 1991; 27:395-409.

61. Wood WF. Photoadaptive responses of the tropical red alga *Eucheuma striatum* Schmitz (Gigartinales) to ultra-violet radiation. Aquat Bot 1989; 33:41-51.

62. IPCC, Climate change 1992: Houghton JT, Callender BA, Varney SK, eds. The Supplement Report to the IPCC Scientific Assessment. Cambridge: Cambridge University Press, 1992.

63. Weaver AJ. The oceans and global warming. Nature 1993; 364:192-193.

64. Bowes G. Facing the inevitable: plants and increasing atmospheric $CO_2$. Ann Rev Plant Physiol Plant Mol Biol 1993; 44:309-332.

65. Melillo JM, McGuire AD, Kicklighter DW et al. Global climate change and terrestrial net primary production. Nature 1993; 363:234-240.

66. Häder D-P, Worrest RC, Kumar HD et al. Effects of increased solar ultraviolet radiation on aquatic ecosystems. Ambio 1995; 24:174-180.

67. Lampitt RS, Hillier WR, Challenor PG. Seasonal and diel variation in the open ocean concentration of marine snow aggregates. Nature 1993; 362:737-739.

68. Toggweiler JR. Carbon overconsumption. Nature 1993; 363:210-211.

69. Nixon SW. Physical energy inputs and the comparative ecology of lake and marine ecosystems. Limnol Oceanogr 1988; 33: 1005-1025.

70. Gucinski H, Lackey RT, Spence EC. Fisheries. Bulletin Amer Fish Soc 1990; 15:33-38.

71. USEPA (U.S. Environmental Protection Agency). An assessment of the effects of ultraviolet-B radiation on aquatic organisms. In: Assessing the Risks of Trace Gases That Can Modify the Stratosphere, EPA 400/1-87/001C. 1987:1-33.

72. Damkaer DM. Possible influence of solar UV radiation in the evolution of marine zooplankton. In: Calkins J, ed. The Role of Solar Ultraviolet Radiation in Marine Ecosystems. New York: Plenum Press, 1982: 701-706.

73. Siebeck O, Böhm U. Untersuchungen zur Wirkung der UV-B-Strahlung auf kleine Wassertiere. BPT Bericht, Gesellschaft für Strahlen- und Umweltforschung, München. 1987.

74. Williamson CE, Zagarese HE, Schulze PC et al. The impact of short-term exposure to UV-B radiation on zooplankton communities in north temperate lakes. J Plankton Res 1994; 16.

75. Bothwell ML, Sherbot DMJ, Pollock CM. Ecosystem response to solar ultraviolet-B radiation: Influence of trophic level interactions. Science 1994; 256:97-100.

76. Hunter JR, Kaupp SE, Taylor JH. Assessment of effects of UV radiation on marine fish larvae. In: Calkins J, ed. The Role of Solar Ultraviolet Radiation in Marine Ecosystems. New York: Plenum Press, 1982: 459-497.

77. Damkaer DM, Dey DB. UV damage and photo-reactivation potentials of larval shrimp, *Pandalus platyceros*, and adult euphausiids, *Thysanoessa raschii*. Oecologia 1983; 60:169-175.

78. El Sayed SZ. Fragile life under the ozone hole. Natural History 1988; 97:73-80.

79. Maragos JE. A study of the ecology of Hawaiian reef corals. Ph.D. thesis, Univ. of Hawaii. 1972.

80. Jokiel PL, York HR. Solar ultraviolet photobiology of the reef coral *Pocillopora damicornis* and symbiotic zooxanthellae. Bull Mar Sci 1982; 32:301-315.

81. Shick JM, Lesser MP, Stochaj WR. Ultraviolet radiation and photooxidative stress in zooxanthelate anthozoa: the sea anemone *Phyllodiscus semoni* and the octocoral *Clavularia* sp. Symbio 1991; 10:145-173.

82. Gleason DF, Wellington GM. Ultraviolet radiation and coral bleaching. Nature 1993; 365:836-838.

83. Little EE, Fabacher DL. Comparative sensitivity of rainbow trout and two threatened salmonids. Appache Trout and Lahontan Cutthroat Trout, to UV-B radiation. Arch Hydrobiol 1994; Beiheft 43:217-226.

84. Visscher PT, Culbertson CW, Oremland RS. Degradation of trifluoroacetate in oxic and anoxic sediments. Nature 1994; 369: 729-731.

85. Groeneveld AHC, de Kok HAM, van den Berg G. The toxicity of sodium trifluoro-acetate to the alga *Selenastrum capricornutum*. AFEAS Report No. 56635/52/92. 1992.

86. Chumley FG. Workshop on the environmental fate of trifluoroacetic acid (TFA). 3-4 March 1994, Miami Beach, FL. AFEAS Administrative Organization: SPA-AFEAS Inc, Washington 1994.

# CHAPTER 4

# INSTRUMENTATION AND METHODOLOGY FOR ULTRAVIOLET RADIATION MEASUREMENTS IN AQUATIC ENVIRONMENTS

John H. Morrow and Charles R. Booth

## INTRODUCTION

The potential for increasing UV irradiance resulting from ozone depletion has fueled interest in obtaining accurate in-water measurements of ultraviolet (UV) irradiance. Stratospheric ozone absorbs UV wavelengths strongly, resulting in an incident surface spectrum that decreases very rapidly with decreasing wavelength below 340 nm. Relative to visible wavelengths, the UV region of the spectrum is reduced dramatically when propagated through even the clearest ocean waters because of the combined influences of water and particulate absorption and scattering (Fig. 4.1). In the water column, the flux in the UV very quickly becomes a small signal that must be measured in the presence of a much larger visible component. In addition, the flux of UV in natural waters may be influenced greatly by features of the air-water interface, such as changes in surface elevation and focusing and defocusing effects near the interface (Fig. 4.2) caused by waves. Finally, the shape of the spectrum in the UV is radically different from typical calibration sources, resulting in increased potential for error in both the amount and spectral distribution of the flux.[1]

In general, an instrument that accurately measures the flux of UV in air will not work as well when submerged, and a number of elements

The Effects of Ozone Depletion on Aquatic Ecosystems, edited by Donat-P. Häder.

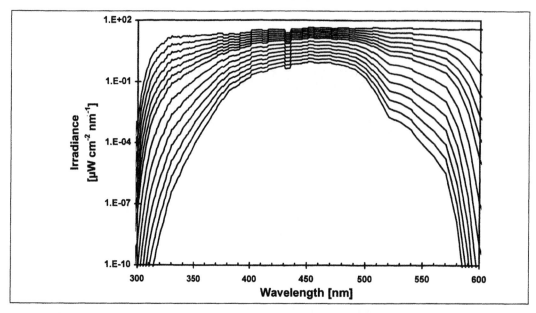

*Fig. 4.1. Theoretical change in spectral distribution for a known irradiance from 300-600 nm in clear ocean waters (modified from Smith and Baker[24,25]). The semilog plot shows modeled irradiance from the surface to 100 m at 10 m intervals and from 100-200 m at 20 m intervals. The dynamic range from 490-350 nm exceeds four orders-of-magnitude in the surface waters. This difference is caused by the optical properties of relatively clear water. If other constituents are added, such as higher concentrations of chlorophyll or dissolved organic matter, the depth of penetration of sunlight is reduced.*

must be optimized to produce instruments for use under water. These include changing the design of the collector to accommodate the refractive index of water, careful specification and quality control of the wavelength selector/detector systems, addition of significant out-of-band blocking to minimize spectral leakage, systematic monitoring of the temperature of the system during deployments, and careful control of dark readings.

Difficulty in obtaining UV irradiance measurements in situ was the subject of a review article[2] concerning photoprocesses in natural waters. In practice, the measurement is not simple to make and readily available instruments are few. Since the early work of Jerlov[3] in the 1950s using a liquid filtered selenium cell, a number of different approaches have been used to measure the flux of UV. In the absence of

specialized instrumentation, much early work was performed using broadband sensors such as the Robertson-Berger (RB) meter, designed for terrestrial studies of human erythema.[4] The fact that until recently commercial UV instrumentation has been relatively unavailable has given rise to a number of ingenious alternative detectors, such as the use of thermopiles,[5] film actinometers,[6] and custom made filter-photodetector instruments.[7,8] There is also increasing interest in the use of biological dosimeters[9] to measure UV effects directly.

Much of the research conducted has emphasized the importance of greater spectral detail, and several different types of radiometers have been used in aquatic environments. In the late 1970s, Smith et al[10,11] measured the spectral penetration of UV (280-340 nm) using an instrument

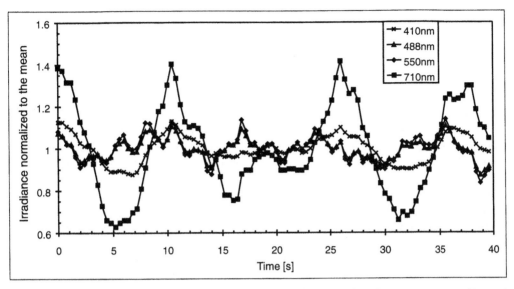

Fig. 4.2. The effect of surface waves on the irradiance flux at several wavelengths over a 40 second interval taken with a MER-1048 spectroradiometer on a mooring off the coast of La Jolla, California.[26] The data are normalized to the mean to emphasize that the effect has a spectral as well as temporal consequence. These fluctuations are caused by changes in the shape and elevation of surface waves,[27,28] and could conceivably impact the measurement depending on the type and time constant of the instrument being used.

equipped with a double monochromator and a UV sensitive photomultiplier tube (PMT). Currently, the LI-1800UW (LI-COR) submersible instrument using a holographic single grating monochromator, order-sorting filter wheel, and silicon photodetector is available commercially for use to depths of 200 m.[12,13] A different approach to obtaining spectral information is used in the PUV-500 Profiling UV radiometer (Biospherical Instruments Inc.).[14,15] The PUV is a battery operated system containing four UV filter-photodetectors centered at 305, 320, 340, and 380 nm as well as PAR (400-700 nm) which are scanned electronically at about two data frames per second. This system may be deployed to 200 m. Still another approach to in-water measurements has been taken by Optronic Laboratories in the OL-740 and 750 series of scanning spectroradiometers. In these instruments, the flux is coupled into a scanning double monochromator through a submersible collector attached to a fiber optic bundle.[15-17] Detailed spectral information concerning irradiance profiles in the Southern Ocean was obtained in support of Icecolors '90[18] and other programs using the Light and Ultraviolet Submersible Spectroradiometer (LUVSS) system developed by Smith et al.[19] Deployed using an remotely operated vehicle (ROV) to reduce interference from the ship, the LUVSS divides the spectrum into two regions (250-380 nm at 0.2 nm resolution and 350-700 nm at 0.8 nm resolution) to address the imbalance in the spectrum between the visible and UV.

Operation and deployment of several types of commercially available UV radiometers was the subject of a recent instrument intercomparison published by Kirk et al[15] and, although manufacturers have introduced improved models subsequent to the intercomparison, there are as yet no completely new additions. Perry Technologies and Optronic Laboratories have recently announced the production and delivery of an underwater housing for the OL-754 which includes a data acquisition module, pressure transducer, and lowering frame. There is no doubt that interest in UV research is sparking production of new, soon-to-be-released instruments which are not included here. Appendix 4.1 briefly presents an overview culled from the literature of the manufacturers and contact information. Although not historically considered submersible, UV instruments from Yankee Environmental Systems and Solar Light Company are included for reference.

Erroneous measurement of the underwater light field can result from the improper use of high quality instruments or from the use of instruments ill suited to work under water. For example, an instrument properly designed with long-wavelength blocking for terrestrial use may respond improperly to increasing long-wavelength flux from outside the desired measurement band as it descends into the water column. This results in overestimation of the actual UV, and, when combined with in situ measurements, underestimation of the consequences of UV exposure. These biased measurements can also affect more traditional laboratory experiments. For example, accurately measured filters are used to attenuate a well defined artificial source to simulate conditions of enhanced UV.[20] When biased in-water measurements are used to interpret these laboratory studies, the in situ consequences of UV exposure will be overestimated. Results such as these will confound all biological studies, such as biological weighting functions, radiation amplification models, and photodegradation and photoprotectant processes.

## SPECTRAL LEAKAGE

The output from a radiometer, V, results from the convolution of the responsivity of the instrument, $R(\lambda)$, with an irradiance spectrum, $E(\lambda)$, over all wavelengths:

$$V = \int_{\lambda=0}^{\infty} E(\lambda)R(\lambda)d\lambda \qquad (1)$$

In practice, the accuracy of the reported irradiance depends upon both the procedure used in calibration and the ability of the sensor to resist the erroneous addition of irradiance from sources that are out of the expected response band (spectral leakage). In the case of spectroradiometers equipped with scanning monochromators, the center wavelength is selected using the monochromator which acts much like a filter.

In the operation of a radiometer, it is often forgotten that all wavelengths of the irradiance spectrum are incident on the sensor, not simply the irradiance from a narrow nominal bandwidth. The absolute spectral responsivity of a sensor, $R(\lambda)$, as a function of wavelength, $\lambda$, is considered to be the spectral product of the response of the detector, $R_{det}(\lambda)$, with the spectral transmission of all optical components, $T_{filt}(\lambda)$, and the diffuser, $T_{diff}(\lambda)$:

$$R(\lambda) = C_r \, R_{det}(\lambda)T_{filt}(\lambda)T_{diff}(\lambda) \qquad (2)$$

$T_{filt}$ in a filter radiometer includes the filter function of any components before the detector, but may equally represent the slit bandwidth and scattering function of a monochromator in a scanning system. $R_{det}(\lambda)$ is frequently determined in practice by irradiating a bare detector with a nearly monochromatic source. Although the relative quantum flux is characterized in this way, the absolute intensity applied to the detector may not be well known due to geometric uncertainties. $C_r$ is a constant that corrects for this geometric uncertainty.

The measured spectra (Fig. 4.3) of the main components of a sensor may be used to model the behavior of the instrument to a known irradiance using Equations 1 and 2. A low-pass filter design was chosen for the 305 nm sensor to maximize sensitivity to the shortest wavelengths influenced by changes in ozone concentration. Given an approach to modeling the vertical

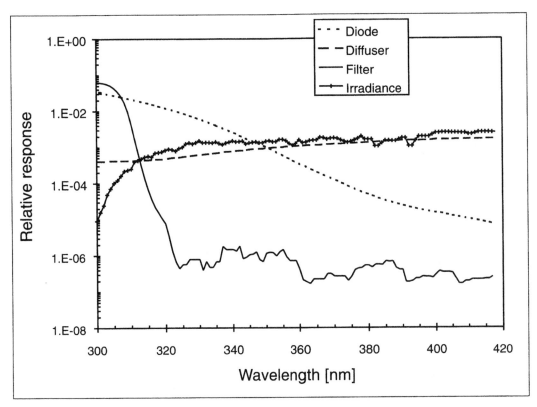

*Fig. 4.3. Main components of the 305 nm sensor in the PUV-500. The response is principally determined by the filter, which is hard-coated and especially formulated to preclude out-of-band leakage in this application. Spectral features above 325 nm are artifacts of the noise floor on the spectrophotometer used to characterize the filter. Additional blocking is achieved via a solar blind photodiode. The GER-23 Teflon diffuser is approximately neutral.*

distribution of spectral irradiance such as is illustrated in Figure 4.1, Equations 1 and 2 may be used to model vertical profiles of reported irradiance, including features such as spectral leakage and shifting of the center wavelength of the sensor as the spectral distribution changes with depth.

To illustrate these effects, a measured irradiance at Ushuaia, Argentina, under clear skies and 32° solar zenith angle from the NSF UV Spectroradiometer Monitoring Network[21] was propagated through a water column using a modified model based on a concentration of chlorophyll and dissolved organic material (DOM) similar to Smith and Baker.[10,11] Underwater, the spectral shape of the irradiance field becomes even more sharply sloped than in air due to attenuation of irradiance by water and its constituents. Where z is depth and

$k(\lambda)$ is the diffuse attenuation coefficient:

$$E(\lambda,z) = R(\lambda)E(\lambda,0)e^{-k(\lambda)z} \qquad (3)$$

In this example (Fig. 4.4), the modeled properties of the water included 5 µg l$^{-1}$ of chlorophyll and 0.5 µg l$^{-1}$ dissolved organic matter. At a depth of 7.6 m in this simulation, the 305 nm signal "reported" from the sensor is down six decades from the signal just beneath the surface. There is no evidence of spurious signals from outside the passband (spectral leakage), but the center wavelength of the detector has shifted one nanometer to the red. This shift results from the change in the spectral distribution of the irradiance as shorter wavelengths are removed with increasing depth.

The results shown in Fig. 4.4 can be contrasted with a simulation made using the same light regime but a different

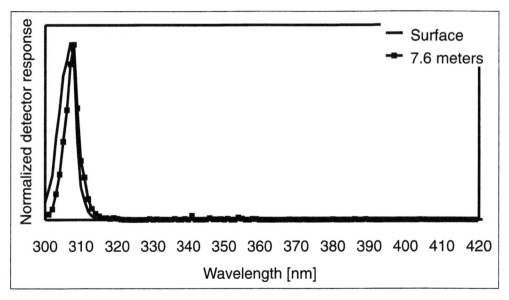

*Fig. 4.4. Modeled response of the 305 nm detector at the surface and at depth. The plots are the product of the incident irradiance and the responsivity of the detector (Eq. 1) at two depths. The shift in the peak wavelength of the sensor is clearly evident. It is also noteworthy that there is no significant prediction of long-wavelength leakage with this sensor (even beyond the plotted range in this figure).*

305 nm detector configuration. Figure 4.5 shows the theoretical result using an 8 nm FWHM detector with a stray light specification of 10-4. The out-of-band filtering is provided by 3 mm of UG-11 filter glass which passes UV wavelengths but blocks the visible region. At the surface, the result from the detector combination strongly resembles that shown in Figure 4.4 and will give an excellent result. There are major differences in the predicted behavior at depth, however: the center wavelength in-band for the detector has shifted three nanometers, but more importantly, greater than 40% of the signal results from wavelengths outside the band. This spectral leakage results from the combined interaction of the change in irradiance distribution and the UV-A passing characteristics of the UG-11 filter.

The question of leakage is not only important for research in the UV. In the past, several ocean optics investigators,

noting unusually high readings from sensors, removed or ignored data which they attributed to spectral leakage but which has been subsequently identified as resulting from the solar-induced fluorescence of the phytoplankton crop. Similarly, variations in the vertical structure of $k(\lambda)$ from ODEX data[22] that were attributed to spectral leakage have been subsequently identified as Raman scattering. The impact of leakage in UV sensors is to overestimate the penetration of UV into the ocean, leading to overestimates of the impact of ozone depletion on the marine food chain. These examples serve to emphasize the degree of performance required of a UV sensor and the consequences of sensor design and calibration on the direction of ocean optics research.

## ARTIFICIAL SOURCES

Sensors designed for proper operation in the ocean may not work well in the

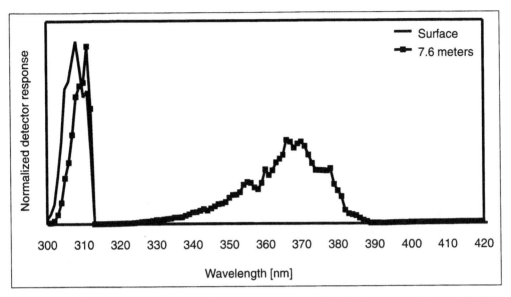

*Fig. 4.5. Modeled response of a 305 nm detector using an 8 nm FWHM detector and 3 mm of UG-11 glass filter. In this simulation, the sensor combination works well at the surface, but the dynamic range difference between 305 and 360-380 nm exceeds the combined stray light and out-of-band blocking at depth, resulting in spectral leakage.*

laboratory. For example, the PUV-500 may produce erroneous results when used to measure artificial sources. These result from potential differences between the spectrum of the source used to calibrate the instrument and the spectral characteristics of the UV lamp. Determination of the "absolute" calibration of a radiometer is frequently conducted by placing the device in front of a Standard of Spectral Irradiance (typically a 1000 watt type FEL lamp) under conditions specified by the National Institute of Standards and Technology (NIST). The output voltage from a sensor is related to the known reference irradiance through a nominal calibration constant, $C_n$:

$$C_n = \frac{V_n - O}{E_l(\lambda_n)} \qquad (4)$$

where $V_n$ is the output voltage of the calibration irradiance, $O$ is the dark offset voltage, $E_l(l_n)$, at the nominal center wavelength of the device, $\lambda_n$. Subsequently, an unknown irradiance may be determined by rearrangement of Equation 4. If the spectral distribution of the calibration source and the source to be measured are relatively flat, then this approach to providing a calibration constant may be adequate. However, in the case where the spectrum is radically different or exhibits significant structure, serious inaccuracies can result from the mismatch between the calibration spectrum and the actual spectrum of the unknown light source. Rather than a filter radiometer, for best results, a scanning spectroradiometer with sufficiently small slits to resolve spectral features in the unknown source should be used to characterize the lamp.

## COSINE COLLECTOR DESIGN

An ideal cosine collector is an aperture which collects all photons transversing the entrance. To work in the ocean, diffusers are often used to approximate this ideal collector. At large zenith angles (large angles of incidence relative to the normal plane of the plate), reflection of light off even the best diffusing materials causes a significant decrease in the flux reaching the photodetector. A wide variety of submersible cosine collector designs have been used, including flat plate, raised plate and variously curved collectors. Largely because of their ease of manufacture, the plate-type collector (Fig. 4.6) is one of the most commonly found in submersible spectroradiometers. The cosine response of the popular raised plate design patterned after the Scripps Spectroradiometer and tested by the Visibility Laboratory (Vislab), Scripps Institution of Oceanography is documented.[23]

A number of factors may contribute to departures from a true cosine response. These include transmission characteristics of the diffusing material at different wavelengths, reflection off the surface at differing angles of incidence, and asymmetric placement of detectors when multiple diodes view a common collector.

As shown in Figure 4.6, the irradiance diffuser of the PUV-500 is formed from a raised trapezoidal quartz piece covered with a thin sheet of vacuum-formed Teflon®, which acts as a diffuser. Photodiodes are arranged in a circular array at the base of the assembly such that all view the same area of the bottom of the diffuser. As with all diffusers, the upper Teflon® surface reflects light at large angles of incidence, significantly underreporting the flux. To mitigate this problem, the sides of the trapezoid are raised to provide a surface with a reduced angle of incidence, thereby increasing the response from larger zenith angles. A cosine collector should not transmit light from angles greater than 90°; and an outer rim is fitted to the irradiance assembly, level to the top of the diffuser to

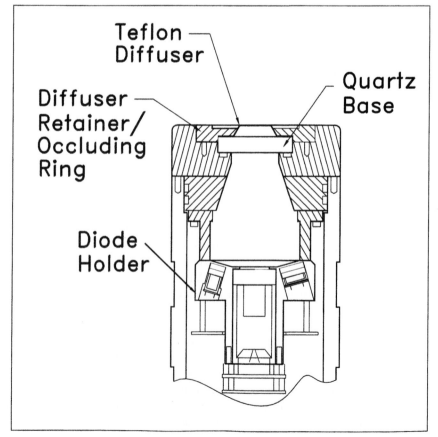

Fig. 4.6. Cross-sectional diagram of the irradiance end piece of the PUV-500.

Teflon Diffuser

Quartz Base

Diffuser Retainer/ Occluding Ring

Diode Holder

act as an occluding ring. In the optimization process, the dimensions of the trapezoid, diffuser retainer, and occluding rings are varied systematically to achieve better cosine response over the spectrum.

In the case of the cosine response of the PUV-500, the instrument was placed on an automated rotating arm in a watertight test tank equipped with nonreflecting sides. A collimated beam was projected to fill the collector at the precise center of rotation of the assembly. Under computer control, the instrument was rotated in 5° increments from +90° to -90° "zenith" angle (ZA) and the response from each channel was recorded relative to the source. In order to test for geometric asymmetries in the response due to the arrangement of the photodiodes, the instrument was turned axially at 10° increments and the cosine response recorded as a function of axial rotation.

The results of just such an analysis for the PUV design show agreement with true cosine of ±10% from +85 to -85° ZA

(Fig. 4.7). The UV-B data shows evidence of some asymmetry in the design, which is most probably a reflection of the placement of the photodiode in the holder.

The impact on the measurement of the design of the cosine collector is subtle. When instruments are deployed in clear waters and small solar zenith angles (SZA), the contribution to the error budget for the measurement from cosine deviation will be small and usually well within the calibration accuracy of the sensor. In contrast, an instrument with poor cosine response when zenith angles exceed 60° may show measurable biases in turbid waters where the signal is diffuse or at higher solar zenith angles. This effect makes comparison of datasets containing data taken at different times of the day or high latitudes much more problematic.

## DARK CORRECTIONS

The output from a photodetector such as those in the PUV-500 is a voltage ($V_n$) which is proportional to the incident

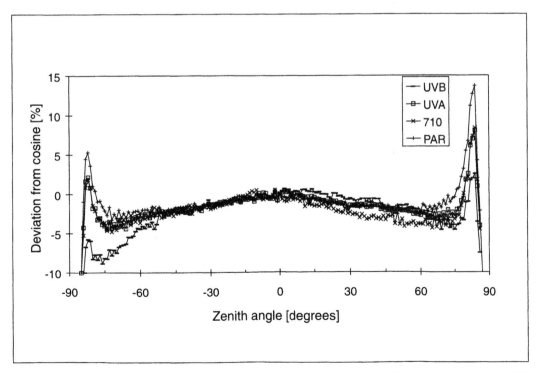

*Fig. 4.7. Percentage deviation from true cosine response for PUV-500 collector. The response is within ±10% of true cosine from ±85° ZA.*

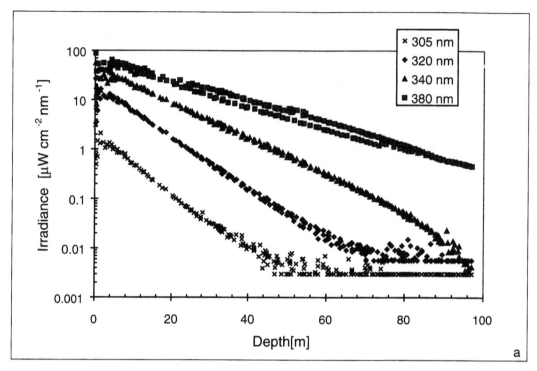

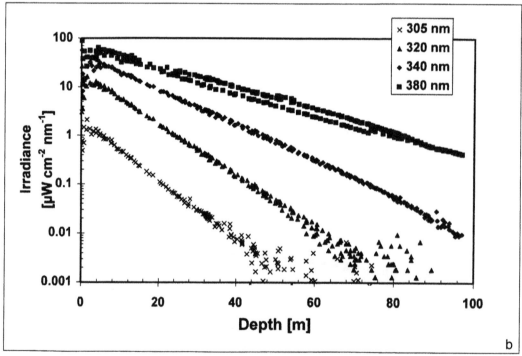

Fig. 4.8. Down- and up-cast from a vertical profile of irradiance for the UV channels of the PUV-500 in the clear equatorial waters off the coast of Nauru (0°32′44″S 166°54′25″E). The water column was isothermal to below 100 m. (a) The change in the trend for the 305 and 320 nm channels below 40 m is an artifact caused by inappropriate values for the zero offset (Eq. 4). (b) The same profiles using measured dark offset voltages. The increasing spread in measured points at the bottom of the sensitivity of the instrument result from single bit shifts in the analog-to-digital converter. The "noise" in the data at the bottom of the sensitivity of the instrument is equally distributed on either side of the trend.

irradiance. This voltage is converted into the appropriate irradiance ($E(\lambda_n)$) by rearrangement of Equation 4. The dark offset is the output voltage from the sensor in the absence of light. By design, the dark offset voltage is typically a small value, on the order of the lowest resolvable signal by the system. This value is not a constant, but varies primarily as a function of the ambient temperature of the electronic components. Thus, the value of the dark voltage will typically vary slightly under field conditions.

Figure 4.8 shows the impact of using a correct zero offset on a typical profile of irradiance versus depth for the four UV channels of the PUV-500. The profiles using the offset measured in the darkroom at Biospherical Instruments tail noticeably and never actually result in near-zero irradiances (Fig. 4.8a). When dark values measured in the field are used (Fig. 4.8b), the profiles appear much more linear and noise at the bottom of the sensitivity of the instrument is distributed more equally across the trend. The 305 nm channel shows measurable penetration of some of the shortest and most damaging wavelengths of UV down to 40 m. The increase in irradiance near the surface occurs as the instrument exits the shadow of the ship. Features near the bottom of the usable range represent the noise floor for each channel. It is important to note that in general the slopes of the profiles for all channels are generally constant throughout the usable range of each sensor, an indication of minimum spectral leakage.

All electrical and optical components have temperature coefficients—even those that "should not." Dark offset values are influenced by temperature, and for the most demanding work, it may be necessary to monitor the temperature of the electronic array board to develop an empirical response change to temperature changes caused by features in the water column.

## CONCLUSION

Accurate and precise measurements of UV in aquatic environments are difficult to make. Instruments for this application are becoming commercially available (see Appendix 4.1). However, instrument designs tend to reflect compromises between a number of competing design goals, such as spectral resolution, sensitivity, stability, calibration, ease of deployment, cost, and data handling. To ensure the quality of the data, it is as important for the UV researcher as it is for the instrument manufacturer to understand the difficulties inherent in the measurement and the strengths and weaknesses of the instruments being used.

## APPENDIX 4.1: COMMONLY AVAILABLE COMMERCIAL INSTRUMENTS

The following appendix contains an alphabetical listing of manufacturers of UV instruments available for use in-water. It is not an exhaustive list, but is the result of references collected from the published literature during the course of research for this publication. No manufacturer has been purposely shunned and not all products are listed. No statement is implied about the fitness of any specific product for an application.

### Biospherical Instruments PUV-500 Profiling UV Radiometer

Battery powered, narrow bandwidth filter radiometer. Custom filters/photodetectors at 305, 320, 340 and 380 nm as well as PAR (400-700 nm), temperature and pressure/depth. Single Teflon® diffuser. Measures UV and visible light at depths up to 200 meters.

*Biospherical Instruments Inc.*
*5340 Riley Street*
*San Diego, CA 92110-2621 USA*
*VOX: 619-686-1888*
*FAX: 619-686-1887*
*email: sales@biospherical.com*
*WWW: http://www.biospherical.com*

## International Light IL1776 Ozone Depletion/Erythemal Radiometer System with SUV240/SPS300/W Super Solar Blind Detector

Narrowband filter detector for measuring underwater (maximum depth 120 feet) irradiance. 297 nm center wavelength using a quartz wide-eye diffuser, narrowband interference filters and a Solar Blind Vacuum Photodiode.

*International Light Inc.*
*17 Graf Road*
*Newburyport, MA 01950-4092 USA*
*VOX: 508-465-5923;*
*FAX: 508-462-0759*

## LI-COR LI-1800UW Underwater Scanning Spectroradiometer

Holographic, single grating monochromator, order sorting filter wheel, cosine corrected silicon detector. Measures irradiance from 300-850 nm or from 300-1100 nm. Operates to depths of 200 meters.

*LI-COR, Inc.*
*P.O. Box 4425*
*Lincoln, NE 68504 USA*
*VOX: 402-467-3576*
*FAX: 402-467-2819*

## Optronic Laboratories OL754 Scanning Spectroradiometer

Double grating monochromator for high resolution measurements of irradiance in the UV region. Can be fitted with an OL IS-470-WP submersible integrating sphere and OL 730-7Q-WP waterproof fiber optic probe for making spectral scans underwater. Recently, OL and Perry Technologies announced a submersible housing system for the 754 called the Subsea Scanning Spectroradiometer System that can be deployed in depths up to 100 meters.

*Optronic Laboratories, Inc.*
*4470 35th Street*
*Orlando, FL 32811 USA*
*VOX: 407-422-3171*
*FAX: 407-648-5412*

## Solar Light Co. Solar UV Biometer

Pyranometer type instrument that measures irradiance near the Erythema Action Spectrum (290-320 nm). Measures biologically effective UV-B to depths of 5 meters.

*Solar Light Co., Inc.*
*721 Oak Lane*
*Philadelphia, PA 19126 USA*
*VOX: 215-927-4206*
*FAX: 215-927-6347*

## Yankee Environmental Systems UV Pyranometer

Uses colored glass optical filters, a UV sensitive fluorescent phosphor to convert UV light to visible light, and a solid state photodetector to measure irradiance from 280-330 nm (model UV-A-1) or from 300-390 nm (model UV-B-1). Primarily used to make atmospheric measurements of UV irradiance, but can be submersed to 1 meter to make underwater measurements.

*Yankee Environmental Systems, Inc.*
*P.O. Box 746*
*Turner Falls, MA 01376 USA*
*VOX: 413-863-0200*
*FAX: 413-863-0255*
*e-mail: yankee@risky.ecs.umass.edu*

## REFERENCES

1. Booth CR, Mestechkina T, Morrow JH. Errors in reporting of solar spectral irradiance using moderate bandwidth radiometers: an experimental investigation. SPIE Ocean Optics XII 1994; 2258:654-663.
2. Zafiriou OC, Joussot-Dubien J, Zeep RG et al. Photochemistry of natural waters. Environ Sci Technol 1984; 18:358A-371A.
3. Jerlov NG. Ultraviolet radiation in the sea. Nature 1950; 166:111.
4. Calkins J. Measurements of the penetration of solar UV-B into various natural waters. Impacts of climatic change on the biosphere. Climatic Impact Assessment Program (CIAP) Monogr 1975; 5:267-296.
5. Flocchini G, Picco P. On the attenuation of solar radiation in the sea. Bolletino di Oceanologia Teorica ed Applicata 1991; IX:15-20.

6. Fleischman EM. The measurement and penetration of ultraviolet radiation into tropical marine water. Limnol Oceanogr 1989; 34:1623-1629.

7. Ryan KG, Beaglehole D. Ultraviolet radiation and bottom-ice algae: laboratory and field studies from McMurdo Sound, Antarctica. Ultraviolet Radiation in Antarctica: Measurements and Biological Effects. In: Weiler CS, Penhale PA, eds. Ultraviolet radiation in Antarctica: measurements and biological effects. Antarctic Research Series, 1994; 62:229-242.

8. Herndl GJ, Muller-Niklas G, Frick J. Major role of ultraviolet-B in controlling bacterioplankton growth in the surface layer of the ocean. Nature 1993; 361:717-719.

9. Karentz D, Lutze LH. Evaluation of biologically harmful ultraviolet radiation in Antarctica with a biological dosimeter designed for aquatic environments. Limnol Oceanogr 1990; 35:549-561.

10. Smith RC, Baker KS. Optical classification of natural waters. Limnol Oceanogr 1979; 23:260-267.

11. Smith RC, Baker KS. Optical properties of the clearest natural waters (200-800 nm). Appl Opt 1981; 20:177-184.

12. Smith RC, Ensminger RL, Austin RW. Ultraviolet submersible spectroradiometer. SPIE Ocean Optics VI 1979; 208:127-140.

13. Smith RC, Baker KS, Holm-Hansen O et al. Photoinhibition of photosynthesis in natural water. Photochem Photobiol 1980; 31:585-592.

14. Behrenfeld M, Hardy J, Gucinski H et al. Effects of ultraviolet-B radiation on primary production along latitudinal transects in the South Pacific Ocean. Marine Environmental Res 1993; 35:349-363.

15. Kirk JTO, Hargreaves BR, Morris DP. Measurements of UV-B radiation in two freshwater lakes: an instrument intercomparison. In: Impact of UV-B radiation on pelagic freshwater ecosystems. Williamson CE, Zagarese HE, eds. Archiv Hydrobiol. Ergebnisse der Limnologie 1994; 43:71-99.

16. Booth CR, Morrow JH, Neuschuler DA. A new profiling spectroradiometer optimized for use in the ultraviolet. SPIE Ocean Optics XI 1992; 1750:354-365.

17. Ilya R, Vasseliev, Prasil O et al. Inhibition of PS II photochemistry by PAR and UV radiation in natural phytoplankton communities. Photosynthesis Research 1994; 42:51-64.

18. Piazena H, Häder D.-P. Penetration of solar UV irradiation in coastal lagoons of the Southern Baltic Sea and its effect on phytoplankton communities. Photochem Photobiol 1994; 60:463-469.

19. Scully NM, Lean DRS. The attenuation of ultraviolet radiation in temperate lakes. In: Impact of UV-B radiation on pelagic freshwater ecosystems. Williamson CE, Zagarese HE, eds. Archiv Hydrobiol. Ergebnisse der Limnologie 1994; 43:135-144.

20. Prezelin BB, Boucher NP, Smith RC. Marine primary production under the influence of the Antarctic ozone hole: Icecolors '90. Ultraviolet Radiation in Antarctica: Measurements and Biological Effects. In: Williamson CE, Zagarese HE, eds. Impact of UV-B radiation on pelagic freshwater ecosystems. Antarctic Research Series 1994; 62:159-186.

21. Smith RC, Prezelin BB, Baker KS. Ultraviolet radiation and phytoplankton biology in Antarctic waters. Science 1992; 255:952-959.

22. Booth CR, Mitchell BG, Holm-Hansen O. Development of a moored spectroradiometer. SBIR Data Report No. 1. Biospherical Instruments Inc. 69 pp. 1987.

23. Stramski D, Booth CR, Mitchell BG. Estimation of downward irradiance attenuation from a single moored instrument. Deep-Sea Research 1992; 39:567-584.

24. Dera J, Stramski D. Maximum effects of sunlight focusing under a wind-disturbed sea surface. Oceanologia 1986; 23:15-42.

25. Neale PJ, Lesser MP, Cullen JJ. Effects of ultraviolet radiation on the photosynthesis of phytoplankton in the vicinity of McMurdo Station, Antarctica. Ultraviolet radiation in Antarctica: measurements and biological effects. Weiler CS, Penhale PA, eds. Antarctic Research Series. 1994; 62:125-142.

26. Booth CR, Lucas TB, Morrow JH. The United States National Science Foundation's polar network for monitoring ultraviolet

radiation. In: Ultraviolet radiation in Antarctica: measurements and biological effects. Weiler CS, Penhale PA, eds. Antarctic Research Series. 1994; 62:17-37.

27. Siegel DA, Booth CR, Dickey TD. Effects of sensor characteristics on the inferred vertical structure of the diffuse attenuation coefficient spectrum. SPIE Ocean Optics VIII 1986; 637:115-123.

28. Smith RC. An underwater spectral irradiance collector. Journal of Marine Research 1969; 27:341-351.

# CHAPTER 5

# PENETRATION OF SOLAR UV AND PAR INTO DIFFERENT WATERS OF THE BALTIC SEA AND REMOTE SENSING OF PHYTOPLANKTON

Helmut Piazena and Donat-P. Häder

## INTRODUCTION

To determine the effects of solar UV and photosynthetic active radiation (PAR) on marine ecosystems and phytoplankton two basic problems have to be analyzed: (1) the penetration of UV and PAR into the water column, and (2) the vertical and horizontal distribution of the affected organisms.

### PENETRATION OF SOLAR RADIATION (OPTICAL PROPERTIES)

Marine waters show large regional and temporal differences in their concentrations of dissolved and particulate substances such as nutrients, chlorophyll *a*, seston, yellow substance and detritus as well as of microorganisms such as phytoplankton and bacteria. These substances and organisms cause a selective spectral attenuation of solar radiation (UV and PAR) penetrating the water column by absorption and scattering which depends on the concentration of each component. Whereas yellow substance is the main absorber of short-wavelength radiation (UV-B and UV-A), seston and detritus attenuate optical radiation by absorption and scattering both in the UV and PAR.[1-3] In contrast, phytoplankton shows different spectral maxima of absorption as well as of fluorescence, depending on the individual content of chlorophyll and accessory pigments.[1-5]

*The Effects of Ozone Depletion on Aquatic Ecosystems*, edited by Donat-P. Häder.

Several systems of classification have been established to characterize the optical properties of marine waters.[1,6] The original Jerlov system of classification divides marine waters into 3 oceanic and 9 coastal types, showing typical differences in their transmission spectra.[7-9]

The basis of this characterization is the analysis of spectral attenuation coefficients which consider both absorption and scattering processes. These coefficients can be derived from irradiance measurements at different depths under the condition of vertical homogeneous distribution of attenuating substances.

Most recent investigations are based on measurements of downward irradiance on a horizontal surface using a cosine corrected receiver.[1,6,10-14] However, due to multiple scattering by water molecules, particles and organisms, not only absorption and forward scattering, but also backscattering processes can contribute significantly to the total irradiance in the water column.[1,14] Thus, in contrast to downward irradiance, data of *scalar* or *spherical measurements* yield more relevant information to characterize systemic effects of solar radiation upon organisms like phytoplankton.[15-24]

## BIO-OPTICAL CHARACTERISTICS

Essential differences between the different water types are found in the UV and in the PAR ranges which can be quantified by analysis of the attenuation coefficients or (equivalently) by the depth of the 0.1% value of surface irradiance. The ratio between the 0.1% depths for UV-B and PAR can be used to estimate damaging effects on phytoplankton by solar UV-B radiation which hits the organisms in the euphotic zone where they dwell in order to optimize their photosynthesis. (Note that the depth of the euphotic zone is defined by the 0.1% depth of PAR.)

A more precise estimation of the effects of solar UV-B radiation on phytoplankton can be based on analysis of the bio-optical characteristics of the water column. These parameters are derived mathematically from spectral irradiance data at different depths weighted by the action spectra of interesting photobiological effects.[10,11,25]

## DETECTION OF PHYTOPLANKTON AND OF OPTICAL CHARACTERISTICS BY BACKSCATTERING MEASUREMENTS

To estimate radiation effects on phytoplankton populations, information about the vertical and horizontal distribution of the organisms is essential. The vertical organism distribution can be measured easily by synchronous sampling of organisms at different depths. But, single measurements are not sufficient to estimate the total number of organisms contained in extended waters. In this case remote sensing by aircraft or satellite is a promising method to monitor both the regional population of phytoplankton, their photosynthetic productivity stimulated by solar PAR, their damage by solar UV-B as well as the optical and bio-optical characteristics of the waters.[1,14]

Remote sensing is based on the determination of spectral backscattering irradiance. However, the backscattered signal not only depends on the concentration of organisms but also on their vertical distribution in the water column, on the water type and on the individual content of chlorophyll and accessory pigments in the species that establish the population.[26]

To evaluate the method of remote sensing for its applicability for monitoring of phytoplankton not only at the near surface layer but also in the euphotic zone, the influence of vertical stratification of the organisms and the water type on the backscattered spectral irradiance at the sea surface is analyzed. Following ground truth, relationships are derived for remote sensing of both optical properties of the waters and the vertical stratification of the phytoplankton.

The investigations reported here were performed in different waters of the Baltic Sea and in the offshore zone of the Mediterranean. The Baltic Sea and its lagoons are typical coastal waters with comparatively high concentrations in both dissolved and particulate matter as well as chlorophyll *a* and phytoplankton modulating the spectral transparency and reflectance. The optical properties show strong horizontal changes due to a limited water exchange with the North Sea considerable freshwater inflow from rivers containing nutrients, suspended and dissolved matter. The lagoons form a series of shallow water bodies with or without water exchange with the open Baltic Sea. In contrast, the waters of the Mediterranean Sea belong to the oceanic type, and show a comparatively high degree of transparency due to low densities of attenuating substances and microorganisms.

## MATERIALS AND METHODS

### STATIONS AND PERIODS OF MEASUREMENTS

The data to analyze optical and bio-optical properties presented here were sampled in the near shore zone (Darßer Ort/peninsula of Zingst, Gullmarsfjorden/Kattegat, Sweden) and in different coastal lagoons of the southern Baltic Sea (Greifswalder Bodden: central lagoon, Vierow/harbor, Gager, Zicker; Strelasund: Barhöft; Barther Bodden: Zingster Strom) during spring and summer 1994. For comparison, additional investigations were performed in the western (Arkona Sea) and central Baltic Sea (station BY8A) during the GOBEX'94 expedition on board the research vessel "Alexander von Humboldt" in September 1994 and in the off shore zone of the Mediterranean Sea near Malaga, Spain in March 1995.

Backscattering and ground truth measurements for remote sensing of phytoplankton and optical properties were performed in different coastal lagoons of the southern Baltic Sea during spring and summer 1994. The lagoons involved in the investigation are: Greifswalder Bodden (stations: FTL, Fahrwassertonne Lauterbach; P, Palmerort; SG and K, Salzboddengrund), Barther Bodden (station: buoy R1) and the lagoon between the islands of Hiddensee and Rügen (station: RS, Rassower Strom). In addition, analogous measurements were performed about one sea mile off the coast of Malaga, Spain during March 1995.

### OPTICAL AND BIO-OPTICAL CHARACTERISTICS

The irradiance data were measured by using a double monochromator spectroradiometer (type 752, Optronic Laboratories, Orlando, FL, USA) equipped with an Ulbricht sphere as optical head for the irradiance measurements above sea surface and a $4\pi$ sensor as receiver for the measurements of scalar irradiance and downward scalar irradiance in the water column developed for the spectroradiometer. The optical spherical entrance of the $4\pi$ sensor is realized by the polished ends of 72 quartz fibers (250 µm diameter) which were oriented radially and distributed isotropically over the surface of a sea water resistant PVC hollow sphere (100 mm diameter) filled with an elastomere. To measure the downward scalar irradiance the lower hemisphere of the $4\pi$ sensor was covered by a thin hemispherical PVC shield painted black on the inside. The quartz fiber bundle had a length of 20 m. The cross section of the other end of the cable matched the entrance slit of the spectroradiometer. The deviation of the spherical response of the $4\pi$ receiver was smaller than $\pm15\%$ (Fig. 5.3). In order to calibrate the spherical sensor, an Ulbricht sphere with an inner diameter of 55 cm was produced internally coated with $BaSO_4$. Before starting each series of measurements the wavelength stability of the spectroradiometer was checked using a mercury calibration lamp (Optronic Laboratories, Orlando, FL, USA). The duration of each spectrometer run between 290 and 800 nm was about 4 min. To exclude errors due to significant changes of the astronomical and meteorological conditions, the series of spectral measurements of the solar irradiance at different depths in the water column were performed between 11 and 13 h local time and under nearly constant atmospheric transparency. The error in the irradiance measurements is about $\pm15\%$ causing uncertainties in the calculated attenuation coefficients of about $\pm20\%$.

Furthermore, the visual depth of the water column was measured by using a Secchi disk after each series of spectrometric measurements as an additional parameter characterizing the optical properties.

## GROUND TRUTH MEASUREMENTS

### a) Spectral backscattering of solar radiation

Simultaneous measurements were performed during noon time under cloudless skies in two ways:

- by using the Optronic spectroradiometer with an Ulbricht sphere and the $4\pi$ sensor to measure the spectral reflectance at the surface and the optical properties in the water column at different stations directly; and
- by using the backscattering data sampled in different wavelength channels by a CZCS sensor installed on board of an aircraft which flew over the investigated waters.

### b) Vertical distribution of phytoplankton

The vertical distribution of phytoplankton in the column of deep waters was estimated by the analysis of 20 water samples of 800 ml, each which were sampled synchronously at equidistant intervals of 32 cm between surface and 6 m depth by electric pumps (type 1004, Comet) submerged into the water during 1 min.[27] In shallow waters the samples were taken between surface and ground in the same way.

The organisms contained in each sample were concentrated by centrifugation (Minifuge RF, Heraus Sepatech, rotor BS 4402/A, 2 min at 2000 rpm) or by tangential flow filtration (Filtron) in a volume of 1 ml. Cell density was determined by automatic image analysis using a video digitizer (PIP 1024 B, Matrox, Quebec, Canada) housed in an AT type computer (486 CPU) running the program COUNTC.[28] Two aliquots of each sample were filled in a quartz cuvette (1 mm deep, Hellma, Mülheim), and five counts were performed for each sample, resulting in 10 measurements of which means and standard deviations were calculated.

## BASIC PARAMETERS, DEFINITIONS AND EQUATIONS

The nomenclature used here agrees with the recommendation by the IAPSO (International Association of the Physical Sciences of the Ocean) Working Group on symbols, units and nomenclature in Physical Oceanography.[1]

### BASIC PARAMETERS TO CHARACTERIZE OPTICAL PROPERTIES

- Spectral irradiance at depth $z_i$ in the water column

$$E_\lambda(z_i) = E_\lambda(z_{i-1}) \cdot e^{-K_\lambda \cdot d} \qquad (1)$$

where $K_\lambda$ denotes the spectral attenuation coefficient and

$$d = \Delta z / \cos \beta \qquad (2)$$

defines the optical effective depth characterizing the optical pathway in the water if the direct component of solar irradiation enters the water surface at an elevation which is different from $90°$ and the diffuse component is not isotropic.

$$\beta = \arc \sin \left[ \sin(90° - \gamma)/\eta \right] \qquad (3)$$

in Eq. 2 denotes the angle of direct solar radiation in the water column ($\gamma$, solar elevation above horizon; $\eta \approx 1.33$, the refractive index of optical radiation at the boundary between atmosphere and sea water resulting from Snellius' law).

$\Delta z = z_i - z_{i-1}$ defines the difference between the depths $z_i$ and $z_{i-1}$. Note that the ratio $d/\Delta z$ increases from 1.00 at $90°$ solar elevation to 1.22 at $40°$ and to 1.49 at $10°$ elevation above the horizon. Thus, the increase in the optical effective depth in comparison with vertical depth, $\Delta z$, is small and mostly negligible during noon time at lower and mid latitudes in summer, but plays a larger role in the morning and evening as well as at high latitudes.

Eq. 1 has to be modified in dependence of "geometry of viewing" to define

- Spectral downward (upward) irradiance $E_{dl}$ ($E_{ul}$) which characterizes the spectral radiant flux on a horizontal area from the upper (lower) hemisphere using $K_{dl}$ ($K_{ul}$) as spectral attenuation coefficients and

- Spectral scalar irradiance $E_{o\lambda}$ describing the spherical integral of the spectral radiant flux at and around a point with $K_{o\lambda}$ as spectral attenuation coefficient.
- Spectral scalar downward (upward) irradiance $E_{od\lambda}$ ($E_{o\lambda}$) with the attenuation coefficients $K_{od\lambda}$ ($K_{ou\lambda}$).

The spectral attenuation coefficient $K_\lambda$ includes the attenuation of spectral irradiance due to absorption as well as forward and backward scattering processes caused by all substances ($\mu$) contained in the water column (including water molecules)

$$K_\lambda = \Sigma_\mu \, k_\lambda(\mu) \qquad (4)$$

($\mu$: water molecules, dissolved organic matter such as yellow substance, microorganisms such as phytoplankton and bacteria as well as particulate inorganic matter and detritus).

Furthermore, $K_\lambda$ depends on the geometry of viewing and is defined by

$$K_\lambda = -(E_\lambda)^{-1} \cdot (\delta E_\lambda/\delta z) \qquad (5)$$

using infinitesimal small quantities. Using data of field measurements $K_\lambda$ can be derived according to

$$K_\lambda = -d^{-1} \cdot \ln (E_\lambda(z_i)/E_\lambda(z_{i-1})) \qquad (6)$$

under the assumption of vertical homogeneity within the depth d. Using Eq. 1

- Spectral transmission of a water column limited by two different depths $z_i$ and $z_{i-1}$ is defined by

$$T_\lambda = E_\lambda(z_i)/E_\lambda(z_{i-1}) = e^{-K_\lambda \cdot d} \qquad (7)$$

For comparability, transmission data are normalized usually to d = $\Delta z/\cos \beta$ = 1.0 m. Transmission mostly is determined from measurements of downward irradiance data. Hence, its data can also be based on data of scalar as well as of scalar downward irradiance.

## BIO-OPTICAL CHARACTERISTICS

To characterize the bio-optical properties of waters three parameter types are used:

1. parameters to determine penetration within broadband ranges such as UV-B, UV-A and PAR
2. percent depth of surface irradiance and
3. parameters of penetration for photobiologically effective radiation.

Characteristics of broadband penetration and penetration of photobiological effective radiation (transmission and attenuation coefficients) can be determined directly by broadband measurements of irradiance

$$E_{\Delta\lambda} = \Sigma_\lambda \, E_\lambda \cdot \delta\lambda \qquad (8)$$

and by measurements or calculations of efficiency

$$\epsilon_{\Delta\lambda} = \Sigma_\lambda \, A_\lambda \cdot E_\lambda \cdot \delta\lambda \qquad (9)$$

at different depths using Eq. 1 or by the arithmetic mean of the attenuation coefficient between the lower ($\lambda_1$) and upper limit ($\lambda_2$) of broadband range (*solar UV-B*: $\lambda_1$= 290 nm, $\lambda_2$ = 315 nm; *UV-A*: $\lambda_1$= 315 nm, $\lambda_2$ = 400 nm; *PAR*: $\lambda_1$= 400 nm, $\lambda_2$ = 700 nm; $A_\lambda$: action spectrum of any effect, $\lambda = \lambda_1 + j \cdot \delta\lambda$, j = 1 ,..., m; $\delta\lambda$ spectral step).

Thus, the attenuation coefficient for broadband irradiance is defined by

$$K_{\Delta\lambda} = -d^{-1} \cdot \ln (E_{\Delta\lambda}(z_i)/E_{\Delta\lambda}(z_{i-1})) \qquad (9a)$$

whereas the attenuation coefficient for photobiological effective radiation is given by

$$K_{\Delta\lambda} = -d^{-1} \cdot \ln (\epsilon_{\Delta\lambda}(z_i)/\epsilon_{\Delta\lambda}(z_{i-1})) \qquad (9b)$$

Note that $K_{\Delta\lambda}$ for both broadband and effective radiation can be calculated easily by using the data of Eq. 6 according to

$$K_{\Delta\lambda} = m^{-1} \cdot \Sigma_\lambda \, K_\lambda \qquad (10)$$

if the spectral limits are chosen to the limits of the broadband range or to the spectral limits of the action spectrum. This way broadband transmission is given by

$$T_{Dl} = E_{Dl}(z_i)/E_{Dl}(z_{i-1}) = e^{-K_{Dl} \diamond d} \quad (11)$$

0.1% depth of solar surface irradiance, often used to characterize the photo-biologically relevant depth of penetration, can be derived from Eqs. 1 and 6 in the case of vertical homogeneity as a function of wavelength and water type according to

$$d_{01, \lambda}(J) = - (K_\lambda(J))^{-1} \cdot \ln 0.001 \ (12)$$

Eq. 12 defines the depth of the euphotic zone if the attenuation coefficient is determined in the PAR range

$$d_{01,PAR}(J) = - (K_{PAR}(J))^{-1} \cdot \ln 0.001 \quad (13)$$

(In contrast, $d_{01}$ exceeds the 1% depth $d_1$ by a factor of 1.5, which earlier defined the euphotic zone in the PAR range.)

## PARAMETERS FOR BACKSCATTERING MEASUREMENTS AND REMOTE SENSING

Ground truth algorithms usually are related to the reflectance ratio

$$X = F_R(\Lambda_1)/F_R(\Lambda_2) \quad (14)$$

of two characteristic bands ($\Lambda_1$ and $\Lambda_2$) according to

$$Y = p \cdot X^q \quad (15)$$

where p and q are empirical coefficients and $F_R$ is the spectral backscattered (upward) signal at the sea surface. These equations were derived to estimate both the optical water characteristics and the concentrations of seston and chlorophyll *a*. Usually they use backscattering information under the assumption of well mixed waters.[2,3,29]

Using the spectral attenuation coefficients defined in Eq. 6, the upwelling solar irradiation at the sea surface which is backscattered by phytoplankton and seston at different depths $z_i$ of the water column of Jerlov type J can be calculated as

$$F_R(J,\lambda,N,F_0,R) = \Sigma_i f_R(z_i) \cdot \Delta z$$
$$= F_0(\lambda) \cdot \Delta z \cdot \Sigma_i (\Sigma_k r_k(z_i,\lambda) \cdot n_k(z_i)) \cdot$$
$$e^{-2K_\lambda(J) \cdot z_i} \quad (16)$$

The spectral upwelling signal at the sea surface $F_R(\lambda)$ depends on the downwelling irradiance at the surface $F_0(\lambda)$, on the attenuation coefficient $K_\lambda$ (J) which is a function of wavelength $\lambda$, on the Jerlov water type J, and on the kind of species (denoted by index k), their spectral individual reflectance $r_k(\lambda)$ and their concentration at every depth $n_k(z_i)$.

The total number of organisms which populate the phytoplankton community in the water column is given by

$$N = \Sigma_i n(z_i) \cdot \Delta z, \qquad i = 0, ..., g \quad (17)$$

where

$$n(z_i) = \Sigma_k n_k(z_i) \quad (18)$$

defines the sum of all kinds of species at the depth $z_i$. (The investigated water column is limited by a defined horizontal area (of horizontal homogeneity) and the vertical extent between surface and the depth $z_g$. $\Delta z$ denotes the length element of vertical distance of sampling.)

The influence of both attenuation of radiation in the water column and of the vertical distribution and concentration of the organisms is calculated for different central wavelengths of the CZCS channels (440, 515, 560 and 680 nm) by analysis of the ratio

$$s(\lambda,z_i) = f_R(\lambda,z_i)/(F_0(\lambda) \cdot R(\lambda))$$
$$= n(z_i) \cdot e^{-2K_\lambda(J) \cdot z_i} \quad (19)$$

which determines the relative spectral signal at the sea surface backscattered from the number of phytoplankton organisms n populating the community at depth $z_i$ and normalized to the spectral values of down-

*Fig. 5.1. The attenuation coefficient (a) and the 0.1% depth of subsurface irradiance (b) as a function of wavelength for different oceanic (O: I, IA, IB, II, III) and coastal water types (C: 1-9) in the original Jerlov system of classification.[1]*

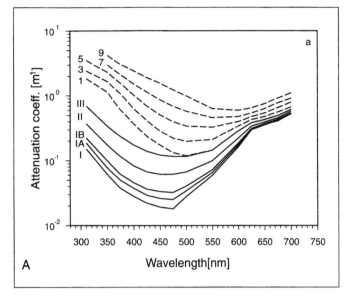

A

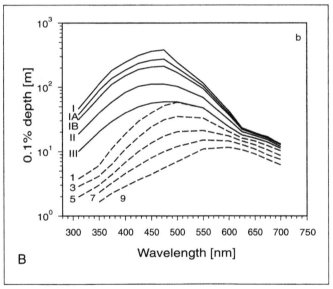

B

welling irradiance $F_0(\lambda)$ and individual reflectance of the organisms $R(\lambda)$. Eq. 19 was derived from Eq. 16 under the simplifying assumptions of:

- uniformity of the spectral reflectance of the phytoplankton community distributed at different depths of the column caused by statistical uniformity of the species composition $[R(\lambda) = \Sigma_i \Sigma_k r_k(\lambda) \cdot n_k(z_i) \cdot \Delta z/N]$;
- up- and downwelling of radiation perpendicular to the surface and neglecting of surface reflection; and
- neglecting multiple backscattering effects caused by phytoplankton at different depths.

Using Eq. 19 the upwelling signals at the sea surface which were backscattered in the column between surface and the depth $z_j$ are calculated by

$$\Delta S(\lambda, z_j) = \Sigma_i \ s(\lambda, z_i) \cdot \Delta z, \quad i = 0, \dots, j \tag{20}$$

whereas the number of organisms contained in this column is given by

$$\Delta N(z_j) = \Sigma_i \ n(z_i) \cdot \Delta z, \quad i = 0, \dots, j \tag{21}$$

$(j < g)$

With reference to the total backscattered signal at the surface

$$S(\lambda) = \Sigma_i \ s(\lambda, z_i) \cdot \Delta z \tag{22}$$

and the total number of phytoplankton organisms N which were sampled between surface and $z_g = 6$ m depth (or ground) the percentage contributions backscattered between surface and depth $z_j$ ( $|z_j| \leq |z_g|$ ) are given by

$$\Delta S(\lambda, z_j)/S(\lambda) \cdot 100 \text{ and } \Delta N(z_j)/N \cdot 100 \tag{23}$$

Fig. 5.2. Spectral distribution of scalar solar irradiance at a series of depths in different marine waters during local noon and cloudless sky. Stations and visual depths: (a) Mediterranean Sea, 2 km off Malaga (Spain), 15 March 1995, Secchi depth: 12.0 m; (b) Gullmarsfjorden/Kattegat, 27 May 1994, Secchi depth: 5.0 m; Opposite page: (c) Barhöft/harbor (Strelasund, Baltic Sea), 12 August 1995, Secchi depth: 1.50 m; (d) Zingster Strom (Barther Bodden, Baltic Sea), 10 August 1995, Secchi depth: 0.55 m.

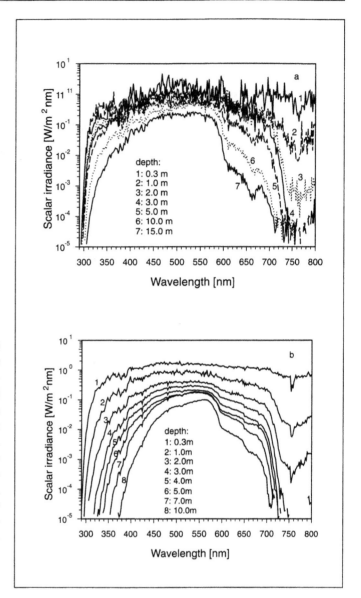

## THE JERLOV SYSTEM OF CLASSIFICATION OF OPTICAL WATER TYPES

Due to different contents of dissolved and particulate additions like yellow substance, chlorophyll $a$ and seston, marine waters have been classified into 5 oceanic (O) and 9 coastal (C) optical types (J) in the Jerlov system,[1,7-9] showing typical differences in the transmission spectra. Thus, using Eq. 7 the attenuation coefficient K depends on wavelength and water type according to

$$K_\lambda(J) = -d^{-1} \cdot \ln T_\lambda(J)$$
(24)

which characterizes the optical properties of the water column completely in the case of vertical homogeneity.

Figure 5.1a shows the spectra of the attenuation coefficient between 300 and 700 nm for different oceanic and coastal Jerlov water types. It has to be noted that the functions decrease nearly linearly with wavelength between 300 and 450 nm on the semilogarithmic plot mainly due to the absorption of short-wavelength radiation by yellow substances.

The spectral distribution of the 0.1% depths calculated by using Eq. 12 shows

for increasing Jerlov water type (Fig. 5.1b) a shift to longer wavelengths caused by increasing turbidity and chlorophyll $a$ content. For oceanic waters maximal transmission is in the blue range of the spectrum between 55 m (type OIII) and 380 m (type OI), whereas in coastal waters green light penetrates best with 0.1% depths between 12 m (type C9) and 55 m (type C1). In the UV range the 0.1% depths increase nearly linearly with wavelength on a semilogarithmic plot between 1.07 m at 310 nm and 2.87 m at 400 nm for type C9 and between 42 m at 310 nm and about 200 m at 400 nm for type OI.

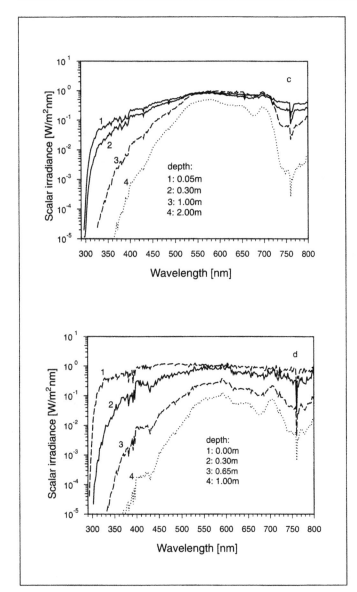

caused by different contents of

- *yellow substance* which is the main absorber in the UV range
- *chlorophyll a* with absorption maxima at 440 and 680 nm
- *phycobiliproteins* which absorb between 550 and 640 nm as well as
- *seston* and *detritus* which attenuates by absorption and scattering in the UV as well as in the PAR range.

In contrast to the waters of the Baltic Sea which contain comparatively high concentrations of attenuating substances, the decrease with depth in the UV and blue range of solar irradiance is small in the waters of the Mediterranean due to small concentrations of yellow substance and seston. But a large decrease was found between 580 and 660 nm in the spectra measured off Malaga and at Gullmarsfjorden/Kategatt which are probably caused by the presence of cyanobacteria.

In addition, the Malaga data derived from the near surface layers are modified by water waves which focus and scatter radiation. Whereas these changes of irradiance do not influence photobiological processes which depend on the absorbed dose over comparatively large time scales, processes which respond to fast changes in the irradiance can be affected.

## RESULTS

### OPTICAL AND BIO-OPTICAL PROPERTIES

#### Spectral data

*Scalar solar irradiance*

Figure 5.2 shows spectra of scalar solar irradiance measured at different depths in the waters of the Baltic Sea and the Mediterranean, which differ due to different modifications with depth. These spectral decreases of irradiance are mainly

*Spectral attenuation coefficients*

Using these data and results of further measurements of scalar irradiance performed in different lagoons of the Baltic Sea, spectral attenuation coefficients for scalar irradiance $K_{o\lambda}$ were calculated between different depths according to Eq. 6. Figures 5.3 and 5.4 show the mean attenuation

Fig. 5.3. The spectral attenuation coefficient of scalar irradiance at different stations of the Baltic Sea and the Mediterranean in comparison with the attenuation coefficients of the Jerlov water types C1-C9 and OIII.
Stations: (a) Greifswalder Bodden/central lagoon, 26 April 1994 (1); Vierow/harbor, 27 April 1994 (2); Gullmarsfjorden/Kattegat, 27 May 1994 (3); Barhöft/Harbor, 9 July 1995 (4); (b) Darßer Ort/Baltic Sea, 10 July 1994 (5); Gager, 2 May 1994 (6); Zicker, 5 May 1994 (7); Zinger Strom, 10 August 1995 (8) and Mediterranean Sea, 2 km off Malaga (Spain), 15 March 1995 (9).

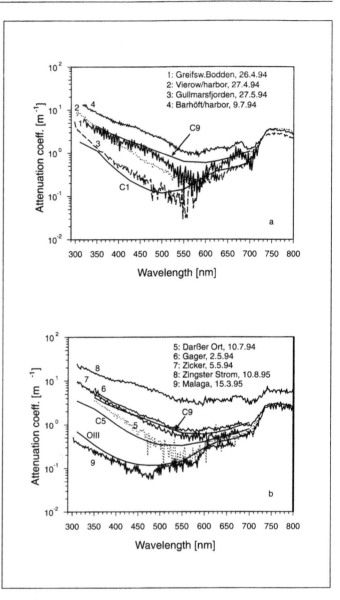

coefficients in comparison to the spectral attenuation coefficients of different Jerlov water types whereas Figure 5.5 analyzes changes of attenuation coefficient with depth.

The spectral data in Figure 5.3 indicate water types which equal or exceed type C9 in the Jerlov system. The spectra differ from each other due to different concentrations of yellow substance, chlorophyll $a$ and seston. But their slopes in the short-wavelength range between 290 and about 550 nm which are caused mainly by yellow substance are semilogarithmic and similar to each other, indicating a similarity in the chemical composition of the yellow substances in different Baltic Sea waters. In contrast to the spectra 2, 3, 5 and 7, the spectra 1, 4, 6 and 8 show additional local increases of the attenuation coefficient between 400 and 500 nm and of the maximum at 480 nm due to absorption by chlorophyll $a$. Furthermore, in contrast to spectrum 5, the spectra 6 and 7 are influenced by higher seston concentrations which cause increases in the attenuation coefficient at wavelengths between 450 and 650 nm by absorption and scattering of radiation at the particles.

The waters of the open Baltic Sea contain lower concentrations of attenuating substances than the Bodden lagoons which result in higher transparency. Figure 5.4 shows mean attenuation coefficients for scalar solar UV irradiance measured during the GOBEX'94 expedition which slope nearly linear on a semilogarithmic scale similar to the coastal types C1-C5 in the Jerlov system.

In comparison to the data found in the Baltic Sea, the attenuation coefficients derived for the Mediterranean waters near Malaga are small and fit between the attenuation coefficients of the types OII and

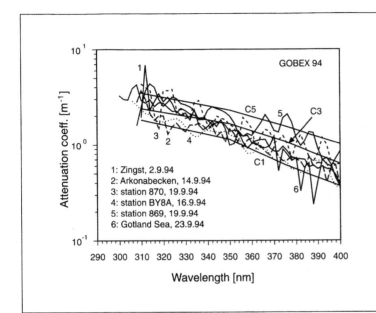

*Fig. 5.4. The spectral attenuation coefficient of scalar UV irradiance at different stations of the open and central Baltic Sea in comparison with the attenuation coefficients of the Jerlov water types C1-C5.*

**Table 5.1. Visual depths (Secchi) and regression coefficients a and b of Eq. 25 for different stations in the Baltic Sea and in the Mediterranean near Malaga, Spain**

| Station | Date | D [m] | a [nm⁻¹] | b | r |
|---|---|---|---|---|---|
| Zingster Strom | 16.06.94 | 0.25 | −0.004496 | 2.7092 | −0.71 |
| Barhöft | 09.07.94 | 0.80 | −0.005377 | 2.8234 | −0.93 |
| Barhöft | 20.06.94 | 1.30 | −0.004004 | 2.0550 | −0.90 |
| Greifswalder Bodden | 26.04.94 | 2.00 | −0.005064 | 2.3276 | −0.84 |
| Darßer Ort | 10.07.94 | 3.50 | −0.006917 | 3.0017 | −0.87 |
| Gullmarsfjorden | 27.05.94 | 5.00 | −0.008907 | 3.2499 | −0.99 |
| Vierow | 27.04.94 | 2.35 | −0.007876 | 3.3456 | −0.99 |
| Gager | 02.05.94 | 1.80 | −0.006402 | 3.0255 | −0.98 |
| Zicker | 05.05.94 | 3.50 | −0.006365 | 2.9464 | −0.99 |
| Arkonabecken | 14.09.94 | 6.00 | −0.007778 | 2.7961 | −0.95 |
| BY8A | 16.09.94 | 6.50 | −0.009183 | 3.1850 | −0.63 |
| Zingster Strom | 10.08.95 | 0.55 | −0.003187 | 2.3096 | −0.99 |
| Malaga | 15.03.95 | 12.00 | −0.004745 | 1.0523 | −0.98 |

r, correlation coefficient

OIII in the Jerlov system. On a semilogarithmic scale the slope of the attenuation coefficients can be approximated with high accuracy by a regression line according to

$$K_{0\lambda}(J) = 10^{(a \cdot \lambda + b)} \qquad (25)$$

($K_\lambda$ = [m⁻¹], $\lambda$ = [nm]) between 300 and 550 nm (UV, violet and partially also blue/green range) for coastal waters (Baltic Sea data) and between 300 and 470 nm for oceanic waters (Malaga data). The coeffi-

cients a and b of Eq. 25 and the measured data of visual depth D are combined in Table 5.1.

In the case of vertical stratification of the absorbing substances the attenuation coefficients change with depth in the water column. Four different examples are combined in Figures 5.5a-d. Figure 5.5a shows nearly constant optical properties between the surface and 2 m depth, whereas the attenuation coefficient increases

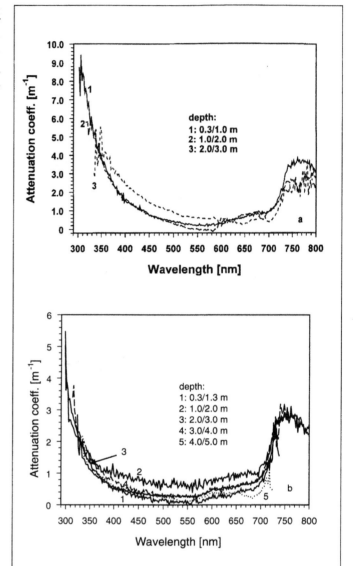

below a depth of 2 m. In contrast, the attenuation coefficients in Figure 5.5b indicate
an optical thick layer between
two optical thin layers, one at
the surface and the other at a
depth below 2 m. The third
example (Fig. 5.5c) shows a
concentration of attenuating
substances in the near surface
layer and more transparent
layers below. Figure 5.5d indicates vertical homogeneity
of yellow substance concentration in the waters near Malaga
due to equal attenuation coefficients in the UV range but
a higher content of scattering
particles in the layer between
5 and 10 m depth.

Often measurements of
penetration of solar irradiance
are performed by using sensors
which receive the downward
component only. But to determine systemic effects of solar
radiation upon spherical organisms such as
phytoplankton, radiation backscattered
from the layers below has to be included
in the irradiance measurement if its
amount contributes significantly. Figure
5.6 compares attenuation coefficients for
scalar and for downward scalar irradiance
which were derived from measurements in
the Baltic Sea as well as in the Mediterranean. Two attenuating processes compete
with each other: absorption and scattering.
If absorption dominates, the difference between scalar and downward scalar irradiance is small or negligible. In contrast, if
scattering processes mainly determine the

attenuation of radiation the differences can
be significant.

The attenuation coefficients determined
for the waters of the Baltic Sea clearly show
smaller coefficients for scalar irradiance in
comparison with downward scalar irradiance (Fig. 5.6a-c). These differences are
caused by the influence of internal backscattering processes which increase with
depth in comparison with absorption and
forward scattering. Due to increases in effectiveness for scattering processes with decreasing wavelengths these effects occur
especially in the UV and blue/green wavelength range. Furthermore, the data in Fig-

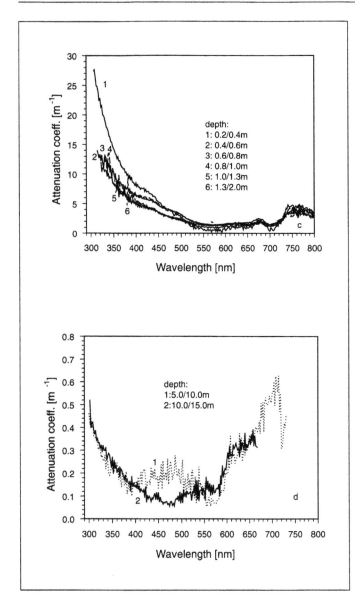

coefficient for scalar irradiance became negligible between 460 and 630 nm, whereas the attenuation coefficient for downward scalar irradiance remained measurable.

In contrast with the results shown for coastal waters (Figs. 5.6a-c), negligible differences in the UV range and only small differences between 400 and 500 nm were found between the attenuation coefficients of scalar and downward scalar irradiance in the offshore waters near Malaga (type: OII~OIII; Fig. 5.6d) due to small concentrations of yellow substance and particles.

To quantify the differences between the attenuation coefficients for scalar and downward scalar irradiance in the short-wavelength range mean slopes and the ratios of the two were calculated for different Baltic Sea waters. Figures 5.7a-d show the attenuation coefficients for scalar irradiance which are (depending on the water type) smaller by 10-50% than the data of downward scalar irradiance.

### 0.1% depths of spectral solar surface irradiance

The 0.1% depths of surface irradiance were calculated according to Eq. 12 by using data of mean spectral attenuation coefficients for scalar irradiance (Figs. 5.8 and 5.9). Their values increase in the short-wavelength range between 300 and about 550 nm for Baltic Sea waters and up to about 480 nm for Mediterranean waters (Fig. 5.8). The increases can be approximated by semilogarithmic functions from the data of Table 5.1 using Eqs. 12 and 25. Furthermore, the 0.1% depths for Baltic Sea waters are characterized by:

• maxima between 6 and 60 m which shift to longer wavelengths from about

ures 5.6a-c show the dependence of backscattering effects on the concentration of scattering particles. Whereas the difference between attenuation coefficients for scalar and downward scalar irradiance is small in comparatively clear waters (Figs. 5.6a and b, water types: C2) it increases significantly with turbidity (Fig. 5.6c, water type: C5).

In addition to internal backscattering, ground reflectance may cause significant decreases of the attenuation coefficient for scalar irradiance, shown in Figure 5.6a. The measurements were performed in the water column over white sand at 4.2 m depth. Due to ground reflectance, the attenuation

*Fig. 5.6. The attenuation coefficient of scalar irradiance (1) and of downward scalar irradiance (2) as a function of wavelength. Stations and conditions: (a) Zingst/Baltic Sea, 2 September 1994, Secchi depth: 2.6 m, depths of measurement: 0.3/ 1.3 m; (b) Arkonabecken, 14 September 1994, Secchi depth: 6.0 m, depths of measurement: 0.3/1.3 m; (opposite page) (c) Zingst/Baltic Sea, 11 August 1995, Secchi depth: 4.2 m, depths of measurement: 2.0/ 3.0 m; (d) offshore waters of the Mediterranean Sea near Malaga (Spain), 15 March 1995, Secchi depth: 14.0 m, depths of measurement: 5.0/10.0 m. Measurements at local noon and cloudless sky.*

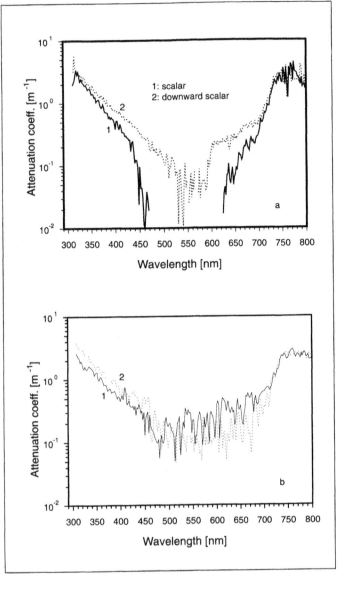

500 nm (green) to about 570 nm (yellow) for decreases in transparency; and

- local minima around 680 nm due to chlorophyll *a* absorption.

In contrast, the maximum of penetration was found between 450 and 480 nm (blue) in the Mediterranean Sea showing the 0.1% depth at about 100 m. For comparison, Figure 5.8b contains the spectra of the 0.1% depths for different coastal water types and for the oceanic type OIII of the Jerlov system. Figure 5.9 suggests values between 1.6 and 3.6 m at 310 nm and between 5.5 and 16 m at 400 nm for the open Baltic Sea (water types C1-C5).

In contrast to the data for the most turbid coastal water type C9 in the Jerlov system, 0.1% depths with smaller values (below 1.0 m at 300 nm, below 2.9 m at 400 nm and below 12 m at the wavelength of maximal transmission) were found in the inner lagoons with little water exchange with the open Baltic Sea (stations: Zingster Strom, Barhöft) as well as in the near shore zones of the half-open lagoon Greifswalder Bodden (stations: Vierow, Zicker, Gager).

These data indicate water types of higher turbidity, beyond the Jerlov system of classification (Fig. 5.8). On the other hand, the 0.1% depths show that phytoplankton organisms which populate the top layers of the water column[25] may be affected by solar UV-B irradiation even in these turbid waters.

## CLASSIFICATION OF DIFFERENT BALTIC SEA WATERS (EXTENSION OF THE JERLOV SYSTEM FOR TURBID COASTAL WATERS)

The composition and concentration of attenuating substances as well as their ver-

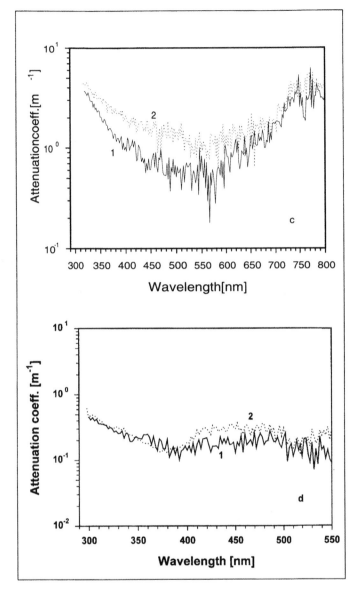

ter type given at different wavelengths between 300 and 550 nm (Fig. 5.10) marine waters can be classified by using only one spectral value of the attenuation coefficient in this wavelength range according to

$$J = m_\lambda \cdot lg \; K_{o\lambda} + n_\lambda \quad (26)$$

The coefficients m and n are listed in Table 5.2 for coastal and oceanic waters using wavelengths between 350 and 550 nm. The spectral values $K_{o\lambda}$ of the attenuation coefficient used in Eq. 26 can easily be estimated from monochromatic or narrow band irradiance measurements at different depths. The absolute deviation of the calculated water types from each other which were determined by using different wavelength is less or equal $\pm$ 0.5 classes.

Furthermore, this method allows the extension of the Jerlov system of classification for coastal waters of a turbidity higher than of type C9 which were found in some cases. The whole spectrum of the attenuation coefficient of these waters between 300 and 550 nm can be estimated by Eq. 25 using the semilogarithmic slope of these spectra as well as the coefficients a and b of each water type. Figures 5.11a and 5.11b show linear changes of the coefficients a and b with water type according to

$$a = p_a \cdot J - q_a \quad (27)$$

and

$$b = p_b \cdot J - q_b \quad (28)$$

The coefficients of Eq. 27 and 28 listed in Table 5.3 differ from each other to a

tical distribution mainly determine the optical properties in the water column and the penetration of solar irradiation. In the case of vertical inhomogeneity of the added substances the attenuation coefficient changes with depth (cf. Fig. 5.5) causing deviations from the semilogarithmic slope of the vertical decrease of solar irradiance. For vertical homogeneity or small vertical changes in the composition and concentration of attenuating substances the different optical properties of the waters can be characterized by different water types. Based on the semilogarithmic slope of the attenuation coefficient with the Jerlov wa-

*Fig. 5.7. Spectral distribution of the attenuation coefficients of scalar irradiance (1), of downward scalar irradiance (2) and of their ratio (3) in the short-wavelength range. Stations and conditions: (a) Zingst/Baltic Sea, 2 September 1994, Secchi depth: 2.6 m, depths of measurement: 0.3/ 1.3 m; (b) Arkonabecken, 14 September 1994, Secchi depth: 6.0 m, depths of measurement: 0.3/1.3 m; (opposite page) (c) Gotland Sea, 23 September 1994, Secchi depth: 9.0 m, depths of measurement: 0.3/ 1.3 m; (d) Zingst/Baltic Sea, 11 August 1995, Secchi depth: 4.2 m, depths of measurement: 2.0/3.0 m.*

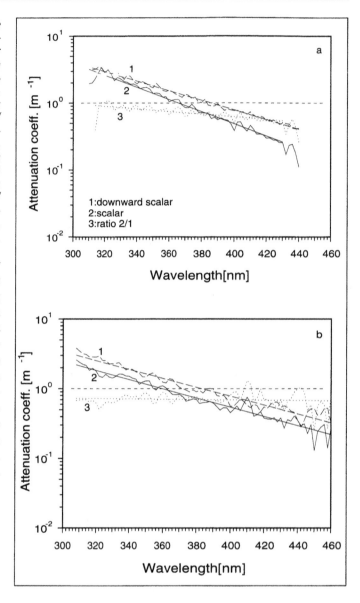

smaller degree for the Jerlov and for the Baltic Sea data at a given water type, causing differences of less than ±0.001 nm⁻¹ for coefficient a and less than ±0.4 for the coefficient b. With respect to the deviation of the Baltic Sea data in Figure 5.11, the spectra of the attenuation coefficient for coastal waters of type C9 to C20 shown in Figure 5.12 were calculated for both the Jerlov and the Baltic Sea data.

## ESTIMATION OF OPTICAL PROPERTIES BY SECCHI DEPTH MEASUREMENTS

### Visual (Secchi) depth and optical properties in the UV and blue/green range

The attenuation coefficients derived for scalar irradiance between 300 and 550 nm are highly correlated with the values of the visual (Secchi) depth (D) which can be measured easily. Figure 5.13a shows that the attenuation coefficients are semilogarithmically dependent on the wavelength slope with increasing visual depth. This connection is given by regression analysis according to

$$K_{o\lambda}(D) = 10^{(a_v \cdot D + b_v)} \qquad (29)$$

(D = [m], correlation coefficients r = -0.95

...-0.89). The coefficients $a_v$ and $b_v$ of Eq. 29 can be calculated for each wavelength between 300 and 550 nm by

$$a_v = -7.061 \cdot 10^{-4} \cdot \lambda + 0.0842 \qquad (30)$$

$$b_v = -4.466 \cdot 10^{-3} \cdot \lambda + 2.6114 \qquad (31)$$

($\lambda$ = [nm], $a_v$ = [m⁻¹]). Its data vary between $a_v$ = -0.128 m⁻¹ and $b_v$ = 1.272 at $\lambda$ = 300 nm and $a_v$ = -0.304 m⁻¹ and $b_v$ = 0.155 at 550 nm. By analogy with this

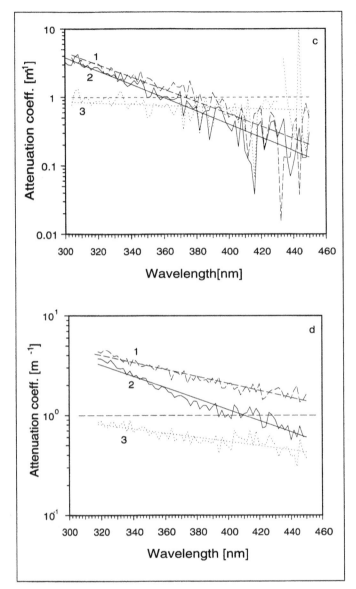

and

$$f_v = 4.454 \cdot 10^{-3} \cdot \lambda - 1.7661 \quad (34)$$

$(\lambda = [nm])$

### Estimation of Jerlov water type and optical properties by visual depth

The data of Figure 5.13 show a high correlation between the attenuation coefficients and visual depth, which often is used for first estimation of bio-optical properties. Thus, Eq. 25 also allows the simple estimation of the spectral optical properties of the water between 300 and 550 nm by data of visual (Secchi) depth between 0.2 and 6.5 m using the dependence of its parameters a and b on visual depth. Both parameters change linearly with visual depth D according to

$$a = -7.075 \cdot 10^{-4} \cdot D - 0.0045$$
$$(r = -0.85) \quad (35)$$

$$b = 0.085 \cdot D + 2.606$$
$$(r = -0.45) \quad (36)$$

Furthermore, the Jerlov water type of Baltic Sea waters can be estimated with an accuracy of ±1 class from visual depth measurements using the data of Figure 5.14.

connection also the 0.1% depth of irradiance is highly correlated with the visual depth (Fig. 5.13b). Its values show linear increases with visual depth on a semilogarithmical scale and can be estimated from data of the visual depth between 300 and 550 nm by using Eqs. 12 and 29 or directly according to

$$d_{01,\lambda}(D) = 10^{(g_v \cdot D + f_v)} \quad (32)$$

with

$$g_v = 7.093 \cdot 10^{-4} \cdot \lambda - 0.08555$$

$$(33)$$

### BROADBAND AND BIO-OPTICAL DATA

Microorganisms like phytoplankton populating the waters are in a difficult situation. On the one hand, they need solar PAR (photosynthetic active radiation, 400-700 nm) to perform their photosynthesis as well as solar RER (repair effective radiation, 390-470 nm) which is effective for stimulating their repair mechanisms for

UV-caused damages. Thus, using passive (by buoyancy) or active movements the organisms try to move to a depth in the water column characterized by optimal irradiance of PAR and RER. On the other hand, the organisms receive solar UV radiation which penetrates into the euphotic zone of the water column and may cause damage in different ways.[11,14,17-24,30]

Using the data from spectral irradiance measurements in the water column as discussed previously, scalar irradiance was determined, according to Eq. 8, as a function of depth for solar UV-B reaching the Earth's surface (290-315 nm), UV-A (315-400 nm), PAR and RER. In addition, efficiencies for different damaging effects caused by solar UV-B and UV-A as well as PAR were calculated by weighting scalar irradiance data with action spectra, shown in Figure 5.15 according to Eq. 9. The data were normalized with their values given at subsurface to compare the

slopes of irradiances with depth calculated for different spectral ranges, photobiological effects and waters of coastal and oceanic type. The relative irradiance changes with depth in dependence of spectral range and water type (Fig. 5.16). Linear slopes with depth were found in the case of vertical homogeneity (Figs. 5.16a-c) whereas stratifications of attenuating substances cause changes of the vertical gradient of radiation (Figs. 5.16d-f). In contrast to PAR, which penetrates the water column best, UV-B radiation shows the strongest decrease with depth. The relative efficiencies for DNA damage[31] and for photoinhibition of motility of *Euglena gracilis*[32] slope similarly to that of UV-B due to the similarity between the spectral range of solar UV-B reaching the Earth's surface and the effective ranges of these effects.

The photoinhibition of photosynthesis in *Nodularia spumigena*[21,30] is mainly caused by UV-A radiation. Thus, the slope of the

**Table 5.2. Coefficients $m_\lambda$ and $n_\lambda$ of Eq. 26 for estimation of the Jerlov water type of coastal and oceanic waters using (narrow band) wavelengths**

**a. Coastal waters**

| $\lambda$ [nm] | $m_\lambda$ | $n_\lambda$ |
|---|---|---|
| 350 | 15.794 | -0.694 |
| 380 | 11.116 | 3.600 |
| 400 | 9.963 | 5.113 |
| 440 | 8.670 | 7.083 |
| 450 | 8.381 | 7.450 |

**b. Oceanic waters**

| $\lambda$ [nm] | $K_g$ [m-1] | $K \leq K_g$ | | $K \geq K_g$ | |
|---|---|---|---|---|---|
| | | $m_\lambda$ | $n_\lambda$ | $m_\lambda$ | $n_\lambda$ |
| 350 | 0.0998 | 9.626 | 12.644 | 3.724 | 6.762 |
| 380 | 0.0567 | 9.834 | 15.155 | 3.357 | 7.196 |
| 400 | 0.0460 | 9.436 | 15.547 | 3.445 | 7.646 |
| 440 | 0.0348 | 8.581 | 15.467 | 3.509 | 8.143 |
| 450 | 0.0336 | 8.184 | 15.015 | 3.561 | 8.268 |

**Table 5.3. Coefficients p and q of Eqs. 27 and 28 for Jerlov water types and Baltic Sea data**

| Data | $p_a$ | $q_a$ | $p_b$ | $q_b$ |
|---|---|---|---|---|
| Jerlov | $4.232 \cdot 10^{-4}$ | 0.00822 | -0.071 | 2.800 |
| Baltic Sea | $3.361 \cdot 10^{-4}$ | 0.00900 | -0.034 | 3.103 |
| Jerlov + Baltic Sea | $2.900 \cdot 10^{-4}$ | 0.00829 | -0.017 | 2.818 |

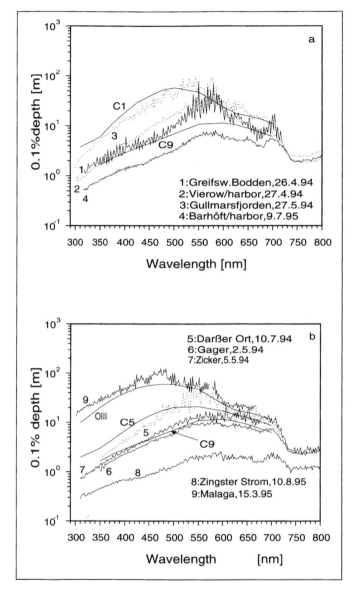

Fig. 5.8. The 0.1% depth of subsurface scalar irradiance at different stations of the Baltic Sea and the Mediterranean in comparison with the attenuation coefficients of the Jerlov water types C1-C9 and OIII. (For stations, see Figs. 5.3a and b.

(Figs. 5.16c,e,f). In contrast, turbid waters strongly attenuate UV radiation, violet and blue light which causes a slope of the efficiency for RER which is smaller in comparison to that of UV-A, but stronger than that of PAR (Figs. 5.16a,b,d).

Using a semilogarithmic scale the attenuation coefficients (0.1% depths) for broadband and weighted solar scalar radiation penetrating coastal waters show a linear increase (decrease) with water type (Figs. 5.17a-c and d-f). The slopes can be estimated according to

$$K_{\Delta\lambda} = 10^{\,(a_{\Delta\lambda} \cdot J + b_{\Delta\lambda})} \quad (37)$$

and

$$d_{01,\Delta\lambda} = 10^{\,(-a_{\Delta\lambda} \cdot J + c_{\Delta\lambda})} (38)$$

(coefficients see Table 5.4).

In addition, the ratios of the 0.1% depths between UV irradiance as well as further spectral ranges and between PAR were calculated for different coastal waters using the data in Figure 5.17d-f. These ratios allow the estimation of the part of the euphotic zone which is influenced down to 0.1% of subsurface irradiance by transmission of different ranges of the solar spectrum. Thus, this knowledge is basic information to discuss the importance of different photobiological processes such as damage by solar UV, photoinhibition and repair effects in aquatic microorganisms which populate the euphotic zone to perform photosynthesis.

efficiency of photoinhibition in *Nodularia spumigena* was found to be similar to that of UV-A.

As an example of an effect stimulated by both UV-B and UV-A, the slope of erythema efficiency shows similarity with that of UV-B at the near surface layer and with that of UV-A in deeper layers due to the stronger attenuation of UV-B in the water. Furthermore, due to the stimulation of repair mechanisms with radiation between 390 and 470 nm the efficiency of RER is similar to PAR in oceanic and comparatively clear coastal waters which are highly transparent for blue light

Fig. 5.9. The 0.1% depth of
the subsurface scalar UV irradiance at different stations of
the open and central Baltic
Sea in comparison with the
attenuation coefficients of the
Jerlov water types C1-C5.

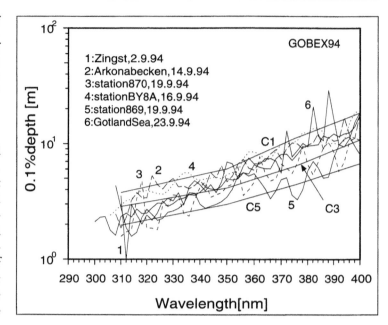

The calculated
0.1% depth ratios (Fig.
5.17g-i) show that
UV-B radiation as well
as effective radiation (efficiencies for DNA
damage and photo-
inhibition of motility of
*Euglena gracilis*) transmit
up to 15% of the
euphotic zone if the
coastal water is com-
paratively clear (type C1), whereas the
transmission decreases linearly on a semilogarithmic scale with increasing turbidity. Thus, for water type C15, a mean
transmission by UV-B of only about 5%
of the euphotic zone was found.

UV-A irradiance and the efficiency of
photoinhibition of *Nodularia spumigena,*
which is affected by both UV-B and UV-A,
penetrate through about 35% of the euphotic zone in water type C1 and about
20% in water type C15, also showing a
semilogarithmic slope with water type.

In contrast to UV-B and UV-A, repair
effective radiation and effective irradiance
for photoinhibition of chloroplast reactions[33] transmit up to 70% of the euphotic
zone in water type C1. The 0.1% depths
of PAR and of the efficiency of photoinhibition decrease similarly with increasing water type. Thus, the 0.1% depth ratio is independent of the water type. In
contrast, the attenuation of repair effective
radiation was found to be stronger than the
attenuation of PAR with water type. Thus,
the transmission of the euphotic zone by
RER decreases to about 40% at water type
C15.

The ratios between the 0.1% depths
for UV-B, UV-A, RER and the efficiencies involved here and between the 0.1%
depth of PAR can be quantified in

dependence of water types according to

$$R = 10^{(-a_R \cdot J + b_R)} \tag{39}$$

(coefficients see Table 5.5).

The results show that the UV-protec-
tion of the organisms which populate the
euphotic zone by attenuating substances
contained in the water column becomes
more effective as the turbidity increases.
On the other hand, the risk of damaging
effects caused by UV radiation increases
especially in comparatively clear waters due
to small or negligible concentration of UV
protective substances such as yellow substance. Thus, especially oceanic waters show
a high transmission of UV radiation which
even can exceed the 0.1% depth of long-
wavelength range of PAR. Figure 5.18
shows the 0.1% depth ratio calculated from
spectral transmission data at 310, 450 and
650 nm for different oceanic water types
in the Jerlov system of classification.[7-9]
Whereas UV-radiation at 310 nm exceeds
the penetration of long-wavelength photosynthetic radiation at 650 nm by up to a
factor of about 2.5, its penetration is only
about 12-18% of the 0.1% depth of photosynthetic active radiation at 450 nm.
Thus, these differences in the ratios of UV
penetration into the euphotic zone contribute to ecological selection of species in

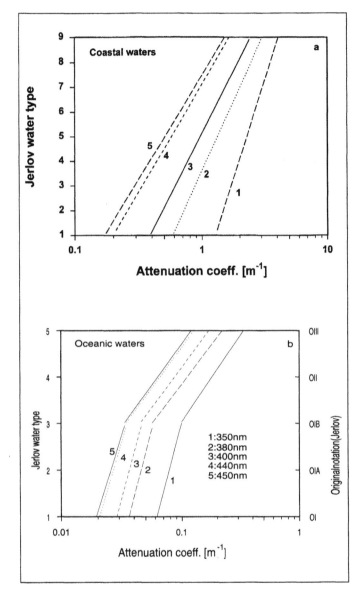

Fig. 5.10. The Jerlov water type (J) for coastal waters (a) and for oceanic waters (b) as a function of the attenuation coefficient of scalar irradiance at 350 nm (1), 380 nm (2), 400 nm (3), 440 nm (4) and 450 nm (5).

irradiance and the efficiencies of photoinhibition and DNA damage. The slopes can be estimated according to

$$K_{\Delta\lambda} = 10^{({}^a D, \Delta\lambda} \cdot {}^D + {}^b D, \Delta\lambda}) \quad (40)$$

(coefficients see Table 5.6).

## SPECTRAL BACKSCATTERING AND PASSIVE REMOTE SENSING

To detect the phytoplankton concentration and optical properties using the CZCS, suitable channels were chosen with bands centered at wavelengths $\lambda_1 = 443$ nm (channel 1)—or $\lambda_1 = 520$ nm (channel 3) if the 443 nm signal is too low—and at $\lambda_2 = 550$ nm (channel 4) for chlorophyll concentrations above 1.5 mg/$m^3$ in oceanic waters.[34,35] For coastal waters which usually contain high concentrations of yellow substance, optimal channel ratios are found at $\lambda_1$ = 520 and 550 nm and at $\lambda_2$ = 670 nm (channel 6).[3] (All channels show half-bandwidths of 20 nm.)

The applicability of phytoplankton detection by using the CZCS is limited by:
• high concentrations of yellow substance which absorbs short-wavelength radiation;
• high concentrations of accessory pigments (phycobiliproteins), which absorb radiation around 550, 620 and 640 nm;[29]
• high bandwidths of the optical channels which prevent the discrimination of background signals, and
• vertical stratification of the organisms during low winds by active or passive

coastal waters which are more transparent for - PAR and in oceanic waters showing a higher transparency for short-wavelength PAR in comparison with UV-B transparency.

Due to high correlation coefficients between spectral optical characteristics and visual depth also the characteristics of broadband irradiance as well as efficiencies of photobiological effects can be estimated by simple measurements of the visual depth. Figure 5.19 shows the slopes of the attenuation coefficients and the 0.1% depths with visual depth (Secchi) for PAR, UV-B, UV-A as well as for repair effective

*Fig. 5.11. The coefficients a (a) and b (b) of Equation 25 as a function of the Jerlov water type for Baltic Sea data and data published by Jerlov.*

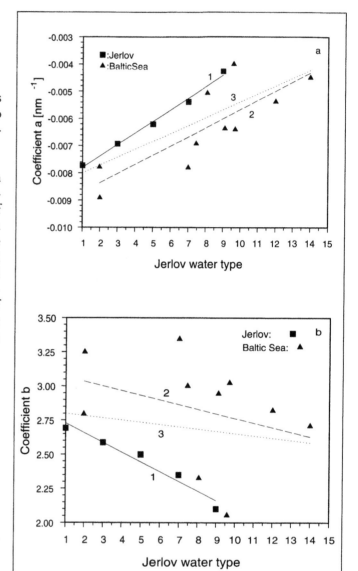

(by buoyancy) movements in dependence of PAR to optimize the photosynthetic rate.

In the following section the influence of vertical stratification of organisms and of water type on both backscattered signals at the sea surface and on ground truth relationships is discussed by using data which were derived from different lagoons of the southern Baltic Sea and for comparison also from waters of the Mediterranean Sea.

## VERTICAL DISTRIBUTION OF ORGANISMS, OPTICAL PROPERTIES AND SPECTRAL BACKSCATTERING

### Optical properties of the investigated waters

The results reported here were calculated by using data from direct measurements of both phytoplankton concentration and of spectral irradiance at different depths in the column with a double monochromator spectroradiometer.[26]

The water types listed in Table 5.5 indicate that the waters selected for the investigation are extremely turbid due to high concentrations of dissolved organic matter and seston. Thus, comparatively high attenuation coefficients were found especially in the short- and long-wavelength ranges of the optical solar spectrum (visual depths (Secchi) between 0.5 and 1.9 m and water type numbers between C9 and C13 in the extended Jerlov system). In contrast, the waters of the Mediterranean Sea near Malaga were identified as typical oceanic with low concentrations of yellow substance and seston, visual depth of 14 m and water type OIII in the Jerlov system.

Hence, the attenuation coefficient does not only depend on yellow substances and particles but also on the absorption and scattering caused by organisms contained in the column. The changes of the attenuation coefficients by an inhomogeneous distribution of phytoplankton in the column were investigated for the Malaga data. Figure 5.20 shows that the attenuation coefficients at 440, 510 and 680 nm essentially increase with the concentration of

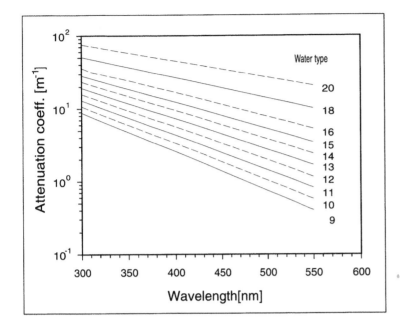

Fig. 5.12. The attenuation coefficient of scalar irradiance between 300 and 550 nm for the coastal water types C9-C20 in the extended Jerlov system of classification.

Table 5.4. Coefficients $a_{\Delta\lambda}$, $b_{\Delta\lambda}$ and $c_{\Delta\lambda}$ of Eqs. 37 and 38 to estimate attenuation coefficients and 0.1% depths of broadband and weighted scalar irradiance at coastal waters as a function of Jerlov water type

| Broadband/weighted range | $a_{\Delta\lambda}$ | $b_{\Delta\lambda}$ | $c_{\Delta\lambda}$ | $|r|$ |
|---|---|---|---|---|
| PAR | 0.0644 | – 0.6616 | 1.5010 | 0.84 |
| UV-B | 0.0618 | 0.3855 | 0.4538 | 0.94 |
| UV-A | 0.0757 | – 0.1679 | 1.0073 | 0.94 |
| PIH (J&K) | 0.0610 | – 0.5305 | 1.3700 | 0.84 |
| RER | 0.0826 | – 0.5044 | 1.3437 | 0.94 |
| DNA damage | 0.0718 | 0.2768 | 0.5626 | 0.94 |
| PIH (*Nodularia spumigena*) | 0.0767 | – 0.1477 | 0.9870 | 0.93 |
| PIH (*Euglena gracilis*) | 0.0628 | 0.2765 | 0.5629 | 0.74 |

Table 5.5. Optical parameters of the investigated waters

| Station | Date | J | D [m] | $K_{om}$ [m$^{-1}$] | | | |
|---|---|---|---|---|---|---|---|
| | | | | 440 nm | 510 nm | 560 nm | 680 nm |
| FTL | 11.5.94 | C9 | 1.9 | 1.573 | 0.625 | 0.359 | 1.148 |
| P | 11.5.94 | C10 | 1.6 | 2.093 | 0.873 | 0.516 | 1.262 |
| SG | 11.5.94 | C10 | 1.4 | 2.093 | 0.873 | 0.516 | 1.345 |
| R1 | 7.7.94 | C12.5 | 0.6 | 4.326 | 2.045 | 1.305 | 1.735 |
| RS | 11.7.94 | C11 | 1.3 | 2.788 | 1.220 | 0.743 | 1.389 |
| K | 12.7.94 | C13 | 0.5 | 4.940 | 2.384 | 1.540 | 1.791 |
| Malaga | 15.3.95 | OIII | 14.0 | 0.095 | 0.120 | 0.130 | 0.400 |

D, visual depth (Secchi); $K_{om}$, mean spectral attenuation coefficient of the water column

*Fig. 5.13. The attenuation coefficient of scalar irradiance (a) and the 0.1% depth of subsurface irradiance between 300 and 550 nm (b) in dependence of visual depth (Secchi) for the Baltic Sea stations listed in Table 5.1.*

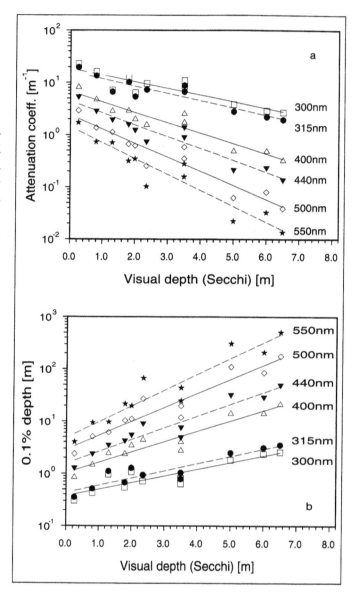

phytoplankton due to the absorption by chlorophylls and carotenoids. (For illustration, their changes are approximated by linear functions.) In contrast, the attenuation coefficient at 560 nm also shown in Figure 5.20 does not depend on the organism concentration because of the absence of accessory pigments such as phycoerythrin in the species populating these waters. Thus, the attenuation coefficients at 440, 510 and 680 nm change essentially with depth in dependence of the vertical changes of phytoplankton concentration (Fig. 5.21).

## Phytoplankton in the water column
## and backscattering of solar irradiation

Three different types of vertical distribution of the phytoplankton community in the water column were found during noon time and low winds in the investigated waters in the Baltic Sea and in the Mediterranean. The stratification of the organisms was caused by active or passive movements to optimize their photosynthesis and show
- maxima of phytoplankton concentration at different depths for Baltic Sea waters of type C9-C13 (Figs. 5.22a-d,f)
- concentration of the organisms at the surface (Fig. 5.22e)
- an increase of phytoplankton concentration with depth, including two minima at 3 and 4 m depth which was measured in the Mediterranean Sea near Malaga (Fig. 5.22g).

Using these data of vertical stratification of phytoplankton, the modification of the relative intensity of the spectral signals at the sea surface which were backscattered from the phytoplankton organisms at different depths $z_i$ was determined by calculation of the ratios $s(\lambda, z_i)$ defined in Eq. 19 (and shown in Fig. 5.22a-g for the wavelengths 440, 515, 560 and 680 nm). These spectral backscattering signals at the surface strongly depend on wavelength, vertical distribution of the organisms and water type. It is shown that the deeper the organisms are in the col-

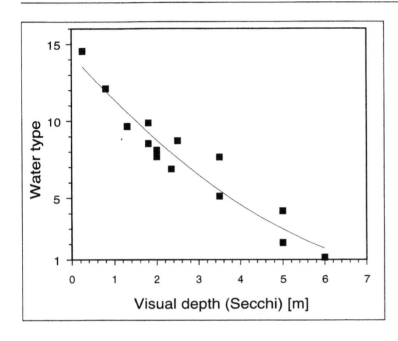

Fig. 5.14. The Jerlov type of Baltic Sea waters in dependence of visual depth (Secchi).

**Table 5.6. Ratio of the organism number between depth of maximal concentration $z_{max}$ and surface and of the spectral backscattered signals at 440, 515, 560 and 680 nm for different coastal lagoon waters of the Baltic Sea**

| Station | J | $z_{max}$ [m] | $n(z_{max})/n(z_0)$ | $s(\lambda,z_{max})/s(\lambda,z_0)$ | | | |
|---------|-----|------|-------|---------|---------|---------|---------|
| | | | | 440 nm | 515 nm | 560 nm | 680 nm |
| FTL | C9 | 1.8 | 3.50 | 0.012 | 0.369 | 0.961 | 0.056 |
| P | C10 | 1.2 | 1.26 | 0.008 | 0.155 | 0.365 | 0.061 |
| SG | C10 | 3.6 | 15.00 | < 0.001 | 0.028 | 0.365 | 0.001 |
| SG | C10 | 4.2 | 13.00 | < 0.001 | 0.009 | 0.171 | < 0.001 |
| R1 | C12.5 | 2.1 | 2.87 | < 0.001 | < 0.001 | 0.012 | 0.002 |
| K | C13 | 0.3 | 4.60 | 0.238 | 1.100 | 1.828 | 1.571 |

umn and the higher the Jerlov water type, the lower are the backscattered signals in the Baltic Sea. For example, the ratio of the organism number between the depth of maximal concentration ($z_{max}$) and the near surface range ($z_0$) given by $n(z_{max})/n(z_0)$ is compared with the ratio of the spectral backscattered signals $s(\lambda,z_{max})/s(\lambda,z_0)$ in Table 5.6. The data indicate that even high concentrations of organisms only contribute on the order of a few percent or less to the backscattered signal due to its attenuation in the column if the organisms populate deeper layers of turbid waters.

Furthermore, Figures 5.22a-f show a strong attenuation of the backscattered signals especially in the short-wavelength range (wavelengths shorter than 560 nm) which is due to absorption not only by the organisms but also by the usually high concentration of yellow substance in the Baltic Sea waters. In contrast, the waters of the Mediterranean Sea typically contain only a small concentration of yellow substance. Thus, the attenuation of the backscattered radiation at 440 and 510 nm in Figure 5.22g is comparatively small. Because of the spectral attenuation in the specific water type and the vertical stratification of the organisms, the backscattered signal at the surface reflects only part of the total content of organisms in the water

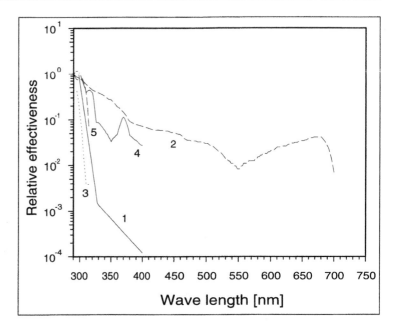

Fig. 5.15. Action spectra for the stimulation of erythema (CIE definition) (1), photoinhibition[33] (2), DNA damage (3), PIH of photosynthesis in Nodularia spumigena (4) and PIH of motility in Euglena gracilis (5). Action spectra are expressed in relative units, normalized to 290 nm.

column. Essentially, only the phytoplankton in the upper part of the water column contribute to the backscattered signal at the surface, whereas organisms in deeper layers do not.

Using Eqs. 17 and 22 this loss of information is analyzed by calculation of the ratios between the sum of the spectral upwelling signals $S(\lambda)$ which are backscattered between surface $z_0$ and $z_i = z_g = 6$ m depth (or ground) and the total content of phytoplankton organisms $N$ in the investigated waters. The ratios $S(\lambda)/N$ listed in Table 5.7 show

• values of only a few percent (especially in the 440 nm and in the 680 nm channel) if the organisms populate deeper layers. Furthermore, these values strongly depend on the water type

• values on the order of 30-45% in the case of near surface stratification of the organisms

and for the measurements in the Mediterranean Sea:

• values between 36 and 46% at 440, 515 and 560 nm and of about 10% at 680 nm.

In order to determine the depth of the upper part of the water column, which is characterized by significant contributions of the contained organisms to the back-

scattered signal at the surface, the percentage ratio $\Delta S(\lambda,z_i)/S(\lambda)$ (Eq. 20) was calculated as a function of depth and compared with the relative number of organisms $\Delta N(z_i)/N$ (Figs. 5.23a-g). In contrast to the relative number of organisms the spectral signals backscattered by the organisms between surface and depth $z_i$ converge to 100% in the Baltic Sea lagoons already at depths which are much smaller than the depth of sampling $z_g$. Figures 5.22a-f indicate that the depth of convergence $z_c$ is a function of both water type and vertical distribution of the organisms. In contrast, depth $z_c$ is much closer to $z_g$ for 440, 515 and 560 nm in the Malaga data due to small effects of attenuation.

From these data the depths $z_{90}$ were determined which define the 90% values of the ratios $\Delta S(\lambda,z_i)/S(\lambda)$ and $\Delta N(z_i)/N$ (Table 5.8). Thus, signals from deeper layers cause only contributions to the total signal at the surface on the order of 10% or less which defines the signal-to-noise resolution of a good optical receiver. The data listed in Table 5.8 show that in the turbid waters of the Baltic Sea lagoons only the organisms concentrated in the surface layer essentially contribute to the signals at the surface which are used to detect the content of organisms in the column.

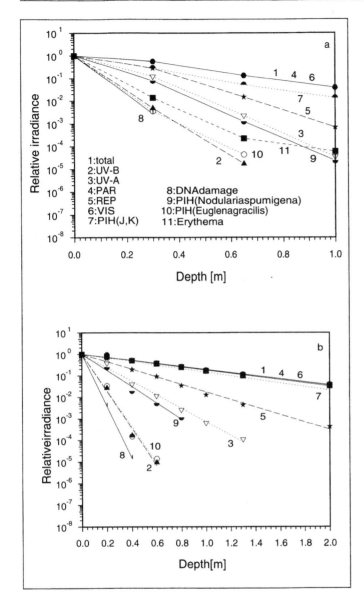

Fig. 5.16. Relative broadband scalar irradiance at several ranges and relative efficiencies of several photobiological effects as a function of depth at different stations. Data are derived from spectral measurements at different depths, cloudless sky and at local noon. Broadband ranges: 1: total optical range (290-800 nm), 2: UV-B, 3: UV-A, 4: PAR, 5: RER, 6: VIS (visible range, 380-720 nm). Photobiological effects: 7: PIH[33], 8: DNA damage, 9: PIH of photosynthesis in Nodularia spumigena, 10: PIH of motility in Euglena gracilis and 11: erythema. Stations: (a) Zingster Strom, 10 August 1995; (b) Barhöft/harbor, 9 July 1994; (following two pages) (c) BY8A/Baltic Sea, 16 September 1994; (d) Darßer Ort, 10 July 1994; (e) Gullmarsfjorden/Kattegat, 27 May 1994 and (f) the offshore waters of the Mediterranean Sea near Malaga (Spain), 15 March 1995.

In contrast, the depths $z_{90}$ of both the ratios $\Delta N/N$ and $\Delta S/S$ are on the same order for the Malaga data. Hence, the increase of the organism number with depth is only detectable as a signal which is attenuated by the transmitted water column. Thus, the ratio $\Delta S(\lambda, z_g)/\Delta N(z_g)$ listed in Table 5.7 is only between 35 and 45% in the relevant channels for remote sensing.

The results show that channels in the short-wavelength range (440 and 510 nm) as well as channels in the long-wavelength range (680 nm) are suitable for remote sensing of phytoplankton only in oceanic waters which contain small concentrations of added absorbing and scattering substances. Remote sensing of phytoplankton may be based on the absorption of both accessory pigments and chlorophyll *a* in these waters. Hence, the choice of the most suitable channel ratio depends on the specific absorbing substances of the species which populate the water. As shown in Figure 5.20, the attenuation coefficient at 680 nm depends much more on the concentration of organisms than the attenuation coefficient at 440 and 515 nm in the Mediterranean Sea waters. Thus, the

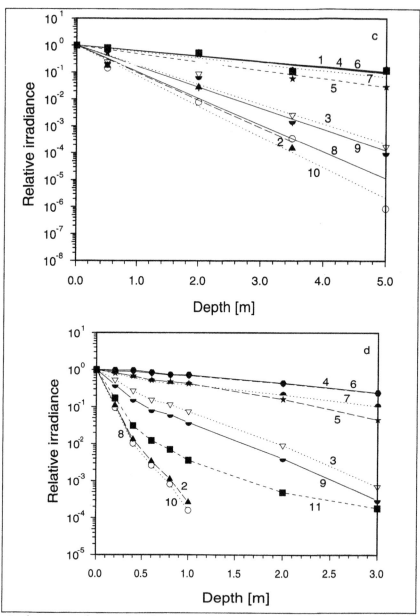

channel ratio between 680 and 560 nm gives a better indication than the ratio between 440 and 560 nm or between 515 and 560 nm of the content of phytoplankton in these waters in March 1995.

In contrast, comparatively high concentrations of yellow substance prevent the use of channels in the short-wavelength range for remote sensing of phytoplankton in Baltic Sea waters. Thus, in agreement with the results by Siegel and co-workers,[3] only the ratio of backscattered signals between long-wavelength radiation (670-680 nm) and 550-560 nm seems to be suitable to

determine the concentration of organisms in these waters using their chlorophyll *a* absorption.

## GROUND TRUTH RELATIONSHIPS FOR OPTICAL CHARACTERISTICS AND PHYTOPLANKTON CONTENT

The ground truth relationships were derived from simultaneously measured data of the vertical profile of phytoplankton concentration and of optical properties as well as of data sampled by the CZCS on board of an aircraft. The measurements were performed in the Baltic Sea lagoons

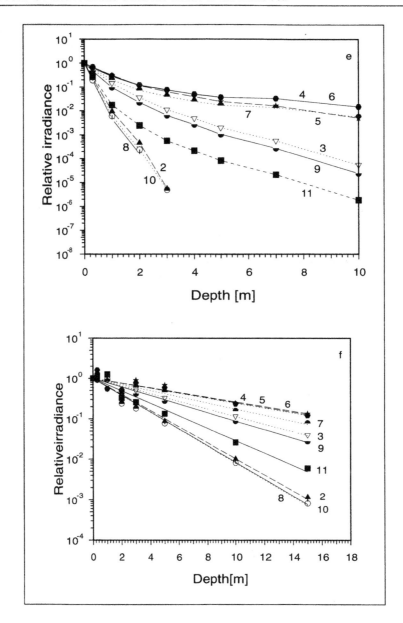

Greifswalder Bodden, Barther Bodden and at the waters between the islands of Hiddensee and Rügen during spring and summer 1994. The results are shown in the Tables 5.9 and 5.11 as well as in Figures 5.24-26.

In contrast to Eq. 15, three types of equations were derived to define ground truth relationships:

$$C_s = 10^{(p \cdot x + q)} \qquad (41)$$

for the near surface concentration of organisms $C_s$

$$\Delta N(z_i) = 10^{(p(z_i) \cdot x + q(z_i))} \qquad (42)$$

for the number of organisms the water column between surface and depth $z_i$ (surface area of the column = 1 m²) and

$$Y = p \cdot x + q \qquad (43)$$

for optical and bio-optical characteristics. The coefficients p and q are listed in Tables 5.9 and 5.11.

In general, high correlation coefficients were found between $C_s$, $\Delta N$ and Y and between the channel ratios 1/4 and 3/6. In contrast, the use of the channel ratio 4/6 results in insignificant relations only, either

**Table 5.7. Intensity of the spectral backscattering signals at the surface $\Delta S(\lambda, z_g)$ integrated between surface and $z_i = z_g = 6$ m depth (or ground) in comparison to the phytoplankton content $\Delta N(z_g)$ in the column**

| | $\Delta S(\lambda, z_g)/\Delta N(z_g)$ [%] | | | |
| Station | 440 nm | 515 nm | 560 nm | 680 nm |
|---|---|---|---|---|
| FTL | 7.83 | 14.83 | 24.89 | 9.28 |
| P | 10.06 | 16.34 | 23.07 | 13.08 |
| SG | 2.04 | 4.72 | 8.89 | 2.94 |
| R1 | 9.53 | 11.99 | 15.23 | 12.97 |
| RS | 31.45 | 36.45 | 42.26 | 35.34 |
| K | 4.07 | 7.54 | 11.57 | 9.98 |
| Malaga | 45.93 | 38.40 | 35.86 | 9.81 |

**Table 5.8. Depth $z_{90}$ of the ratios $\Delta N/N$ and $\Delta S/S$ (for stations see Table 5.1)**

| | $z_{90}(\Delta N/N)$ [m] | $z_{90}(\Delta S/S)$ [m] | | | |
| Station | – | 440 nm | 515 nm | 560 nm | 680 nm |
|---|---|---|---|---|---|
| FTL | 4.65 | 0.20 | 1.70 | 2.80 | 0.60 |
| P | 5.00 | 0.20 | 0.85 | 1.45 | 0.40 |
| SG | 4.85 | 0.20 | 1.40 | 3.00 | 0.80 |
| R1 | 2.20 | ≈ 0.00 | 0.23 | 0.55 | 0.33 |
| RS | 2.45 | ≈ 0.00 | 0.25 | 0.65 | 0.20 |
| K | 2.35 | 0.15 | 0.30 | 0.55 | 0.45 |
| Malaga | 6.30 | 6.15 | 6.10 | 6.05 | 3.30 |

due to equivalent influence of the back-scattering particles and organisms upon the reflectance at the channels 4 and 6 or due to absence of characteristic information.

## Detection of phytoplankton

To demonstrate the effect of vertical stratification of organisms on the ground truth relationships, the data were analyzed using increasing steps of the depth of column with and without organism stratification (Tables 5.9a-c). Vertical homogeneous distribution of organisms was found in 10 of 17 cases, whereas maxima in the near surface layer occurred in 2 cases, and a concentration of organisms in the deeper part of the investigated column was seen in five cases.

As expected, the highest values of the correlation coefficient were found during homogeneous organism distribution in the column (r ≈ - 0.82 ... - 0.78, Table 5.9b),

whereas vertical stratification of phytoplankton causes comparatively small values (r ≈ - 0.62 ... - 0.52, Table 5.9c). Furthermore, the correlation coefficient increases if not only the surface concentration of organisms, but also the organisms contained in the upper part of the column are involved in the analysis (Table 5.9a). Thus, it is shown that not only phytoplankton contained in the near surface water, but also the organisms which populate the first decimeters below the surface essentially contribute to the backscattered signal. In contrast, organisms which populate deeper layers become more and more invisible for remote sensing due to strong attenuation of their backscattered signals in the transmitted column. Consequently, the correlation coefficient in generally decreases the more, the more the organisms are stratified. However, even if the organisms are distributed homogeneously in the

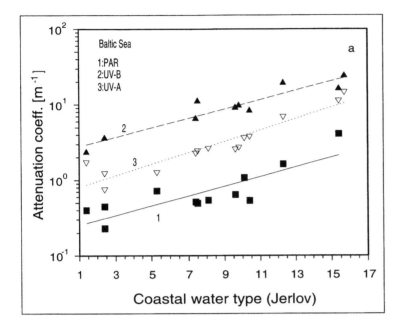

Fig. 5.17. The attenuation coefficient of scalar irradiance (a-c) and the 0.1% depth of subsurface scalar irradiance (d-f) for several broadband ranges and photobiological efficiencies as a function of the (coastal) Jerlov water type of several Baltic Sea waters. (g-i) Ratio between the 0.1% depth of scalar irradiance at several broadband ranges and efficiencies and between the 0.1% depth of PAR in dependence of the (coastal) Jerlov water type (Baltic Sea waters). See following four pages for Figures 5.17b-i.

water column only a part of the population is observable directly. In the case of negligible deviations from vertical homogeneity, however, the ground truth relationships can be extrapolated to the whole population contained in the column. For illustration, Figures 5.24a-c show ground truth relationships which were derived by correlation of organism numbers contained in the first 32 cm of the column as well as in the whole analyzed column (0-2.84 m) with the intensity ratio of spectral backscattered signals measured above the surface (Tables 5.9a-c). As expected, the slopes with depth of both functions are nearly equivalent in the case of vertical homogeneity. In this case the extrapolation from the directly visible part of the population to the whole population is possible using $p(z_i) \approx n_p \approx$ const. and a simple modification of the parameter $q(z_i)$ in Eq. 42 by a linear function according to

$$q(z_i) = m_q \cdot z_i + n_q \qquad (44)$$

(The variation of the coefficients $n_p$, $m_q$ and $n_q$ derived for the total of analyzed data and for the data representing vertical homogeneous distribution of organisms (Fig. 5.24) is listed in Table 5.10 for vertical extension of the water column be-

tween surface and 2.84 m depth. (Note, however, that the coefficients p and q as well as the coefficients of Eq. 44 are only valid for the data analyzed here.)

In general, these results show that the ground truth for phytoplankton detection not only requires the analysis of the species composition and their vertical distribution, but also on the vertical extension of the populated zone. Ignorance of these effects may cause errors of the determined number of organisms by up to several orders of magnitude. For example, in the case of the data analyzed here differences the estimated number of organisms contained in the water between 20 and 200% are caused if stratification of the organisms and of vertical extension of the population in the column are neglected.

## Optical and bio-optical characteristics

In contrast with the influence of vertical inhomogeneities of phytoplankton concentration upon the detection of the total content of organisms, the influence of vertical stratification of attenuating substances upon the determination of the mean optical and bio-optical characteristics of the column by remote sensing was found to be small. The ground truth relationships derived for the mean attenuation coefficients and for the mean 0.1% depth of penetration of solar UV-B, UV-A and PAR as well as for visual depth (Secchi) and Jerlov water type are shown in Figures 5.25 and 5.26 for different channel ratios of the CZCS (coefficients of Eq. 43 see Table 5.11).

## SUMMARY AND CONCLUSIONS

The results show that not only the content of phytoplankton but also its influence by solar UV and PAR are observable in both oceanic as well as in comparatively turbid waters by remote sensing. However, there are substantial difficulties which are caused both by the modulation of the backscattering signals by attenuating substances in the transmitted water column and by the vertical distribution of the organisms. Vertical inhomogeneities of the phytoplankton concentration cause vertical differences of the spectral attenuation coefficient, especially in the relevant channels of optical detection. Thus, the attenuation of the backscattered signals from different depths is not only a function of the transmitted water layer but also of the vertical distribution of absorbing organisms and of attenuating substances.

Depending on water type and stratification of the organisms, only a small part of the phytoplankton population is usually detectable directly by backscattering measurements in turbid waters like those of the Baltic Sea lagoons. Thus, the signals received from backscattered measurements only cover the near surface concentration of organisms which may differ substantially from the total content in the column. This

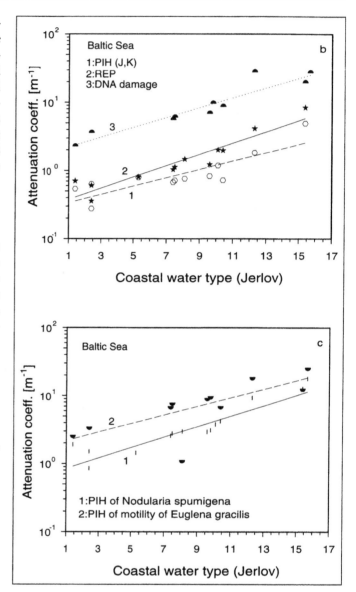

means that phytoplankton organisms which populate deeper layers of the water column are invisible for remote sensing.

If the organisms are well mixed in the column by strong winds, the total content of organisms can be estimated by an additional determination of the water type which is also possible by comparing suitable optical channels (for example 440 nm and 560 nm). In contrast, the high transparency of oceanic waters allows the direct detection of phytoplankton by remote sensing even at deeper layers. Hence, the quantitative determination of the content of

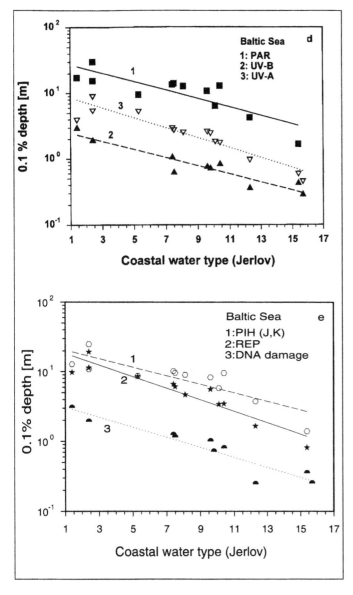

species differ from each other in their content of chlorophyll and accessory pigments. Consequently, the remote sensing method has to be adapted to the special conditions in the investigated waters by a simultaneous ground truth measurement.

The quantitative determination of the content of organisms and the estimation of photobiological processes in the organisms, such as photosynthesis and solar UV-B damage, requires additional information on water type, species composition and vertical distribution and extension in the column. In order to obtain information on optical properties of the water and vertical distribution of the population, multichannel backscattering measurements with bandwidths smaller than 20 nm are necessary which allow one to distinguish between phytoplankton and background information. In addition, calibration of the remote sensing data is necessary by ground truth which considers the individual optical properties of the species in the water.

organisms requires additional information on their vertical distribution and on the attenuation of the backscattered signals which can be estimated from multispectral measurements which are performed, e.g., by the SeaWiFS instrument.

Thus, the choice of suitable optical channels not only depends on the composition of species which establish the phytoplankton population but also on the water type. Different types and contents of absorbing and scattering substances attenuate the backscattered signals in dependence of the wavelength. Furthermore, different

## ACKNOWLEDGMENTS

The authors are grateful to the Bundesministerium for Forschung und Technologie (01 LB 9292/0) and to the Commission for the European Community (EVSV-CT 91-0026). We thank U. Vietinghoff (Universität Rostock, Institut für Biophysik), H. Baudler (Universität Rostock, Fachbereich Biologie, Biologische Station Zingst), L. Meyer-Reil (Ernst-Moritz-Arndt-Universität Greifswald, Institut für Ökologie), E. Hagen (Institut für Ostseeforschung, Rostock-Warnemünde), H.-J. Schönfeldt (Universität Leipzig, Maritimes

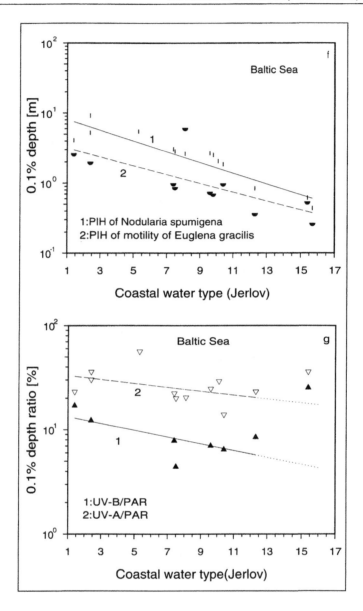

Observatorium Zingst), F. Lopez-Figueroa (Universidad de Malaga, Facultad de Sciencias, Departamento de Ecologia) for generous support of the investigation.

REFERENCES

1. Høerslev NK. Optical properties of sea water. Landoldt-Bjornstein New Series V/ 3a, 1986.

2. Doerfer R. Imaging spectroscopy for detection of chlorophyll and suspended matter. In: Torselli, Bodechtel, eds. Imaging Spectroscopy: Fundamentals and Prospective Applications. Dordrecht, Netherlands: Kluwer Academic Press, 1992:215-257.

3. Siegel H, Gerth M, Beckert M. The variation of optical properties in the Baltic Sea and algorithms for the application of remote sensing data. Ocean Optics 1994; XII 2258:894-905.

4. Heath OVS. Die Physiologie der Photosynthese. New York: Thieme, Stuttgart, 1972.

5. Lawlor DW. Photosynthese. New York: Thieme, Stuttgart, 1990.

6. Smith RC, Baker KS. Optical classification of natural waters. Limnol Oceanogr 1978; 23:260-267.

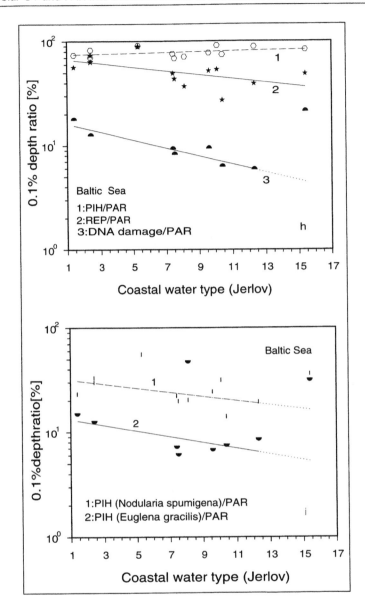

7. Jerlov NG. Ultraviolet radiation in the sea. Nature 1950; 166:111.

8. Jerlov NG. Light—General introduction. In: Kinne O, ed. Marine Ecology, vol 1. London, New York: Wiley, 1970:95-102.

9. Jerlov NG. Marine Optics. Elsevier, Amsterdam, 1976.

10. Smith RC, Baker KS. Penetration of UV-B and biologically effective dose-rates in natural waters. Photochem Photobiol 1979; 29:311-323.

11. Smith RC, Baker KS, Holm-Hansen O et al. Photoinhibition of photosynthesis in natural waters. Photochem Photobiol 1980; 31:585-592.

12. Smith RC, Baker KS. Optical properties of the clearest natural waters (200-800 nm). Applied Optics 1981; 20:177-184.

13. Smith RC, Tyler JE. Transmission of solar radiation into natural waters. In: Smith RC, ed. Photochemical and Photobiological Reviews, vol. 1, Plenum Press, London, New York, 1976:117-155.

14. Kirk JTO. Light and photosynthesis in aquatic ecosystems, 2nd ed., University Press, Cambridge, 1994

15. Häder D-P, Rhiel E, Wehrmeyer W. Ecological consequences of photomovement and

**Table 5.9. Phytoplankton**

| CZCS-Channel ratio | | | | | | | | | |
|---|---|---|---|---|---|---|---|---|---|
| | 1/4 | | | 3/6 | | | 4/6 | | |
| $z_i$ | p | q | r | p | q | r | p | q | r |
| **a. Phytoplankton–all data** | | | | | | | | | |
| [m] | – | – | – | – | – | – | – | – | – |
| 0 | -2.542 | 2.266 | -0.70 | -1.689 | 2.560 | -0.64 | 0.007 | 0.438 | 0.00 |
| 0.32 | -2.690 | 1.931 | -0.79 | -1.906 | 2.389 | -0.76 | -0.488 | 0.750 | -0.15 |
| 0.64 | -2.688 | 2.222 | -0.78 | -1.951 | 2.738 | -0.78 | -0.582 | 1.186 | -0.18 |
| 0.96 | -2.643 | 2.359 | -0.77 | -1.863 | 2.787 | -0.77 | -0.531 | 1.278 | -0.16 |
| 1.26 | -2.662 | 2.497 | -0.78 | -1.925 | 3.000 | -0.77 | -0.502 | 1.359 | -0.15 |
| 1.58 | -2.708 | 2.629 | -0.78 | -1.943 | 3.121 | 0.76 | -0.475 | 1.416 | -0.14 |
| 1.89 | -2.727 | 2.722 | -0.78 | -1.959 | 3.221 | -0.76 | -0.457 | 1.469 | -0.14 |
| 2.21 | -2.754 | 2.808 | -0.77 | -1.988 | 3.324 | -0.76 | -0.457 | 1.535 | -0.13 |
| 2.52 | -2.790 | 2.900 | -0.77 | -2.015 | 3.424 | -0.76 | -0.452 | 1.594 | -0.13 |
| 2.84 | -2.672 | 2.855 | -0.76 | -1.952 | 3.379 | -0.76 | -0.723 | 2.018 | -0.22 |
| **b. Phytoplankton–vertical homogeneous distribution** | | | | | | | | | |
| [m] | – | – | – | – | – | – | – | – | – |
| 0 | -2.915 | 2.571 | -0.85 | -1.962 | 2.934 | -0.81 | -0.742 | 1.502 | -0.24 |
| 0.32 | -3.157 | 2.240 | -0.86 | -2.026 | 2.503 | -0.78 | -0.720 | 0.954 | -0.22 |
| 0.64 | -3.148 | 2.530 | -0.86 | -2.061 | 2.845 | -0.80 | -0.792 | 1.361 | -0.24 |
| 0.96 | -3.064 | 2.653 | -0.85 | -1.916 | 2.825 | -0.79 | -0.793 | 1.550 | -0.24 |
| 1.26 | -3.059 | 2.780 | -0.85 | -2.023 | 3.114 | -0.80 | -0.765 | 1.637 | -0.24 |
| 1.58 | -3.103 | 2.915 | -0.86 | -2.044 | 3.242 | -0.80 | -0.762 | 1.735 | -0.23 |
| 1.89 | -3.107 | 2.994 | -0.85 | -2.058 | 3.360 | -0.80 | -0.753 | 1.796 | -0.23 |
| 2.21 | -3.103 | 3.048 | -0.84 | -2.073 | 3.413 | -0.79 | -0.767 | 1.875 | -0.23 |
| 2.52 | -3.111 | 3.109 | -0.84 | -2.081 | 3.479 | -0.79 | -0.781 | 1.951 | -0.23 |
| 2.84 | -3.072 | 3.131 | -0.83 | -2.074 | 3.521 | -0.79 | -0.786 | 2.010 | -0.24 |
| **c. Phytoplankton–vertical stratified organisms** | | | | | | | | | |
| [m] | – | – | – | – | – | – | – | – | – |
| 0 | -2.049 | 1.912 | -0.48 | -1.184 | 1.949 | -0.35 | 1.565 | -1.768 | 0.41 |
| 0.32 | -1.491 | 1.211 | -0.55 | -1.308 | 1.758 | -0.60 | 0.309 | -0.225 | 0.13 |
| 0.64 | -1.530 | 1.525 | -0.53 | -1.428 | 2.185 | -0.62 | 0.162 | 0.283 | 0.06 |
| 0.96 | -1.679 | 1.767 | -0.54 | -1.507 | 2.421 | -0.61 | 0.265 | 0.273 | 0.09 |
| 1.26 | -1.809 | 1.964 | -0.55 | -1.601 | 2.643 | -0.62 | 0.275 | 0.371 | 0.09 |
| 1.58 | -1.887 | 2.112 | -0.56 | -1.631 | 2.774 | -0.61 | 0.345 | 0.364 | 0.11 |
| 1.89 | -1.911 | 2.213 | -0.56 | -1.633 | 2.863 | -0.61 | 0.387 | 0.387 | 0.13 |
| 2.21 | -1.919 | 2.300 | -0.56 | -1.632 | 2.943 | -0.59 | 0.440 | 0.389 | 0.14 |
| 2.52 | -1.931 | 2.389 | -0.55 | -1.633 | 3.026 | -0.58 | 0.505 | 0.372 | 0.16 |
| 2.84 | -1.709 | 2.261 | -0.52 | -1.553 | 2.940 | -0.61 | -0.089 | 1.256 | -0.03 |

**Table 5.10. Changes of the coefficients $n_p$, $m_q$ and $n_q$ in a water column between surface and 2.84 m depth for the channel ratios 1/4 and 3/6**

| CZCS-channel ratio | $n_p$ | $m_q$ | $n_q$ |
|---|---|---|---|
| 1/4 | – 3.13 ... -2.66 | 0.335 ... 0.364 | 1.973 ... 2.294 |
| 3/6 | – 1.99 ... -1.89 | 0.388 ... 0.392 | 2.430 ... 2.524 |

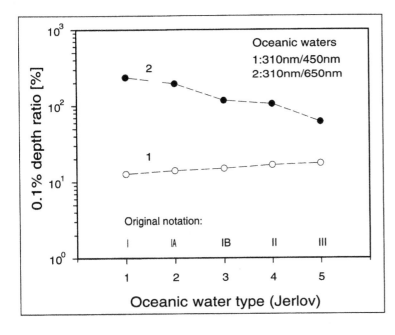

Fig. 5.18. Ratio between the 0.1% depth of subsurface irradiance at 310 nm and between the 0.1% depth at 450 nm (1) as well as at 650 nm (2) as a function of oceanic water type in the original Jerlov system of classification (data determined from Jerlov[1]).

## Table 5.11. Optical and bio-optical characteristics (Y)

**CZCS-Channel ratio**

| | 1/4 | | | 3/6 | | | 4/6 | | |
|---|---|---|---|---|---|---|---|---|---|
| | p | q | r | p | q | r | p | q | r |
| J | −6.453 | 15.713 | −0.78 | −5.670 | 18.114 | −0.83 | −1.603 | 13.627 | −0.20 |
| D | 2.471 | −0.639 | 0.83 | 2.054 | −1.417 | 0.83 | 0.414 | 0.460 | 0.15 |
| $d_{01}$(PAR) | 3.916 | 3.146 | 0.84 | 3.257 | 1.911 | 0.85 | 0.718 | 4.794 | 0.16 |
| $d_{01}$(UV-B) | 0.369 | 0.226 | 0.84 | 0.306 | 0.111 | 0.85 | 0.068 | 0.381 | 0.16 |
| $d_{01}$(UV-A) | 1.124 | 0.524 | 0.84 | 0.935 | 0.169 | 0.85 | 0.212 | 0.989 | 0.17 |
| K(PAR) | −0.779 | 1.734 | −0.81 | −0.648 | 1.980 | −0.82 | −0.119 | 1.370 | −0.13 |
| K(UV-B) | −10.85 | 22.093 | −0.81 | −9.013 | 25.502 | −0.82 | −0.162 | 16.966 | −0.13 |
| K(UV-A) | −4.531 | 8.556 | −0.81 | −3.771 | 9.988 | −0.82 | −0.672 | 6.411 | −0.13 |

photobleaching in the marine flagellate *Cryptomonas maculata*. FEMS Microbiol Ecol 1988; 53:9-18.

16. Häder D-P, Griebenow K. Orientation of the green flagellate, *Euglena gracilis*, in a vertical column of water. FEMS Microbiol Ecol 1989; 53:159-167.

17. Häder D-P, Worrest RC, Kumar HD. Aquatic ecosystems. UNEP Environmental Effects Panel Report. Nairobi, Kenya: United Nations Environmental Programme, 1989:39-48.

18. Häder D-P, Worrest RC, Kumar HD. Aquatic ecosystems. UNEP Environmental Effects Panel Report. Nairobi, Kenya:

United Nations Environmental Programme, 1991:33-40.

19. Häder D-P, Reinecke, E. Phototactic and polarotactic responses of the photosynthetic flagellate *Euglena gracilis*. Acta Protozool 1991; 30:13-18.

20. Häder D-P. Effects of enhanced solar ultraviolet radiation on aquatic ecosystems. In: Tevini M, ed. UV-B Radiation and Ozone Depletion. Boca Raton, Ann Arbor, London, Tokyo: Lewis Publishers, 1993: 155-192.

21. Häder D-P, Worrest RC, Kumar HD et al. Effects of increased solar ultraviolet radiation on aquatic ecosystems. UNEP Envi-

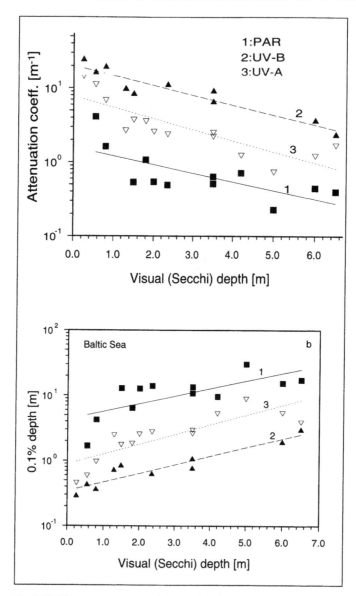

Fig. 5.19. The attenuation coefficient of scalar irradiance (a,c,e) and the 0.1% depth of subsurface scalar irradiance (b,d,f) for several broadband ranges and photobiological efficiencies as a function of visual depth (Secchi). See following two pages for Figures 5.19c-f.

ronmental Effects of Ozone Depletion. Nairobi, Kenya: United Nations Environment Programme, 1994:65-77.

22. Nultsch W, Agel G. Fluence rate and wavelength dependence of photobleaching in the cyanobacterium *Anabaena variabilis*. Arch Microbiol 1986; 144:268-271.

23. Smith RC. Ozone, middle ultraviolet radiation and the aquatic environment. Photochem Photobiol 1989; 50:459-468.

24. Voitek MA. Addressing the biological effects of decreasing ozone in the Antarctic environment. Ambio 1990; 19:52-61.

25. Piazena H, Häder D-P. Penetration of solar UV irradiation in coastal lagoons of the southern Baltic Sea and its effect on phytoplankton communities. Photochem Photobiol 1994; 60:463-469.

26. Piazena H, Häder D-P. Vertical distribution of phytoplankton in coastal waters and

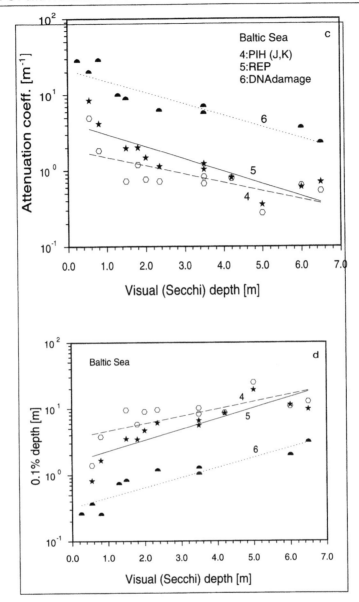

its detection by backscattering measurements. Photochem Photobiol 1995; 62: 1027-1034.

27. Häder D-P. Novel method to determine vertical distributions of phytoplankton in marine water columns. Env Exp Bot 1995; 35:547-555.

28. Häder D-P, Vogel K. Simultaneous tracking of flagellates in real time by image analysis. J Math Biol 1991; 30:63-72.

29. Aiken J, Moore GF, Holligan PM. Remote sensing of oceanic biology in relation to global climate change. J Phycol 1992; 28:579-590.

30. Häder D-P, Worrest RC, Kumar HD et al. Effects of increased solar ultraviolet radiation on aquatic ecosystems. Ambio 1995; 24:174-180.

31. Setlov RB. Proc Nat Acad Sci USA 1974; 71:3363-3366.

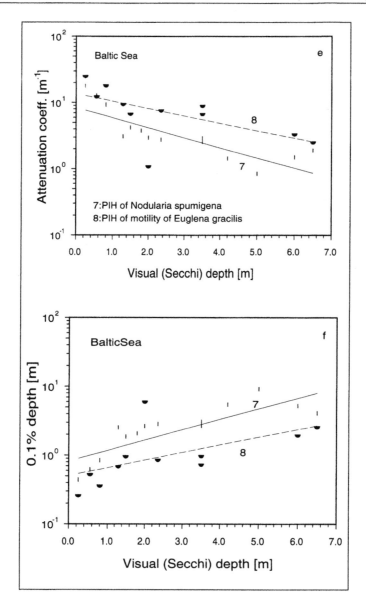

32. Häder D-P, Liu SM. Motility and gravitactic orientation of the flagellate *Euglena gracilis*, impaired by artificial and solar UV-B radiation. Current Microbiol 1990; 21: 161-168.
33. Jones LW, Kok B. Photoinhibition of chloroplast reactions. Plant Physiol 1966; 41:1037-1043.
34. Clark DK. Phytoplankton pigment algorithms for the NIMBUS-7 CZCS. In: Gower JFR, ed. Oceanography from Space. New York: Plenum Press, 1981:227-237.

35. Gordon HR, Morel AY. Remote assessment of ocean colour for interpretation of satellite visible imagery. Lecture Notes on Coastal and Estuarine Studies 4. Berlin: Springer, 1983.

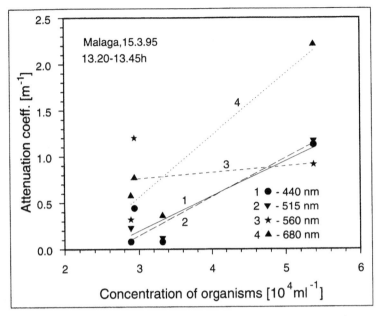

*Fig. 5.20. Attenuation coefficients at 440, 515, 560 and 680 nm in depen-dence of the concentration of phytoplankton in the Mediterranean Sea near Malaga (Spain) on 15 March 1995.*

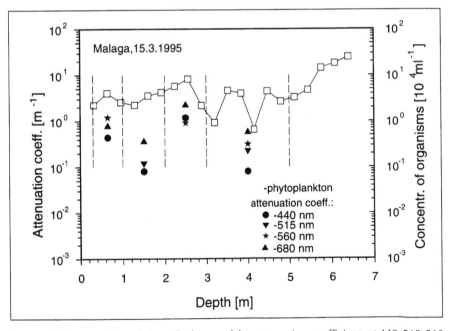

*Fig. 5.21. Concentration of phytoplankton and the attenuation coefficients at 440, 515, 560 and 680 nm as a function of depth in the nearshore waters off Malaga on 15 March 1995 (Jerlov water type: OIII; determination of the attenuation coefficients from irradiance measurements at 0.3, 1.0, 2.0, 3.0, and 5.0 m depth).*

Fig. 5.22. Concentration of phytoplankton (dashed lines) and the relative backscattered signals (dotted lines) at the wavelengths 440, 515, 560 and 680 nm in dependence of depth ($z_i$) for the Baltic Sea lagoon stations FTL, 11.5.94 (a); P, 11.5.94 (b); SG, 11.5.94 (c); R1, 7.7.94 (d); (opposite page) RS, 11.7.94 (e); K, 12.7.94 (f) and for the near shore waters off Malaga (Spain) on 15 March 1995 (g).

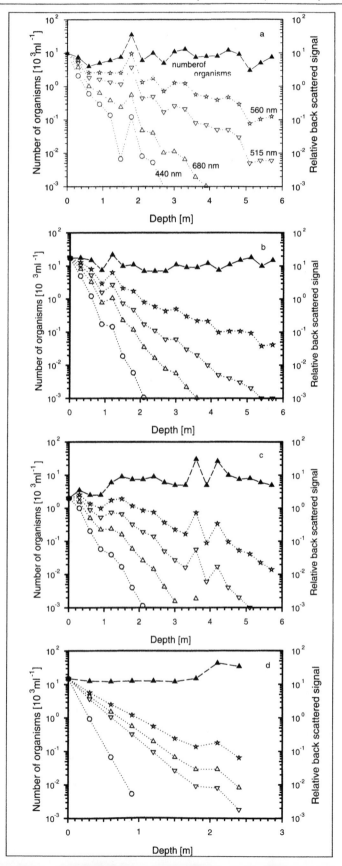

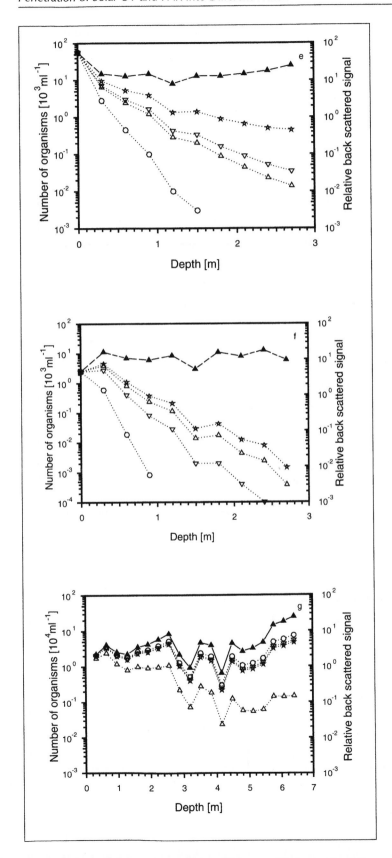

Fig. 5.23. (This and opposite page) Relative increase of the organism number ($\Delta N(z_i)/N$) and of the relative backscattered signals ($\Delta S(\lambda, z_i)/S(\lambda)$) with depth ($z_i$) at the wavelengths 440, 515, 560 and 680 nm (for the stations noted in Fig. 5.22).

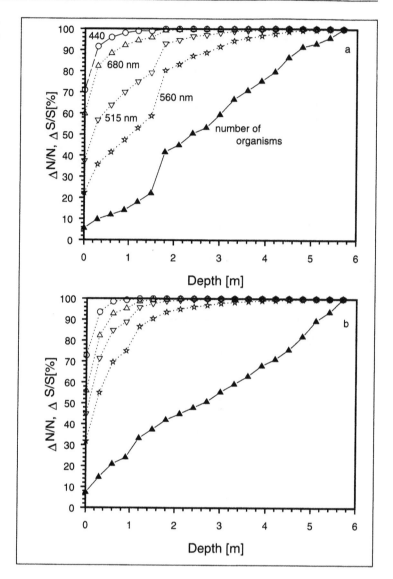

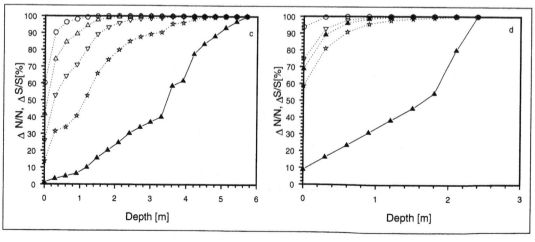

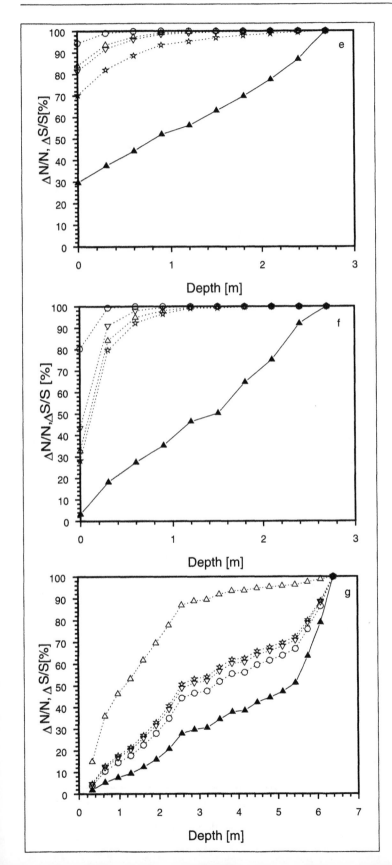

Fig. 5.24. The number of phytoplankton organisms contained in the water column of 1 m² horizontal extension and 0-0.32 m depth (solid symbols, solid lines) as well as 0-2.84 m depth (open symbols, dashed lines) as a function of the reflectance ratio of different spectral channels of the CZCS with and without respect to vertical stratifications of the organisms. Experimental areas and conditions: lagoon between the islands of Rügen and Hiddensee, 11 July 1994 (upward triangles), Barther Bodden 7 July 1994 (downward triangles) and Greifswalder Bodden, 12 July 1994 (stars). Measurements at cloudless sky between 11 and 14 h CEST; Reflectance ratios between the CZCS channels 1/4 (a, this page), 3/6 (b, opposite page), 4/6 (c, next page). Analysis of (all) data without respect to organism stratifications (1), of vertical homogeneous distribution (2) and of vertical stratifications of organisms (3).

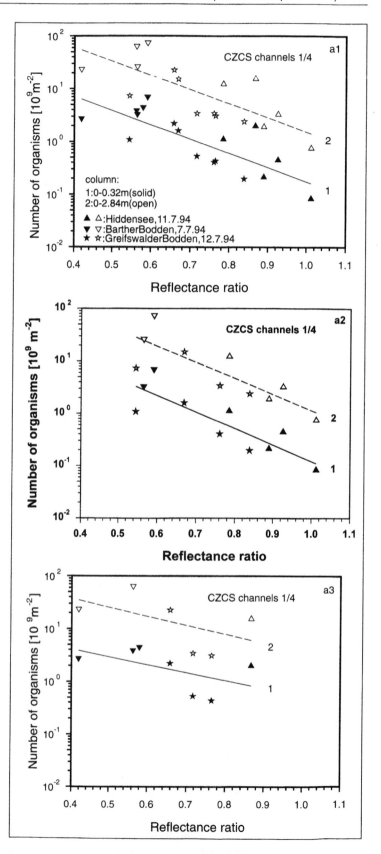

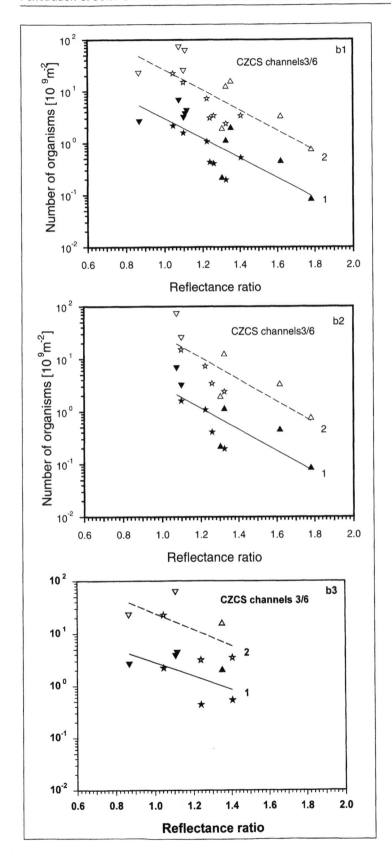

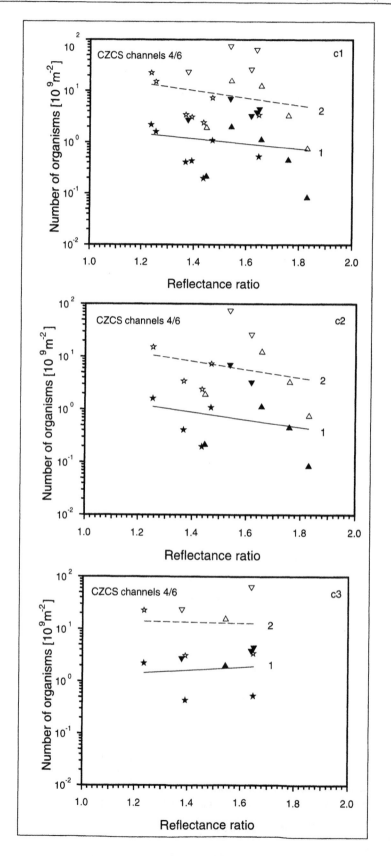

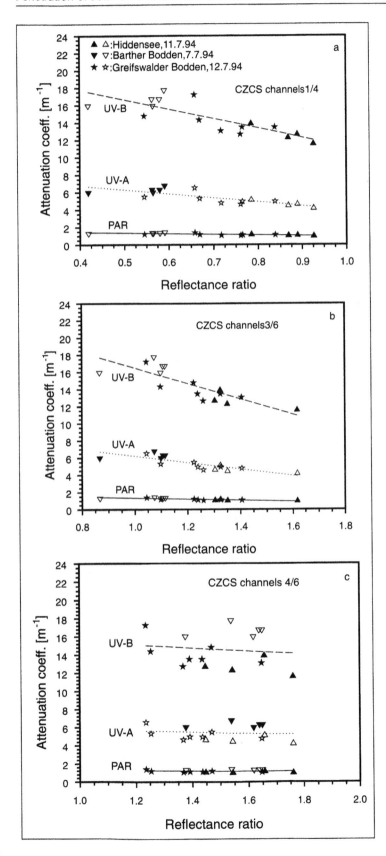

Fig. 5.25. The attenuation coefficient of solar irradiance (a-c) and the 0.1% depth of subsurface irradiance (d-f) in the ranges of UV-B, UV-A and PAR as a function of the reflectance ratio of different spectral channels of the CZCS. Experimental areas and conditions as noted in Fig. 5.24. Reflectance ratios between the CZCS channels 1/4 (a,d), 3/6 (b,e) and 4/6 (c,f). See next page for Figures 5.25d, e and f.

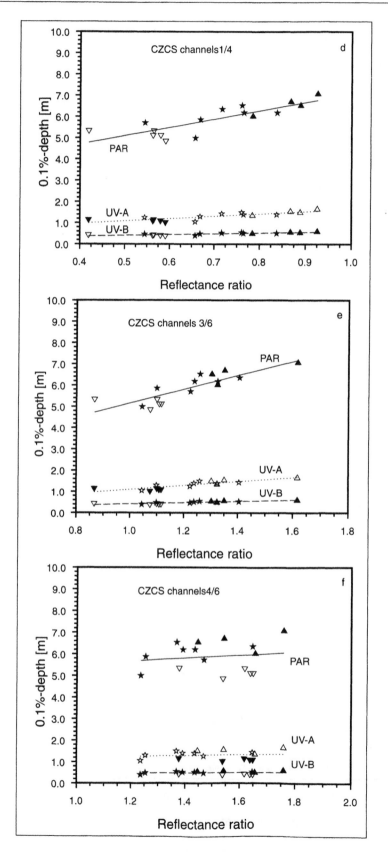

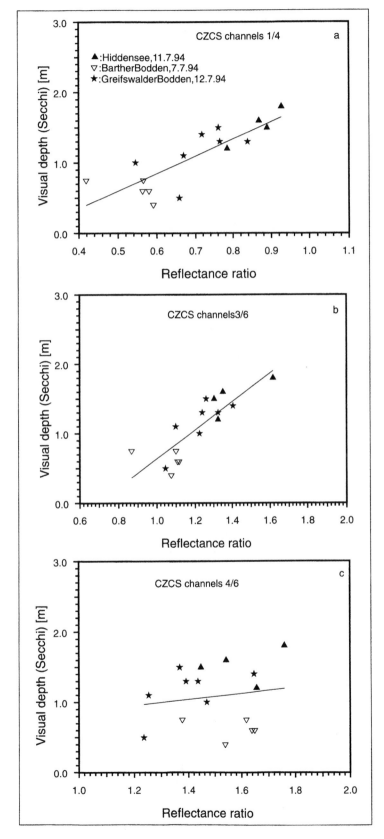

*Fig. 5.26. The visual depth (Secchi) (a-c) and the Jerlov water type for coastal waters (d-f) as a function of the reflectance ratio of different spectral channels of the CZCS. Experimental areas and conditions as noted in Fig. 5.24. Reflectance ratios between the CZCS channels 1/4 (a,d), 3/6 (b,e) and 4/6 (c,f). See next page for Figs. 5.26d, e and f.*

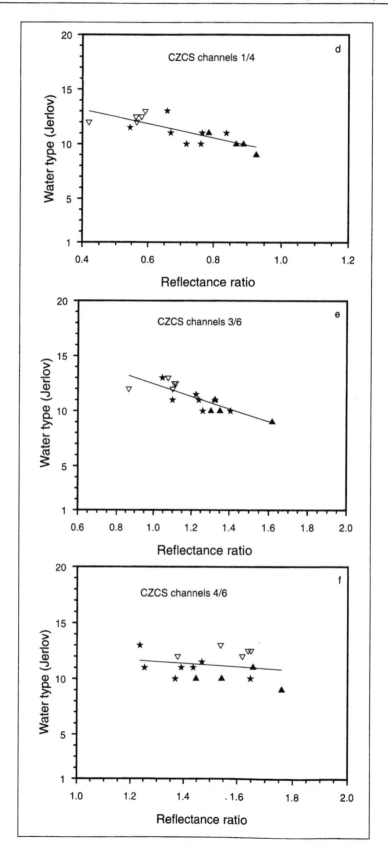

# CHAPTER 6

# Biological Weighting Functions for Describing the Effects of Ultraviolet Radiation on Aquatic Systems

John J. Cullen and Patrick J. Neale

## INTRODUCTION

To evaluate the effects of ultraviolet radiation (UV; 280-400 nm) on aquatic ecosystems, and to predict the potential influence of enhanced ultraviolet-B radiation (UV-B; 280-320 nm)[†] from ozone depletion, it is necessary to observe the effects of natural and experimentally-altered UV on different biological and chemical systems, and to quantify measured responses as functions of irradiance and time. Because biological and chemical effects of UV are strongly dependent on wavelength, spectral weighting functions must be applied to relate experimental responses quantitatively to UV exposure, if the results are to have any use for predicting effects in nature.[1-4] This is reality, not opinion. The fact is that if any biological or chemical effect is expressed as a function of radiant exposure for regimes with different spectral irradiance (i.e., different depths in the water column, times of day or year, or different

---

[†] The Commission Internationale de l'Éclairage (CIE) defines UV-B as 280-315 nm and ultraviolet-A (UV-A) as 315 - 400 nm. Here, we will follow the preference of many aquatic scientists and accept 320 nm as the boundary between UV-B and UV-A. With respect to biological processes, there are no consistent discontinuities in response at either 315 nm or 320 nm.

*The Effects of Ozone Depletion on Aquatic Ecosystems*, edited by Donat-P. Häder.
© 1997 R.G. Landes Company.

artificial sources of UV), a weighting function is used, either explicitly or implicitly. An appropriate weighting function must be applied, therefore, or at least the uncertainties associated with the chosen weighting function should be appreciated. Instead of reviewing the development of operating principles (reviewed in refs. 1-7), we will illustrate them with real data and results of model calculations.

We discuss spectral weighting functions (SWFs) in general and biological weighting functions (BWFs) in particular. In the past, BWFs were called action spectra.[1-3] Subsequently, the terms have been used interchangeably. There are, however, important differences between action spectra and BWFs that can and should be identified. We propose that "action spectrum" should be applied to SWFs that are determined through experimental exposures of samples to monochromatic radiation at different wavelengths (because interactions between wavelengths are unimportant);[3] the term "biological weighting function" should be reserved for SWFs that are determined through polychromatic exposures,[1,2] in explicit recognition of the interactions between wavelengths that characterize complex biological responses to UV. We will focus on biological responses to UV (particularly the inhibition of photosynthesis) because the interactions between wavelengths and the potential for time-dependence make them much more complicated (but not necessarily more ecologically important) than many photochemical processes that can be characterized by action spectra.

## QUANTIFYING RADIANT EXPOSURE

When algal photosynthesis ($P$) is described as a function of irradiance, exposure is often measured and expressed as photosynthetically active radiation (PAR; 400-700 nm, W m$^{-2}$ or μmol quanta m$^{-2}$ s$^{-1}$). A crude spectral weighting function is used to quantify effective irradiance: wavelengths <400 nm and >700 nm are excluded because utilization of those wave-

lengths for photosynthesis is considered to be negligible. Consequently, $P$ vs. PAR relationships are justifiably insensitive to variability in near-infrared relative to PAR. It is well recognized, however, that PAR is not the best measure of the irradiance available for aquatic photosynthesis: for most algal assemblages, blue and red wavelengths are utilized much more efficiently than green wavelengths, but the ratios of those wavebands vary with solar angle and atmospheric conditions, and they change with depth in the water column.[8] It is thus useful to weight spectral irradiance with a function describing the spectral response of algal photosynthesis, thereby quantifying photosynthetically utilizable radiation (PUR).[9] Nonetheless, it is common to see aquatic photosynthesis described as a function of PAR. If both $P$ and PAR are measured in situ, the relationships between $P$ and PAR are accurate, but not directly applicable to other environments where ratios of PUR:PAR might be different. If $P$ vs. PAR is measured in incubators that attenuate light with neutral density screens (i.e. the ratio of PUR:PAR is relatively constant in the incubator), application to natural water columns can be biased[10] to the extent that PUR:PAR is different in nature and variable with depth.[11]

The important points are: (1) A relatively crude weighting function (PAR) can be used to quantify irradiance available for photosynthesis—it excludes wavelengths that we know are generally unimportant in nature; and (2) the PAR weighting function is insensitive to spectral variability within the 400-700 nm waveband, and thus PAR alone cannot be used to predict important modes of variation in photosynthetic efficiency with location and depth.

A comparable situation exists with respect to biological responses to UV. Numerous effects have been quantified as functions of broad-band measures of UV (i.e. UV-B and/or UV-A), and a few have been characterized with spectral weighting functions. However, there is little appreciation of how much predictability a researcher sacrifices by quantifying a biological response as a function of some broadband measure of UV, rather than as a function of appropriately weighted radiant exposure. We will explore some of the consequences here.

## BIOLOGICAL WEIGHTING FUNCTIONS

To quantify biologically effective irradiance ($E^*$, dimensionless), photon fluence at each wavelength ($E(\lambda)$; W m$^{-2}$ nm$^{-1}$) must be weighted according to its biological effectiveness ($\varepsilon(\lambda)$; (W m$^{-2}$)$^{-1}$):[2,3,14,16,17]

$$E^* = \sum_{\lambda=280\,\text{nm}}^{700\,\text{nm}} \varepsilon(\lambda)\cdot E(\lambda)\cdot \Delta\lambda \qquad (1)$$

The coefficients, $\varepsilon(\lambda)$, constitute a BWF. An alternative approach is to define weighted irradiance, $E_{Beff}$, with units (W m$^{-2}$)$_{Beff}$, in which case the coefficients, $\varepsilon(\lambda)$, are dimensionless. The critical requirement, and the raison d'être for a BWF, is that it quantifies radiant exposure so that the effect of the exposure is a function of $E^*$ (or $E_{Beff}$), regardless of the shape of the irradiance spectrum.[18] We choose to determine dimensionless $E^*$ because it directly quantifies the effect if the form of the

response function is known (i.e. if UV irradiance reduces photosynthesis proportional to $1/(1 + E^*)$,[14] then $E^* = 1$ corresponds to 50% inhibition). Functions that are normalized (e.g. to 1.0 at 300 nm; Fig. 6.1) can predict *relative* changes in the biological effect as a function of changes in spectral irradiance, but the absolute magnitude of the biological effect can be calculated only if it is quantified for at least one biologically weighted radiant exposure.

The key role of BWFs has been well described in several papers that should be required reading for anyone interested in quantifying the biological effects of UV.[1-3,18] Spectral weighting functions have been determined for many processes,[19] such as damage to DNA,[15,16] inhibition of motility in an alga,[20] inhibition of photosynthesis in higher plants,[1,2] and in marine phytoplankton.[13,14,21-24] The spectra have different shapes (Fig. 6.1), and many extend well into the UV-A (320-400 nm). Clearly, neither a measurement of UV-B nor irradiance weighted by any one SWF can quantify accurately the influence of environmental radiation on different biological processes.

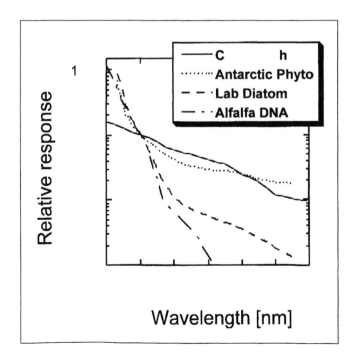

**Relative response**

**Wavelength [nm]**

*Fig. 6.1. Biological weighting functions. For this comparison, relative response has been normalized to 1.0 at 300 nm: (—) inhibition of partial photosynthetic reactions in chloroplasts;[12] (···) inhibition of photosynthesis during 60 min incubations of natural Antarctic phytoplankton grown in outdoor culture (ref.13, in which incubation time was mistakenly reported as 30 min); (- - -) inhibition of photosynthesis during 45 min incubations of the diatom Phaeodactylum in culture;[14] and (— · —) the action spectrum for pyrimidine dimer induction in alfalfa seedlings.[15]*

## THE IMPORTANCE OF A GOOD WEIGHTING FUNCTION

Weighted irradiance spectra show which wavelengths are most biologically effective, and comparisons between spectra emphasize the key wavebands (Fig. 6.2). If the wrong weighting function is used to quantify irradiance, then comparisons between radiant exposures are compromised.[18] For example, if one considers damage to DNA in alfalfa,[15] a 50% depletion of ozone corresponds to a 57% increase in biological damage (example in Fig. 6.2B). If biologically effective irradiance for inhibiting photosynthesis in a diatom is considered,[14] the relative increase is estimated to be 24% (Fig. 6.2C), whereas a broad-band UV-B meter (Fig. 6.3) might report an increase of 146% associated with the same ozone depletion (Fig. 6.2D). These comparisons demonstrate that if the weighting function for a biological process is not known, the

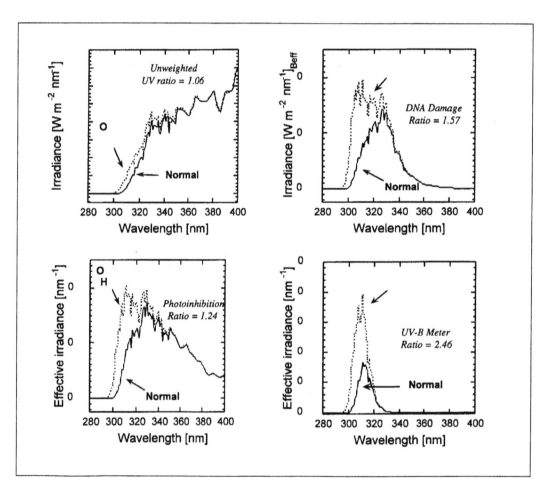

Fig. 6.2. The calculated effects of ca. 50% ozone depletion associated with the Antarctic Ozone Hole. (A) Measurements of solar irradiance[25] were made at McMurdo Station, Antarctica (78°S) on 28 October 1990 (Ozone Hole, 175 Dobson Units [DU]) and 10 November 1990 (Normal, 350 DU). To correct for cloud cover, the spectrum from 28 October was multiplied by a factor of 1.53 to match that of 10 November for integrated irradiance, 350-400 nm; (B) the same measurements, weighted by the spectrum for DNA damage in alfalfa seedlings[15] normalized to 1.0 at 300 nm; (C) weighted by the BWF for inhibition of photosynthesis in Phaeodactylum;[14] and (D) weighted by the response of a hypothetical broad-band UV-B meter (see Fig. 6.3). Relative increases in weighted irradiance (unweighted UV, W m$^{-2}$, in A) associated with ozone depletion are reported as ratios for each measure. Note that the weighted irradiance spectra show what the biological or radiometric system "sees" and integrates (see ref. 1).

biological effect associated with a change in spectral irradiance is uncertain at best. This uncertainty might be a big problem, because appropriate weighting functions are not known for many biological responses, and also because many researchers are able to quantify irradiance only with an inexpensive, broad-band meter that in practice applies a weighting function determined solely by its spectral sensitivity (Fig. 6.3).

How important is the spectral weighting? In some cases, inappropriate representation of biologically effective irradiance can lead to big errors in the prediction of UV effects in nature. To explore the influence of spectral weighting on the quantification of biologically effective UV, we have prepared a rather intricate and perhaps intimidating comparison (Table 6.1). We will discuss subsets of the data to illustrate important principles of environmental radiometry.

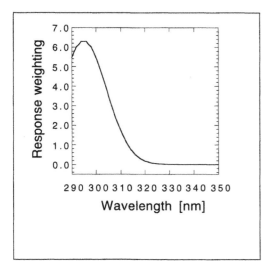

*Fig. 6.3. The response (dimensionless weightings, yielding a broad-band measurement in W m⁻²) of a hypothetical UV-B meter. The response of this meter at environmentally relevant wavelengths (>295 nm) approximates the SWF for sunburn, but some might consider the readings to represent UV-B. The relatively inexpensive meter would report total UV-B irradiance (i.e. W m⁻², 290-320 nm) accurately for mid-day clear-sky solar irradiance at 45° latitude at the equinox. The 1:1 relationship between this measure and unweighted UV-B irradiance would vary for other locations, times, depths, and sources of irradiance (see Table 6.1).*

## ULTRAVIOLET-B AS A MEASURE OF BIOLOGICALLY DAMAGING IRRADIANCE

When a relevant BWF is unavailable, it would seem appropriate to express a measured biological effect as a function of unweighted UV,[29] or UV-B and UV-A considered separately.[30] Exposure to UV-B can be measured accurately with a scanning spectroradiometer,[25,30] or calculated from a solar radiation model.[26,31-33] By relating a biological effect quantitatively to UV-B (W m⁻²), a biological weighting function is applied, with a relative weight of 1.0 for 280-320 nm, and a weight of zero for other wavelengths. Is this an acceptable simplification?

In the lower panel of Table 6.1, different measures of damaging irradiance are normalized to the value for clear-sky mid-day solar irradiance at 45° latitude at the equinox ("mid-latitude standard"; Table 6.1, column 1). If unweighted UV-B (W m⁻²) is a good measure of biologically effective irradiance for a particular process, then UV-B, relative to the mid-latitude standard, would be nearly the same as appropriately weighted irradiance, relative to the same standard exposure. For the comparisons presented here, unweighted UV-B (row H) is indeed a good measure of DNA-damaging solar irradiance (row K), if one accepts a recently published action spectrum for dimer induction in the DNA of alfalfa seedlings.[15] However, the close correspondence between UV-B (W m⁻²) and DNA-damaging irradiance (W m⁻²_Beff) does not extend to comparisons between artificial sources and solar irradiance: an unfiltered sunlamp providing UV-B fluence essentially equal to 45° latitude at noon (H9) would damage DNA 2.86 times as effectively than the mid-latitude standard exposure (K9). A similar lamp, filtered with cellulose acetate to exclude shorter wavelengths, would emit 19% the UV-B of the mid-latitude standard (H10), but it would be 34% as damaging to alfalfa DNA (K10). An experimenter who uses UV-B fluence to relate sunlamp exposures to

*Table 6.1. Different measures of UV radiation*

| | Modeled Spectra | | | | | | Measured Spectra | | | | Row |
| | | | | | | | McMurdo | | Laboratory | | |
| | 45°N | Equator | 60°S high O₃ | 60°S low O₃ | 60°S low O₃ x 0.6 | 60°S low O₃ 5 m | Normal O₃ | O₃ hole | Bare lamp | Filtered lamp | |
|---|---|---|---|---|---|---|---|---|---|---|---|
| UV-B | 1.62 | 3.30 | 1.42 | 2.44 | 1.46 | 0.44 | 1.27 | 2.31 | 1.61 | 0.31 | A |
| UV-B Meter | 1.62 | 3.86 | 1.33 | 3.46 | 2.07 | 0.56 | 1.00 | 2.46 | 4.90 | 0.74 | B |
| Setlow DNA | 0.046 | 0.161 | 0.032 | 0.232 | 0.139 | 0.032 | 0.016 | 0.077 | 2.129 | 0.175 | C |
| Alfalfa DNA | 0.45 | 0.92 | 0.40 | 0.72 | 0.43 | 0.14 | 0.39 | 0.62 | 1.28 | 0.15 | D |
| Chloroplast | 9.40 | 16.11 | 9.03 | 9.97 | 5.98 | 2.97 | 9.93 | 11.22 | 1.75 | 0.40 | E |
| *Phaeodactylum* | 0.89 | 1.61 | 0.84 | 1.09 | 0.65 | 0.29 | 0.89 | 1.10 | 0.96 | 0.12 | F |
| Antarctic phyto | 0.98 | 1.67 | 0.95 | 1.03 | 0.62 | 0.33 | 1.03 | 1.14 | 0.20 | 0.04 | G |
| | | | | | **Relative to 45°N** | | | | | | |
| UV-B | 1.00 | 2.04 | 0.88 | 1.51 | 0.91 | 0.27 | 0.78 | 1.43 | 0.99 | 0.19 | H |
| UV-B Meter | 1.00 | 2.39 | 0.82 | 2.14 | 1.28 | 0.34 | 0.62 | 1.52 | 3.03 | 0.46 | I |
| Setlow DNA | 1.00 | 3.53 | 0.69 | 5.07 | 3.04 | 0.71 | 0.35 | 1.68 | 46.62 | 3.83 | J |
| Alfalfa DNA | 1.00 | 2.05 | 0.90 | 1.62 | 0.97 | 0.31 | 0.88 | 1.37 | 2.86 | 0.34 | K |
| Chloroplast | 1.00 | 1.71 | 0.96 | 1.06 | 0.64 | 0.32 | 1.06 | 1.19 | 0.19 | 0.04 | L |
| *Phaeodactylum* | 1.00 | 1.81 | 0.94 | 1.22 | 0.73 | 0.32 | 1.00 | 1.24 | 1.08 | 0.13 | M |
| Antarctic phyto | 1.00 | 1.70 | 0.96 | 1.05 | 0.63 | 0.34 | 1.05 | 1.16 | 0.21 | 0.04 | N |
| **Column** | 1 | 2 | 3 | 4 | 5 | 6 | 7 | 8 | 9 | 10 | |

The modeled spectra are from a clear-sky solar irradiance model.[26] Mid-day spectra at 45°N and the equator are simulated for 300 Dobson Units (DU: $10^{-3}$ cm O₃) at the equinox; spectra for 60°S are for high O₃ (340 DU) and low O₃ (140 DU) during the austral spring. Effects of clouds on low-O₃ spectra are approximated with a 40% reduction at all wavelengths. Irradiance at 5 m is estimated using diffuse attenuation coefficients ($m^{-1}$) measured in the Weddell-Scotia Sea.[27] Direct measurements of solar irradiance at McMurdo Station, Antarctica (78°S) are the same as in Figure 6.2. Spectra from sunlamps, unfiltered and filtered through cellulose acetate, are from a laboratory study.[28] In the upper panel, unweighted UV-B (W $m^{-2}$) and readings from a hypothetical UV-B meter (Fig. 6.3) are reported in W $m^{-2}$. Weighted irradiance [(W $m^{-2}$)_Beff] for the Setlow DNA action spectrum,[16] alfalfa DNA spectrum (converted to energy units),[15] and a spectrum for photoinhibition of partial reactions in chloroplasts[12] come from spectra normalized to 1.0 at 300 nm. The weighting functions for photoinhibition in *Phaeodactylum*[14] and for outdoor cultures of natural Antarctic phytoplankton[13] are in units of (W $m^{-2}$)$^{-1}$; thus biologically weighted irradiance ($E^*$) is dimensionless: the reduction of photosynthesis due to photoinhibition is $(1/(1+ E^*))$. To illustrate how the comparison of different exposures is sensitive to the weighting function, each measure in the lower panel is normalized to that modeled for 45°N. In the text, elements of this table are identified by row and column, i.e. K9 = 2.86.

natural solar irradiance would thus over-estimate DNA damage in nature when extrapolating laboratory results. Conse-quently, commonly used laboratory expo-sures, quantified as unweighted UV-B, can-not be related accurately to solar irradiance in nature.

Table 6.1 indicates that for natural exposures in surface waters near mid-day, unweighted UV-B, relative to a mid-lati-tude standard, would be a reasonably good measure of damaging irradiance with re-spect to effects on DNA in a higher plant,[15] but UV-B does not serve as well for representing biological effectiveness ac-cording to other weighting functions. Thus, if the weighting function for a bio-logical process is unknown, one cannot expect the biological effect to be a consis-tent function of UV-B. Indeed, irregulari-ties in the relationship between biological effect and weighted irradiance can be used as a diagnostic of an inappropriate BWF.[18]

## BROAD-BAND MEASURES OF UV-B

Ultraviolet-B irradiance is commonly measured with a broad-band radiometer that applies its own weighting function according to its spectral response (Fig. 6.3). Such an instrument can be calibrated to report UV-B (W m$^{-2}$) accurately for a par-ticular source of irradiance (Table 6.1, cell 1B), but because instrument response is not directly proportional to energy for wavelengths between 280 and 320 nm, the proportionality between the meter reading and UV-B irradiance will change with spectral shape of the source. For example, our hypothetical UV-B meter would over-estimate the increase of UV-B with ozone depletion (B4 vs. B3, B8 vs. B7) because it weights the shorter wavelengths heavily (Fig. 6.2D). The discrepancy between UV-B and that measured by the meter is much greater for comparisons of laboratory exposures with the solar irradiance at mid-latitude (H9 vs. I9, H10 vs. I10). Thus, when a laboratory experiment is compared to a natural exposure on the basis of broad-band UV-B measurements, the comparison should be considered approximate at best,

unless it can be shown that the spectral sensitivity of the biological process closely matches the spectral response of the UV-B sensor. We have not explored in detail the use of broad-band sensors to measure the penetration of UV-B in natural waters (col-umn 6). Kirk et al[32] show clearly that such sensors are unsatisfactory for underwater use because they cannot distinguish be-tween light fields that may have the same measured UV-B, but different spectral dis-tributions, hence different biological effects. Some sensors match the response for erythema[34] so their general use is appro-priate if one wishes to quantify the poten-tial for sunburn, or a process with a closely similar SWF.

The conditions represented in Table 6.1 are arbitrarily chosen, so it is not possible to generalize broadly about how well a broad-band sensor might represent biologically effective irradiance for processes such as damage to DNA and inhibition of photosynthesis. Clearly, errors are intro-duced, and they are worse for SWFs that differ greatly from the response function of the meter.

## DIFFERENT WEIGHTING FUNCTIONS TO DESCRIBE SIMILAR PROCESSES

Weighting functions are available to quantify biologically effective irradiance for several processes,[19] but their generality is not guaranteed. For example, Setlow's[16] generalized DNA damage function is con-sistent with the action spectrum for DNA damage in phages, but for alfalfa seedlings it overestimates the effect of wavelengths <310 nm, presumably because of absorp-tion in the seedlings by cellular material other than DNA.[15] Organisms with other geometries and with UV-absorbing com-pounds[35] would be expected to have dif-ferent action spectra for damage to DNA. Likewise, BWFs for inhibition of photo-synthesis vary in shape (i.e. the slope of $\varepsilon(\lambda)$ vs. wavelength) and in absolute sensi-tivity.[13] Note that relative BWFs normal-ized to 1.0 at a particular wavelength (Fig. 6.1, refs. 21,23,24) can describe dif-ferences in shape between spectra, but not differences in biological effect for a given wavelength.

An important implication of differences between SWFs has been well described in the literature: lower weightings for UV-B (which is influenced by ozone depletion) as compared to UV-A (which is not) reduce predicted impacts of ozone depletion.[15,19] This pattern is reflected in Table 6.1, where the steep Setlow action spectrum for DNA predicts the greatest increases of effect with ozone depletion (Table 6.1; C4 vs. C3 and C8 vs. C7). Comparisons between two similarly-determined BWFs for inhibition of photosynthesis (*Phaeodactylum* grown in the laboratory, and Antarctic phytoplankton in outdoor culture; see Fig. 6.1) show that the predicted effects of ozone depletion would be somewhat higher using the laboratory result (relative increases for ozone hole vs. normal ozone; rows M vs. N). As for other SWFs, the biggest differences between the two photoinhibition BWFs are associated with the comparisons between natural irradiance and laboratory exposures to sunlamps.

How should radiant exposure during experiments be described if the BWF is unknown? Presentation of the experimental irradiance spectrum in absolute units is always useful. Further, we recommend that irradiance be weighted by several published SWFs and that each weighted exposure should be reported. Later, it may be known which spectrum is best for the process being studied.

## TEMPORAL DEPENDENCE AND RECIPROCITY OF UV EFFECTS

We have shown that it is critical to use the correct BWF for describing biologically effective irradiance, but we have not yet discussed how BWFs are determined. It is necessary to consider another issue first: how the time-dependence of UV effects may influence the relationship between radiant exposure and the measured response, i.e. the exposure response curve (ERC[36]). This subject has been addressed

directly,[6,7,18,28,36-39] and will be treated only briefly here.

For a process that is essentially irreversible over the time-scale considered, the measured effect should be a function solely of cumulative exposure (weighted with the appropriate SWF): that is, reciprocity should hold. Reciprocity is satisfied when the effect of a total radiation exposure is independent of the time over which the exposure occurs, i.e. cumulative effect is the same, regardless of exposure rate.[18,28] Reciprocity is assumed when measurements made over different periods of time are plotted together and described analytically as a function of cumulative exposure (biologically effective J m$^{-2}$ or equivalent). When reciprocity fails, plots of effect vs. cumulative exposure lose meaning,[28] and attempts to determine wavelength-dependence are compromised. The assumption of reciprocity, either explicit or implicit, is common to many studies of UV effects,[18,21,30,40,41] but dependence on irradiance has also been explicitly assumed (refs. 42,43, based on refs. 14 and 22, respectively). The assumption can be tested rigorously with controlled exposures at widely different fluence rates over a range of time periods,[28,44] but it has been evaluated only partially in the field,[21,29] where experimental exposures to modified solar irradiance are difficult to control and quantify (see discussion in ref. 6).

Reciprocity for the inhibition of photosynthesis by UV-B radiation has been tested in the laboratory.[28] Cultures of a marine diatom growing at 20°C were exposed to different irradiances of supplementary UV-B for periods of 4 h. The rate of photosynthesis declined in response to UV-B, and within about 30 min reached a rate that was maintained for the remainder of the experiment. The inhibited rate could be described as a hyperbolic function of UV-B irradiance, consistent with a dynamic balance between damage and repair.[37] The shape of the hyperbolic function (i.e. *P*, relative to the uninhibited control = $1/[1 + E^*]$) was insensitive to biological weighting, because supplemental UV-B was varied with neutral density screens so that any weighted irradiance would be equivalent when normalized to full exposure. Because the inhibited rate of photosynthesis depended on exposure rate, it could not be described solely as a function of cumulative exposure. This failure of reciprocity supported the assumption that the inhibition of photosynthesis by UV could be described as a function of biologically weighted irradiance (i.e. the BWF-PI model, with weightings, $\varepsilon(\lambda)$, having units of reciprocal W m$^{-2}$).[14]

Subsequently, we observed a contrasting response: for phytoplankton from open waters of the Antarctic at 0°C, reciprocity was well satisfied for the inhibition of photosynthesis by UV.[27,45] Rates of damage were similar to what was studied in the laboratory, so the difference in kinetic response might be due to sluggish repair associated with lower temperature, and perhaps low growth rates of the deeply-mixed Antarctic phytoplankton. Regardless of the mechanistic basis, this fundamentally different response demanded a fundamentally different model for describing the relationship between UV exposure and effect. Consequently, the BWF$_H$-PI model was developed ($H$ is the symbol for cumulative exposure in J m$^{-2}$), in which the cumulative effect of inhibition is described as the integral of a semilogarithmic survival curve.[46] In this model, biological weightings, $\varepsilon_H(\lambda)$, have units of reciprocal J m$^{-2}$ and are applied to measurements of cumulative exposure.[27]

Because reciprocity can hold under some conditions, and not under others, the assumption of reciprocity must be considered explicitly when relating biological effect to radiant exposure. That is, results must be expressed as functions of irradiance or cumulative exposure, and the choice should be justified—preferably by explicit rejection of the alternate model. If all experimental exposures are to unvarying irradiance treatments for the same duration, however, the question can be deferred: in

a relative sense, irradiance is exactly equivalent to cumulative exposure, and wavelength-dependent biological effectiveness can be determined, if only for that time-scale. To extrapolate to other time scales, one must understand, specify and parameterize the time-dependence.

## LINEARITY OF THE EXPOSURE VS. RESPONSE CURVE

If the reciprocity issue is resolved, and biologically effective irradiance (or cumulative exposure) is weighted with an appropriate BWF, then a biological effect can be related to radiant exposure in a quantitative, generalizable function. Indeed, one way of testing a BWF is to see how well it reduces scatter in the relationship between effect and weighted exposure.[18] In turn, a BWF can be determined statistically by finding the weightings that minimize unexplained variability in the exposure response curve (ERC).[1,14]

The fundamental model that relates effect to weighted exposure is no less important than a BWF for quantifying and generalizing the influence of UV on biological processes. That is, to quantify how well a BWF minimizes scatter in the relationship between response and exposure, one must specify the shape (i.e. functional form) of the ERC.[36] Consider the induction of dimer formation in DNA by UV-B: death ensues when only a small fraction of the DNA has been damaged,[47] long before the rate of damage would decline due to a significant reduction in susceptible base pairs. Consequently, cumulative damage can be accurately described as a linear function of cumulative exposure.[15] Processes that repair DNA[35] will influence the relationship between net damage and cumulative exposure.

Inhibition of photosynthesis is a different story. Commonly, rates of photosynthesis much less than 50% of a PAR-only control are measured for samples incubated several hours at surface irradiance (e.g. refs. 29,41,48). Linearity of cumulative effect (inhibition of photosynthesis relative to

PAR-only control) vs. cumulative exposure (weighted by a BWF or reported as some measure proportional to biologically weighted exposure) is neither expected nor generally observed. Why? (1) if the response is consistent with laboratory results (Fig. 6.4A),[28,37] reciprocity would fail and inhibition will be a hyperbolic function of weighted irradiance (dimensionless $E^*$ or $W\ m^{-2}{}_{Beff}$); and (2) if the process is essentially irreversible over the time-scale of damage and inhibition is a function of effective irradiance times the absorption cross section of undamaged photosynthetic systems, photosynthetic rate will decrease as

a nearly linear function of cumulative exposure (dimensionless $H^*$ or $J\ m^{-2}{}_{Beff}$), only as long as the number of susceptible targets in the photosynthetic apparatus has not significantly decreased. As the latter condition occurs (it would certainly be the case when cumulative photosynthesis is less than 50% relative to a control), the decrease in photosynthesis would become noticeably nonlinear (Fig. 6.4B), consistent with a semilogarithmic survival curve (see ref. 46). It would be instructive to consider these two types of underlying models when relating complex biological processes to radiant exposure, remembering

Fig. 6.4. Kinetics of photosynthesis during photoinhibition according to different models, adjusted to describe similar degrees of inhibition for cumulative photosynthesis over about 8 h: each plot shows instantaneous photosynthesis, relative to an uninhibited control ($P/P_{control}$; solid line), and cumulative photoinhibition, expressed as 1 - cumulative photosynthesis relative to the control (dotted line). (A) $P/P_{control}$ declines to a dynamic balance between damage and repair;[37] (B) $P/P_{control}$ is reduced consistent with a survival curve (see text). Constant irradiance is assumed: thus, time is directly proportional to cumulative exposure.

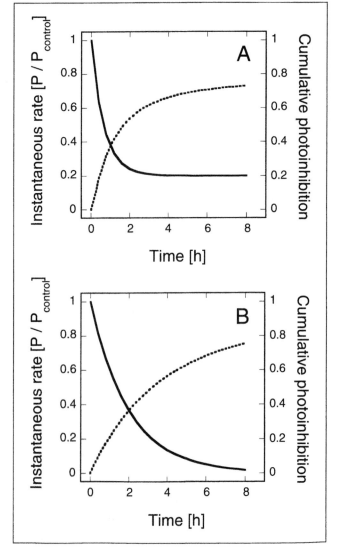

that the shape of the ERC can be distorted if an inappropriate BWF is applied.

Nonlinearity in the relationship between exposure and response should be considered explicitly when estimating the potential influences of ozone depletion. Given an appropriate spectral weighing function and calculations of solar spectral irradiance as a function of column ozone, it is possible to estimate a radiation amplification factor (RAF), a coefficient that quantifies the nonlinear change in effective irradiance as a function of the change in column ozone:[7,19,49]

$$\frac{E_1^*}{E_2^*} = (\frac{\omega_1}{\omega_2})^{-RAF} \qquad (2)$$

where $E_1^*$ and $E_2^*$ represent biologically effective irradiance at column ozone amounts $\omega_1$ and $\omega_2$, respectively. If the biological effect is a nonlinear function of $E^*$, then the ratios of effective irradiance predicted by an RAF will not directly reflect the influence of ozone depletion. In such cases, the relationship between biological effect and column ozone should be described with a total amplification factor that incorporates the nonlinear relationship between $E^*$ and biological effect.[7]

To reiterate: whether one is estimating a BWF given observations of UV-dependent responses, or predicting the influences of altered UV by using a spectral weighting function, it is crucial that biological effect is appropriately related to radiant exposure with an ERC. That is, the time-dependence and linearity of the exposure-response relationship must be known or assumed.

## METHODS FOR DETERMINING BIOLOGICAL WEIGHTING FUNCTIONS

The features necessary for determining environmentally relevant BWFs have been reviewed.[5] It is strongly emphasized that the irradiance treatments should consist of progressively greater amounts of first UV-A, then UV-B, added to a constant background of visible (PAR) irradiance.[2] This polychromatic approach for BWFs

contrasts with the monochromatic approach appropriate for action spectra, which involves measuring the effects of illumination with only a narrow band of radiation, thereby reducing or eliminating counteracting processes that may be stimulated by other wavelengths.[47,50,51] The monochromatic approach can be useful for determining the nature of damage, but it is not satisfactory for representing the net effect in nature.

Rundel[1] tackled the challenge of estimating BWFs from data on photosynthetic responses to polychromatic exposures. In general, the data show an increasingly greater effect (e.g. more severe inhibition of photosynthesis) associated with successively higher ratios of UV:PAR as shorter wavebands of UV were transmitted to the treatment. A simple analysis is to order the results by irradiance of UV exposure (W m$^{-2}$) and to compute the difference in inhibition between successive treatments. Then the biological weight for each waveband is estimated as the differential effect divided by the difference in energy between the treatments that include and exclude that waveband. However, as Rundel pointed out, the estimated weight may be inaccurate if the actual response changes rapidly over the wavelengths for which the treatments differ (the differential bandwidth is usually on the order of 15-20 nm). Moreover, there is no objective method to determine the central wavelength for a weighting. Initial estimates of the biological weighting function for inhibition of photosynthesis by Antarctic phytoplankton were based on this approach,[24,52] and results were used to calculate the relative increase of biologically damaging irradiance associated with Antarctic ozone depletion.

As an alternative to the simple differential method, Rundel advocated assuming *a priori* that the BWF is a general function, e.g. that the natural log of the BWF is a polynomial:

$$\varepsilon(\lambda) = e^{-(a_0 + a_1\lambda + a_2\lambda^2 + ...)} \qquad (3)$$

Many BWFs can be approximated by such a general exponential function (Fig. 6.1; see also refs. 21 and 22). In practice, the parameters for the BWF polynomial ($a_i$) are estimated by nonlinear regression (see ref. 53 and Appendix 6.1). The general function can be modified to introduce various complexities (e.g. thresholds), but the increased complexity should be justified statistically. For example, the fit should initially be performed for a first order exponential function ($a_0$ and $a_1$); higher order functions can be tested, but they should be accepted only if they significantly increase the amount of experimental variance explained. The standard errors of the parameter estimates from the nonlinear regression are used to calculate the uncertainty in estimates of $\varepsilon(\lambda)$.[54]

Many published SWFs decrease monotonically to zero with increasing wavelength. However, most of these spectra have been determined for "simple" processes (e.g. DNA dimerization) using monochromatic treatments. On the other hand, processes in living organisms are usually the net result of competing photo-dependent mechanisms that are all active during polychromatic treatment.[30,48,55] In particular, this means that weights may at times be negative (beneficial effects outweigh damage, e.g. ref. 13; see also refs. 22,48) and may not decrease monotonically to zero with increasing wavelength, as Equation 3 specifies. Choosing a model and justifying it statistically can be difficult under such circumstances, unless one has a large number of treatments or very precise results.

We have developed an alternative method of fitting high resolution BWFs which does not require *a priori* choice of the shape of the weighting function. In this technique (refs. 13,14 and Appendix 6.1), UV spectral irradiance in each treatment is analyzed by Principal Component Analysis (PCA) to generate up to four principal components (essentially, statistically independent shapes defined by weights for each wavelength) which account for nearly 100% of the variance of the treatment

spectra relative to the mean spectrum. Component scores (the relative contribution, $c_i$, of each principal component, $i$, to a given UV spectrum, normalized to PAR) are derived for each of the treatment spectra. The estimation of the BWF then proceeds by nonlinear regression as for the Rundel method, except that the parameters estimate the importance of each spectral component to the measured response, i.e. we estimate coefficients $m_i$ such that

$$E^* = E_{PAR}\left(m_0 + \sum_{i=1}^{4} m_i c_i\right),$$

where $E_{PAR}$ is PAR in W m$^{-2}$, $m_0$ is the coefficient for the contribution to photo-inhibition of PAR plus the mean spectrum, and $m_i$ are the contributions of the spectral components. Again, only as many components are incorporated into the final estimate as can be justified based on variance explained. Finally, once the $m_i$ are estimated by regression, the $\varepsilon(\lambda)$ estimates and their respective statistical uncertainty can be calculated via the original spectral components. Details of the method are given in refs. 13,14, and Appendix 6.1. Pertinent observations are that the calculated BWFs generally explain >90% of the variance in photosynthesis of 72 samples, and that the weighting functions can and do show features more complex than simple exponential slopes.

Accepting that the polychromatic approach is required to determine BWFs for many biological processes, and judging by publications to date, it seems that the Rundel method and the PCA method are the procedures of choice for the statistical estimation of BWFs in aquatic systems. How do they compare? As a guide to the application of the Rundel and PCA methods, we offer a couple of concrete examples. Data for calculation of several BWFs for UV inhibition of photosynthesis in Antarctic phytoplankton were acquired during a cruise to the Weddell-Scotia Confluence (WSC), during the austral spring of 1993. Details of sample collection, experimental procedures and primary photosynthetic data are described in ref. 27. Here we focus on the results from two stations where BWFs were estimated from duplicated sets of 72 measurements of photosynthesis under different spectral irradiance treatments for incubations of 1 h. The BWFs were estimated for the $BWF_H/P$-I model (consistent with the kinetics of Fig. 6.4B) using both the Rundel and PCA method. No significant increase in variance explained ($R^2$) was obtained by using a polynomial with more than linear terms in the exponent for the Rundel method (Eq. 3), or by including more than two principal components in the PCA method. We did not attempt to introduce thresholds. The PCA-fit BWFs are consistent with the overall slope of the Rundel estimated BWFs, with the PCA BWFs within the estimated 95% confidence interval for the Rundel BWFs over the 290-400 nm region most relevant to solar exposure (Fig. 6.5). Both fits are consistent with a 3-4 times greater $\varepsilon_H$ (higher sensitivity to UV inhibition) for phytoplankton from a station with a deeply mixed surface layer (Station P') as compared to a station with a more shallow mixed layer (Station P).[27] Overall, both

methods resulted in a similar total variance explained ($R^2 = 0.90$-$0.94$). Thus, either method is appropriate to fitting BWFs based on the measurements in polychromatic incubations using a large number of spectral treatments. However, the PCA method can capture possible peaks or valleys in the BWF which are not resolved by the general polynomial function. Preliminary results with estuarine dinoflagellates suggest that such features can occur when cells contain high amounts of UV-absorbing compounds (Neale et al, unpublished). The PCA method is more versatile in that it should automatically adjust to any changes in the true shape of spectral response function. For example, the BWF at another WSC station had $\varepsilon_H < 0$ for the near UV-A range ($\lambda > 360$ nm). This could be estimated using the Rundel method by adding an arithmetic offset to Equation 3.

## SUMMARY AND CONCLUSIONS

Relying on examples, real and idealized, rather than on a comprehensive review of the literature, we have illustrated some key considerations for describing the

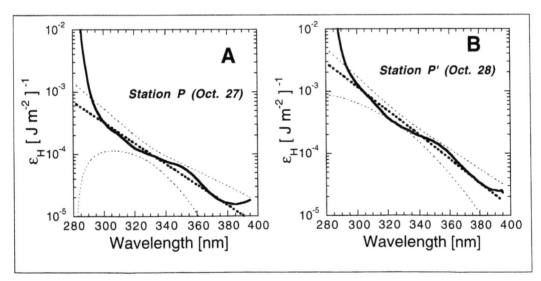

Fig. 6.5. Biological weighting functions for inhibition of photosynthesis, dependent on cumulative exposure ($\varepsilon_H$, [J m$^{-2}$]$^{-1}$; for exposures of 1 h, consistent with the kinetic model in Fig. 6.4B), based on photoinhibition measurements[14] in the Weddell-Scotia Confluence (WSC, Southern Ocean) during austral spring 1993. Statistical fits were made using the Rundel method (dotted lines, ±95% confidence intervals) and the PCA method (solid lines) for phytoplankton samples near the surface at two stations in the WSC: (A) Station P, with a relatively UV-tolerant assemblage; (B) Station P', where a relatively UV-sensitive assemblage was sampled. Full experimental details are given in ref. 27. Calculations are outlined in the Appendix.

influence of ultraviolet radiation on aquatic systems. Several points bear repeating.

1. If an effect of UV is expressed as a function of radiant exposure for re gimes with different spectral irradiance (i.e. different depths in the water column, times of day or year, or different artificial sources of UV), a weighting function is used, either explicitly or implicitiy.

2. The quantitative comparison of different radiant exposures can be very sensitive to the weighting function that is chosen: if the wrong weighting function is used and spectral shape differs strongly between treatments, comparisons can be grossly inaccurate. If the appropriate weighting function is unknown, weighted exposures should be reported according to several published BWFs.

3. Broad-band meters are in many cases unsatisfactory for comparing biologically effective exposures, and they are inappropriate for describing the penetration of biologically effective irradiance in surface waters.

4. To characterize the influence of UV on a biological process, one should determine the degree to which effect is a function of cumulative exposure, independent of exposure rate (or vice-versa), and analyze experimental results accordingly.

5. For some biological processes, such as the inhibition of photosynthesis, the relationship between effect and radiant exposure is unlikely to be linear. If nonlinearity is a feature of biological response, it should be incorporated into descriptive as well as predictive models, and radiation amplification factors should be adjusted to calculate total amplification factors.

6. Polychromatic exposures (PAR, with increments of increasingly shorter wavelengths of UV) should be used to determine environmentally relevant biological weighting func tions. Results can be obtained with an exponential curve-fitting method[1] or through the use of multivariate analysis and curve fitting.[14,27] Both methods can reveal the principal features of BWFs, but the multivariate method can discern more detail.

This discussion is no substitute for the dozens of publications in which the underlying principles have been developed—new experiments and better radiometers will help us to describe better the effects of UV on aquatic systems, but the library remains one of our most valuable research tools.

## ACKNOWLEDGMENTS

Support from NSERC Canada, NASA, and NSF Polar Programs is gratefully acknowledged. Suggestions from D.-P. Häder and an anonymous reviewer are appreciated. CEOTR publication 002.

## APPENDIX 6.1: PROCEDURES FOR CALCULATING BIOLOGICAL WEIGHTING FUNCTIONS

Here we describe how to calculate biological weighting functions (BWFs), using both the exponential weighting method[1] and the (principal component analysis (PCA) method.[2,3] The inhibition of photosynthesis is used as an example, but the principles are essentially the same for calculating any BWF. The procedure is to quantify photosynthesis under different well-measured spectral treatments, using long-pass optical filters to include progressively shorter wavelengths of UV along with PAR; the reduction of photosynthesis associated with inclusion of shorter wavelengths is described quantitatively with spectral weighting coefficients, $e(l)$. It is common to use one PAR-only treatment as a control, and about 4-6 treatments with different cut-off wavelengths for UV. Here, we describe a more general analysis: photosynthesis is measured as the uptake of [14]C-bicarbonate during controlled exposures to temporally constant irradiance in each of 72 spectral treatments: the ratios of UV:PAR are varied with long-pass filters

### Table 6.A1. Symbols and abbreviations

| Symbol | Meaning | Units |
|---|---|---|
| $a_0$ | first coefficient for exponential weighting function | dimensionless |
| $a_1, a_2$ | coefficients for exponential weighting function | $nm^{-1}$ |
| $C$ | proportionality coefficient | $(J\ m^{-2})^{-1}$ |
| $c$ | principal component scores for each case | dimensionless |
| $E(\lambda)$ | spectral irradiance at wavelength $\lambda$ | $W\ m^{-2}\ nm^{-1}$ |
| $E_N(\lambda)$ | normalized spectral irradiance at wavelength $\lambda$ | dimensionless |
| $\overline{E_N}(\lambda)$ | mean normalized spectral irradiance for all treatments | dimensionless |
| $SD[E_N(\lambda)]$ | standard deviation of $E_N(\lambda)$ for all treatments | dimensionless |
| $E_k$ | saturation parameter | $W\ m^{-2}$ |
| $E_{PAR}$ | PAR irradiance | $W\ m^{-2}$ |
| $E^*$ | biologically weighted irradiance | dimensionless |
| $H^*$ | biologically weighted radiant exposure | dimensionless |
| $i$ | case number (1-72) | dimensionless |
| $Inh$ | inhibition term | dimensionless |
| $j$ | wavelength index (1-121) | dimensionless |
| $k$ | principal component number (1-z) | dimensionless |
| $m_0$ | coefficient for influence of mean spectrum + PAR | $(J\ m^{-2})^{-1}$ |
| $m_1, m_2, m_3$ | coefficients for influence of PCs | $(J\ m^{-2})^{-1}$ |
| $n$ | number of treatments (cases) | 72 |
| $P$ | rate of photosynthesis | $mg\ C\ m^{-3}\ h^{-1}$ |
| $P_{avg}$ | average $P$ | $mg\ C\ m^{-3}\ h^{-1}$ |
| $P^B$ | $P$ normalized to chlorophyll $a$ | $mg\ C\ mg\ Chl^{-1}\ h^{-1}$ |
| $P_{pot}$ | potential, uninhibited $P$ at irradiance $E_{PAR}$ | $mg\ C\ m^{-3}\ h^{-1}$ |
| $P_s$ | maximum potential $P$ for saturating $E_{PAR}$ | $mg\ C\ m^{-3}\ h^{-1}$ |
| $P_t$ | $P$ at time $t$ | $mg\ C\ m^{-3}\ h^{-1}$ |
| $q$ | number of PCs that are included in curve-fit | dimensionless |
| $T, \tau$ | time | s |
| $T$ | duration of experiment | s |
| $w_k(\lambda)$ | principal component weights for each wavelength | dimensionless |
| $z$ | number of PCs that describe 99% of spectral variation | dimensionless |
| $\varepsilon$ | biological weighting coefficient | $(W\ m^{-2})^{-1}$ |
| $\varepsilon_H(\lambda)$ | biological weighting coefficient for radiant exposure | $(J\ m^{-2})^{-1}$ |
| $\overline{\varepsilon_H}$ | weighting coefficient for mean spectrum | $(J\ m^{-2})^{-1}$ |
| $\varepsilon_H(\lambda_R)$ | weighting coefficient for reference wavelength | $(J\ m^{-2})^{-1}$ |
| $\varepsilon_H(PAR)$ | weighting coefficient for PAR | $(J\ m^{-2})^{-1}$ |
| $\lambda$ | wavelength | nm |
| $\lambda_R$ | reference wavelength | nm |

Ranges refer to an experiment with 72 spectral treatments and measurements of spectral irradiance at 121 wavelengths from 280 nm to 400 nm.

(8 spectral treatments), and within each spectral treatment, neutral density screens are used to generate 9 treatments with similar UV:PAR, but with a broad range of PAR irradiance.[2] Whether one or many PAR irradiance exposures are used, the objective is to describe the UV-dependent reduction of photosynthetic rate.

### 1. START WITH AN UNDERLYING MODEL

A functional relationship between exposure and response (i.e. an algebraic representation of the exposure response curve) must first be chosen and justified. In its most general form, the inhibition of photosynthesis as a function of biologically effective irradiance ($E^*$, dimensionless) and time ($t$) is:

$$\left(\frac{P}{P_{pot}}\right)_t = f(E^*, t) \qquad (6.A1)$$

where $P_{pot}$ ("pot" for potential) is the un-inhibited rate of photosynthesis, dependent on PAR, and $P$ is the instantaneous rate of photosynthesis (mg C m$^{-3}$ h$^{-1}$) measured under a combination of PAR and UV. For the following examples, we use a model appropriate for the inhibition of photosynthesis in natural phytoplankton from open waters of the Antarctic, where reciprocity was verified for time scales from about 0.5-2 h (see ref. 27). The model (Fig. 6.4B) is

$$\left(\frac{P}{P_{pot}}\right)_t = e^{-H_t^*} \qquad (6.A2)$$

where $H^*$ at any time $\tau$ is biologically weighted irradiance (Eq. 6.A1) integrated through time:

$$H_\tau^* = \int_0^\tau \left(\sum_{\lambda=280nm}^{400nm} \varepsilon_H(\lambda) \cdot E(\lambda) \cdot \Delta\lambda\right) dt$$
$$(6.A3)$$

The weightings [$\varepsilon_H(\lambda)$, reciprocal J m$^{-2}$] define the BWF for inhibition of photosynthesis. The subscript ($H$, for radiant exposure, J m$^{-2}$) distinguishes this dose-dependent weighting from the irradiance-dependent weighting in Equation 6.A1.

Because this model is time-dependent, we must account for cumulative photosynthesis over the duration of the experiment (time = 0 to $T$) by integrating Equation 6.A2,

$$\int_0^T \frac{P_t}{P_{pot}} dt = \frac{(1-e^{-H_t^*})}{H_t^*} \cdot T \qquad (6.A4).$$

Cumulative photosynthesis is then expressed as an average rate over the incubation period, $T$:

$$\frac{P_{avg}}{P_{pot}} = \frac{(1-e^{-H^*})}{H^*} \qquad (6.A5).$$

This is the exposure response curve (ERC). For our experiments, $P_{pot}$ is independent of time because irradiance is constant within treatments.

Given the ERC, our task is to calculate biological weightings for $H^*$ (Eq. 6.A3) comparable to those for $E^*$ in Equation 1 of the main text. The weightings [$\varepsilon_H(\lambda)$] that determine $H$ are incorporated into a model of photosynthesis as a function of PAR, as influenced by biologically weighted UV:

$$P_{avg} = P_s(1-e^{-E_{PAR}/E_k}) \cdot \frac{(1-e^{-H^*})}{H^*}$$
$$\underbrace{1\;44\,2\;4\;43}_{P_{pot}} \quad \underbrace{14\,2\;43}_{P_{avg}} \Big/ P_{pot} \qquad (6.A6).$$

Uninhibited photosynthesis ($P_{pot}$) is a saturation function of $E_{PAR}$, defined by the maximal rate of photosynthesis ($P_s$) and a saturation irradiance ($E_k$).[6] Nonlinear curve-fitting is used to minimize unexplained error in Equation 6.A6, generating estimates of $P_s$, $E_k$, and the BWF coefficients, $\varepsilon_H(\lambda)$. We generally normalize photosynthesis to initial chlorophyll concentration, thereby yielding estimates of $P^B$ (mg C mg Chl$^{-1}$ h$^{-1}$) that are used commonly in models of productivity; the BWF is unaffected by such normalization, and the superscript will be omitted here.

For laboratory cultures at 20°C, we found that inhibition was a function solely of $E^*$ for the time scale considered (0.5-4 h).[4] In contrast to Equation 6.A5, the average rate of photosynthesis was well described by an alternate form of Equation 6.A6, where $P_{avg} / P_{pot} = 1 / (1+ E^*)$, and the BWF was comprised of coefficients [$\varepsilon(\lambda)$] with units of reciprocal W m$^{-2}$.[14]

## 2. EXPERIMENTAL MEASUREMENTS

The analysis requires determinations of $P$ (e.g. from the uptake of $^{14}$C-bicarbonate) and spectral irradiance for each treatment. Spectral irradiance, $E(\lambda)$, must be measured accurately for each of the experimental treatments, with good spectral resolution. For our experiments, $E(\lambda)$ is measured from 280-400 nm at intervals of 1 nm for each of the 72 treatments, and PAR is measured with a broad-band sensor and compared with measures of spectral irradiance from 400-700 nm. It may be impractical to set up such a large number of treatments, but if only a small number of experimental spectra are compared, statistical confidence will suffer.[1]

## 3. STATISTICAL ANALYSIS BY THE RUNDEL METHOD

For each experiment, 72 values of $P$ over time $T$ are fit to corresponding measurements of $E_{PAR}$ and spectral irradiance according to Equation 6.A6, with biologically effective exposure ($H^*$) quantified (Eq. 6.A3) with $\varepsilon_H(\lambda)$ as an exponential function of wavelength (Eq. 6.A3). Nonlinear curve-fitting (we use SAS procedure NLIN) is used to determine the photosynthetic parameters $P_s$ and $E_k$, as well as the coefficients in Equation 3 that determine the exponential dependence of biological effectiveness ($\varepsilon_H$) on wavelength.

a. We assemble a file with $i$ = 1 to 72 cases, each with 123 variables: $P_i$, $E_{PAR_i}$ and spectra of $E(\lambda)_i$ for 121 wavelengths from 280-400 nm. The values are declared as an array, $E(280{:}400)$. In practice, we normalized $E(\lambda)_i$ to $E_{PAR_i}$ for ease of use with other analyses, but the normalization is not included here.

b. The equation for photosynthesis is:

$$P_i = P_s(1 - e^{-E_{PAR_i}/E_k}) \cdot \frac{(1 - e^{-H_i^*})}{H_i^*} \quad (6.A7).$$

where

$$H_i^* = T \cdot [(\varepsilon_H(\text{PAR}) \cdot E_{PAR_i}) +$$
$$\left( \sum_{\lambda=280}^{400} E(\lambda)_i \cdot \varepsilon_H(\lambda_R) \cdot e^{-(a_1 \cdot (\lambda - \lambda_R))} \Delta\lambda )]$$

$$(6.A8).$$

Unlike Rundel, we assign dimensions to our weighting coefficients, so the weighting for any inhibition due to excess PAR, $\varepsilon_H(\text{PAR})$, and the weighting at a

reference wavelength (we use 290 nm) have units of reciprocal J m$^{-2}$. For the easier comparison with Rundel's equation (Eq. 6.A3), we substitute $C \cdot e^{-a_0}$ for the reference weighting, $\varepsilon_H(\lambda_R)$; $C$ is a proportionality constant with a value of 1 (J m$^{-2}$)$^{-1}$:

$$H_i^* = T \cdot [(\varepsilon_H(\text{PAR}) \cdot E_{PAR_i}) +$$
$$( \sum_{\lambda=280}^{400} E(\lambda)_i \cdot C \cdot e^{-(a_0 + a_1 \cdot (\lambda - \lambda_R))}) \Delta\lambda]$$

$$(6.A9).$$

Note that the experimental duration, $T$, is the same for all treatments. As long as the ERC (Eq. 6.A5) is valid, treatments over different time periods $T_i$ could be included in one analysis.

c. In the program, the parameters $P_s$, $E_k$, $a_0$, $a_1$, and $\varepsilon_H(\text{PAR})$ are declared. The model is $P_i = P_{pot_i} \cdot Inh_i$, where $P_{pot_i} = P_s \cdot (1 - e^{-(E_{PAR_i}/E_k)})$. The inhibition term, $Inh_i$, is $(1 - e^{-H_i^*})/H_i^*$. It is defined by evaluating Equation 6.A9 for each case using a do-loop for $\lambda$ = 280 to 400.

d. Estimates for $P_s$, $E_k$, $a_0$, $a_1$, and $\varepsilon_H(\text{PAR})$, along with asymptotic standard errors are obtained. Convergence should be faster if derivatives of the parameters are defined and the Marquardt method is used. The procedure can be repeated with the addition of an additional parameter, $a_2$, for inclusion in the exponent of Equation 6.A9 as $[a_2 \cdot (\lambda - \lambda_P)^2]$. See Equation 6.A3. The higher-order term should be included only if it significantly improves the fit, as evaluated with the $F$-statistic, with 1 and ($n$ - number of parameters) degrees of freedom.

## 4. STATISTICAL ANALYSIS BY THE PCA METHOD

Here, we use Principal Component Analysis (PCA) to determine statistically-independent spectral shapes that are defined by weights for each wavelength. When scored (i.e. weighted) appropriately for each treatment and combined with the mean spectral shape for the experiment, a small number of principal components (PCs) can reproduce the original spectral shape accurately. The advantage is that a few scores for PCs can provide almost as much information as the original 121 values of $E(\lambda)$ for each spectrum. Consequently, we can use nonlinear curve-fitting to determine the degree to which each PC contributes to the biological effect:[13,14] if PCs with large weights for shorter wavelengths contribute strongly to the biological effect, the BWF will likewise have strong weighting at short wavelengths, and vice-versa.

a. We start with the same data as for the Rundel method, above. However, because the analysis focuses on spectral shapes, it is necessary to normalize all values of $E(\lambda)$: $E(\lambda)_i$ is multiplied by $\Delta\lambda$, the wavelength interval for each spectral estimate (1 nm for this example), and the product is divided by $E_{PAR_i}$, generating dimensionless values of $E_N(\lambda)_i$. Normalization to $E(400)$ or irradiance at some other wavelength unaffected by the spectral filters would also be suitable.

b. Using the $n = 72$ normalized spectra, compute means and standard deviations for each of the 121 wavelengths: $\overline{E_N}(\lambda)$ and $SD[E_N(\lambda)]$. Save the file for subsequent computations.

c. Conduct a PCA on the normalized spectra using a multivariate statistical analysis package (e.g. SAS), in which the $j = 1$ to 121 variables are $E_N(\lambda)_i$ and the $i = 1$ to 72 cases are the treatments. This PCA is conducted on the correlation matrix. The analysis yields PCs (sometimes called factors), each with 121 component weights $(w_k(\lambda))$ corresponding to each wavelength, along with information on how much of the aggregate variance in the input spectra is explained by each PC. For each PC, the analysis also yields 72 component scores, $c_i$, which quantify the contribution of that PC to each treatment spectrum. In our experience, the first few PCs ($z$; usually 4) explain >99% of the variance, and only the $k = 1$ to $z$ component scores ($c_{i,k}$)

are retained for subsequent analysis. For this example, we will set $z$ to equal 4.

d. Create a data file with 72 cases and 6 variables: $P_i$, $E_{PAR_i}$ and $c_{i,k}$ for $k - 1$ to $z$.

e. As before, a nonlinear curve-fitting routine will be used to minimize the least-squares error in Equation 6.A7, except that here,

$$H_i^* = T \cdot E_{PAR_i} \cdot (m_0 + \sum_{k=1}^{z} m_k c_{i,k})(6.A10).$$

Note that multiplication by $E_{PAR_i}$ restores the magnitude of each treatment that was removed for the PCA. The coefficient $m_0$ quantifies the influence of the mean spectrum and a residual weight that represents an irradiance-dependent effect that is not dependent on UV: here, that effect is interpreted as PAR-dependent photoinhibition. The parameters of the model are $P_s$, $E_k$, $m_0$, $m_1$, $m_2$, $m_3$, and $m_4$ and the model is $P_i = P_{pot} \cdot Inh_i$, where $P_{pot}$ is defined as in Equation 6.A6, $Inh_i$ is $(1 - e^{-H_i^*})/H_i^*$; and $H_i^*$ is calculated for each case according to Equation 6.A10.

f. The iterative least-squares technique yields estimates for the parameters and their asymptotic errors, along with the amount of variance explained by the model ($R^2$). The model should be run stepwise for $k = 1$ to $z$ PCs, and each successive component (component $q$; $q \leq z$) should be retained only if it significantly increases $R^2$ (use an $F$-test with degrees of freedom = 1, n - total number of parameters).

g. The contributions of the PCs, $m_i$, is transformed back to a spectral weighting, $\varepsilon_H(\lambda)$. For each wavelength:

$$\varepsilon_H(\lambda) = \frac{\sum_{k=1}^{q} m_k \cdot w_k(\lambda)}{SD[E_N(\lambda)]} \qquad (6.A11).$$

h. The standard error of $\varepsilon_H(\lambda)$ is calculated using propagation of the standard errors of the parameters $m_i$, from the curve-fit analysis. For example, if only two PCs can be statistically justified (i.e. $q = 2$):

$$err[\varepsilon_H(\lambda)] = [err(m_1) \cdot w_1(\lambda)^2 +$$

$$(err(m_2) \cdot w_2(\lambda)^2 + 2(r_{1,2} \cdot err(m_1) \cdot$$

$$w_1(\lambda) \cdot err(m_2) \cdot w_2(\lambda)]^{1/2} / SD[E_N(\lambda)]$$

$$(6.A12)$$

where $r_{1,2}$ is the correlation between $m_1$ and $m_2$ from the output of the nonlinear regression analysis. In general for more than two PCs, the standard error is the square root of the matrix product $\mathbf{VRV'}$ divided by $SD[E_N(\lambda)]$, where $\mathbf{V}$ is the vector with elements $[err(m_1 w_1(\lambda), m_2 w_2(\lambda)...)$ and $\mathbf{R}$ is the correlation matrix of the $m$'s. That is, if there are 4 PCs in the fit, $\mathbf{V}$ has 4 elements and $\mathbf{R}$ is a 4 x 4 matrix.

Given the standard error of $\varepsilon_H(\lambda)$, an approximate confidence interval is calculated based on the $t$ statistic with ($n$ cases, total number of parameters) degrees of freedom. These are individual (as opposed to joint) confidence intervals for each of the 121 coefficients, i.e. they are not conditioned on the other values of $\varepsilon_H(\lambda)$. The accuracy of the confidence intervals is dependent on having used the correct underlying model, i.e. Equation 7.A6.

i. Once the values for $\varepsilon_H(\lambda)$ have been calculated, the weighting for the mean treatment spectrum ($\overline{\varepsilon_H}$) is evaluated as

$$\overline{\varepsilon_H} = \sum_{\lambda=280}^{400} \varepsilon_H(\lambda) \cdot \overline{E_N(\lambda)} \qquad (6.A13)$$

where the mean normalized spectral irradiance, $\overline{E_N(\lambda)}$ comes from the initial calculation for the analysis ($\Delta\lambda$ is already included in $\overline{E_N(\lambda)}$: see section 4a). Remember that $m_0$ from the curve-fitting is the contribution of the mean normalized spectrum plus that from PAR: hence the biological weighting coefficient for PAR, $\varepsilon_H(PAR)$ equals $m_0 - \overline{\varepsilon_H}$. Statistical significance of this residual can be assessed by comparison with the standard error of $m_0$.

We have not evaluated the possible consequences of PAR-inhibition being irradiance-dependent, in part because significant inhibition by PAR has seldom been encountered under our experimental conditions.

## REFERENCES

1. Rundel RD. Action spectra and estimation of biologically effective UV radiation. Physiol Plant 1983; 58:360-366.
2. Caldwell MM, Camp LB, Warner CW et al. Action spectra and their key role in assessing biological consequences of solar UV-B radiation change. In: Worrest RC, Caldwell MM, eds. Stratospheric Ozone Reduction, Solar Ultraviolet Radiation and Plant Life. New York: Springer, 1986: 87-111.
3. Coohill TP. Ultraviolet action spectra (280 to 380 nm) and solar effectiveness spectra for higher plants. Photochem Photobiol 1989; 50:451-457.
4. Smith RC, Baker KS. Stratospheric ozone, middle ultraviolet radiation and phytoplankton productivity. Oceanogr Mag 1989; 2:4-10.
5. Coohill TP. Action spectra again? Photochem Photobiol 1991; 54:859-870.
6. Cullen JJ, Neale PJ. Ultraviolet radiation, ozone depletion, and marine photosynthesis. Photosyn Res 1994; 39:303-320.
7. Smith RC, Cullen JJ. Effects of UV radiation on phytoplankton. Rev Geophys 1995; Suppl., U.S. National Report to the IUGG 1991-1994:1211-1223.
8. Kirk JTO. Light and Photosynthesis in Aquatic Ecosystems. 2nd ed. Cambridge: Cambridge University Press, 1994.
9. Morel A. Available, usable, and stored radiant energy in relation to marine photosynthesis. Deep-Sea Res 1978; 25:673-688.
10. Harrison WG, Platt T, Lewis MR. The utility of light-saturation models for estimating marine primary productivity in the field: a comparison with conventional "simulated" in situ methods. Can J Fish Aquat Sci 1985; 42:864-872.
11. Sakshaug E, Johnsen G, Andersen K et al. Modeling of light-dependent algal photosynthesis and growth: experiments with

Barents Sea diatoms *Thalassiosira nor-denskioeldii* and *Chaetoceros furcellatus*. Deep-Sea Res 1991; 38:415-430.

12. Jones LW, Kok B. Photoinhibition of chloroplast reactions. I. Kinetics and action spectra. Plant Physiol 1966; 41:1037-1043.

13. Neale PJ, Lesser MP, Cullen JJ. Effects of ultraviolet radiation on the photosynthesis of phytoplankton in the vicinity of McMurdo Station, Antarctica. In: Weiler CS, Penhale PA, eds. Ultraviolet Radiation in Antarctica: Measurements and Biological Effects. Antarctic Research Series, vol 62. Washington, D.C.: American Geophysical Union, 1994:125-142.

14. Cullen JJ, Neale PJ, Lesser MP. Biological weighting function for the inhibition of phytoplankton photosynthesis by ultraviolet radiation. Science 1992; 258:646-650.

15. Quaite FE, Sutherland BM, Sutherland JC. Action spectrum for DNA damage in alfalfa lowers predicted impact of ozone depletion. Nature 1992; 358:576-578.

16. Setlow RB. The wavelengths in sunlight effective in producing skin cancer: a theoretical analysis. Proc Natl Acad Sci USA 1974; 71(9):3363-3366.

17. Smith RC. Ozone, middle ultraviolet radiation and the aquatic environment. Photochem Photobiol 1989; 50(4):459-468.

18. Smith RC, Baker KS, Holm-Hansen O et al. Photoinhibition of photosynthesis in natural waters. Photochem Photobiol 1980; 31:585-592.

19. Madronich S. Increases in biologically damaging UV-B radiation due to stratospheric ozone reductions: a brief review. Arch Hydrobiol Beih Ergebn Limnol 1994; 43:17-30.

20. Häder D-P, Worrest RC. Effects of enhanced solar ultraviolet radiation on aquatic ecosystems. Photochem Photobiol 1991; 53(5):717-725.

21. Behrenfeld MJ, Chapman JW, Hardy JT et al. Is there a common response to ultraviolet-b radiation by marine phytoplankton? Mar Ecol Prog Ser 1993; 102:59-68.

22. Boucher N, Prézelin BB. An in situ biological weighting function for UV inhibition of phytoplankton carbon fixation in the Southern Ocean. Mar Ecol Prog Ser 1996; (in press).

23. Helbling EW, Villafañe V, Ferrario M et al. Impact of natural ultraviolet radiation on rates of photosynthesis and on specific marine phytoplankton species. Mar Ecol Prog Ser 1992; 80:89-100.

24. Lubin D, Mitchell BG, Frederick JE et al. A contribution toward understanding the biospherical significance of Antarctic ozone depletion. J Geophys Res 1992; 97(D8): 7817-7828.

25. Booth CR, Lucas TB, Morrow JH et al. The United States National Science Foundation's polar network for monitoring ultraviolet radiation. In: Weiler CS, Penhale PA, eds. Ultraviolet Radiation in Antarctica: Measurements and Biological Effects. Antarctic Research Series, vol 62. Washington, D.C.: American Geophysical Union, 1994:17-37.

26. Gregg WW, Carder KL. A simple spectral solar irradiance model for cloudless maritime atmospheres. Limnol Oceanogr 1990; 35(8):1657-1675.

27. Neale PJ, Cullen JJ, Davis RF. Inhibition of marine photosynthesis by ultraviolet radiation: Variable sensitivity of phytoplankton in the Weddell-Scotia Sea during austral spring. Limnol Oceanogr 1996; (in prep).

28. Cullen JJ, Lesser MP. Inhibition of photosynthesis by ultraviolet radiation as a function of dose and dosage rate: results for a marine diatom. Mar Biol 1991; 111: 183-190.

29. Helbling EW, Villafañe V, Holm-Hansen O. Effects of ultraviolet radiation on Antarctic marine phytoplankton photosynthesis with particular attention to the influence of mixing. In: Weiler CS, Penhale PA, eds. Ultraviolet Radiation in Antarctica: Measurements and Biological Effects. Antarctic Research Series, vol 62. Washington, D.C.: American Geophysical Union, 1994: 207-227.

30. Smith RC, Prézelin BB, Baker KS et al. Ozone depletion: ultraviolet radiation and phytoplankton biology in Antarctic waters. Science 1992; 255:952-959.

31. Green AES, Cross KR, Smith LA. Improved

analytic characterization of ultraviolet sky-light. Photochem Photobiol 1980; 31: 59-65.

32. Kirk JTO, Hargreaves BR, Morris DP et al. Measurement of UV-B in two freshwater lakes: an instrument intercomparison. Arch Hydrobiol Beih Ergebn Limnol 1994; 43:71-99.

33. Smith RC, Wan Z, Baker KS. Ozone depletion in Antarctica—Modeling its effect on solar UV irradiance under clear-sky conditions. J Geophys Res 1992; 97:7383-7397.

34. Smith GJ, White MG, Ryan KG. Seasonal trends in erythemal and carcinogenic ultraviolet radiation at mid-southern latitudes 1989-1991. Photochem Photobiol 1993; 57:513-517.

35. Karentz D. Ultraviolet tolerance mechanisms in Antarctic marine organisms. In: Weiler CS, Penhale PA, eds. Ultraviolet Radiation in Antarctica: Measurements and Biological Effects. Antarctic Research Series, vol 62. Washington, D.C.: American Geophysical Union, 1994:93-110.

36. Coohill TP. Exposure response curves action spectra and amplification factors. In: Biggs RH, Joyner MEB, eds. Stratospheric Ozone Depletion / UV-B Radiation in the Biosphere. Berlin: Springer, 1994:57-62. NATO ASI Series; vol 118.

37. Lesser MP, Cullen JJ, Neale PJ. Carbon uptake in a marine diatom during acute exposure to ultraviolet B radiation: relative importance of damage and repair. J Phycol 1994; 30:183-192.

38. Vincent WF, Roy S. Solar ultraviolet-B radiation and aquatic primary production: damage, protection and recovery. Environ Rev 1993; 1:1-12.

39. Smith RC, Baker KS. Assessment of the influence of enhanced UV-B on marine primary productivity. In: Calkins J, ed. The Role of Solar Ultraviolet Radiation in Marine Ecosystems. New York: Plenum Press, 1982:509-537.

40. Behrenfeld M, Hardy J, Gucinski H et al. Effects of ultraviolet-B radiation on primary production along latitudinal transects in the south Pacific Ocean. Mar Environ Res 1993; 35:349-363.

41. Vernet M, Brody EA, Holm-Hansen O et al. The response of Antarctic phytoplankton to ultraviolet radiation: absorption, photosynthesis, and taxonomic composition. In: Weiler CS, Penhale PA, eds. Ultraviolet Radiation in Antarctica: Measurements and Biological Effects. Antarctic Research Series, vol 62. Washington, D.C.: American Geophysical Union, 1994:143-158.

42. Arrigo KR. Impact of ozone depletion on phytoplankton growth in the Southern Ocean: large-scale spatial and temporal variability. Mar Ecol Prog Ser 1994; 114:1-12.

43. Boucher NP, Prézelin BB. Spectral modeling of UV inhibition of in situ Antarctic primary production using a field derived biological weighting function. Photochem Photobiol 1996; 64:407-418.

44. Blakefield MK, Calkins J. Inhibition of phototaxis in *Volvox aureus* by natural and simulated solar ultraviolet light. Photochem Photobiol 1992; 55:867-872.

45. Cullen JJ, Neale PJ, Davis RF et al. Ultraviolet radiation, vertical mixing, and primary productivity in the Antarctic. EOS 1994; 75:200.

46. Harm W. Biological effects of ultraviolet radiation. I.U.P.A.B. Biophysics series, vol. 1. Cambridge Univ. Press:Cambridge, England; 1980; 216.

47. Karentz D, Cleaver JE, Mitchell DL. Cell survival characteristics and molecular responses of Antarctic phytoplankton to ultraviolet-B radiation. J Phycol 1991; 27:326-341.

48. Prézelin BB, Boucher NP, Smith RC. Marine primary production under the influence of the Antarctic ozone hole: icecolors '90. In: Weiler CS, Penhale PA, eds. Ultraviolet Radiation in Antarctica: Measurements and Biological Effects. Antarctic Research Series, vol 62. Washington, D.C.: American Geophysical Union, 1994: 159-186.

49. Booth CR, Madronich S. Radiation amplification factors: improved formulation accounts for large increases in ultraviolet radiation associated with Antarctic ozone depletion. In: Weiler CS, Penhale PA, eds. Ultraviolet Radiation in Antarctica: Mea-

surements and Biological Effects. Antarctic Research Series, vol 62. Washington, D.C.: American Geophysical Union, 1994:39-42.

50. Hirosawa T, Miyachi S. Inactivation of Hill reaction by long-wavelength radiation (UV-A) and its photoreactivation by visible light in the cyanobacterium, *Anacystis nidulans*. Arch Microbiol 1983; 135:98-102.

51. Samuelsson G, Lönneborg A, Rosenqvist E et al. Photoinhibition and reactivation of photosynthesis in the cyanobacterium *Anacystis nidulans*. Plant Physiol 1985; 79:992-995.

52. Mitchell BG. Action spectra of ultraviolet photoinhibition of Antarctic phytoplankton and a model of spectral diffuse attenuation coefficients. In: Mitchell BG, Holm-Hansen O, Sobolev I, eds. Response of Marine Phytoplankton to Natural Variations in UV-B Flux. Washington, D.C.: Chemical Manufacturers Association, 1990: Appendix H.

53. Marquardt DW. An algorithm for least-squares estimation of nonlinear parameters. J Soc Ind Appl Math 1963; 11:431-441.

54. Bevington PR. Data Reduction and Error Analysis for the Physical Sciences. New York: McGraw Hill, 1969.

55. Quesada A, Mouget J-L, Vincent WF. Growth of Antarctic cyanobacteria under ultraviolet radiation: UVA counteracts UVB inhibition. J Phycol 1995; 31:242-248.

# CHAPTER 7

# BIOLOGICAL UV DOSIMETRY

**Gerda Horneck**

## THE BIOLOGICAL EFFECTIVENESS OF SOLAR UV RADIATION

Solar optical radiation is subject to absorption and scattering as it passes through the Earth's atmosphere, which results in a cut-off at wavelengths below 290 nm. It is especially the stratospheric ozone layer that acts as a protective filter, effectively cutting off incoming UV-C (190-280 nm) radiation and greatly reducing the amount of UV-B (280-315 nm) radiation reaching the Earth's surface. Although the UV-C and UV-B regions contribute only 2% of the entire solar irradiance prior to attenuation by the atmosphere,[1] they are mainly responsible for the high lethality of extraterrestrial solar radiation to living organisms. This was demonstrated with spores of *Bacillus subtilis*, which are killed effectively within a few seconds by extraterrestrial solar radiation (>190 nm), whereas at the Earth's surface about a thousand times longer exposure times are required to obtain the same effect (Fig. 7.1).[2] The reasons for this high lethality of extraterrestrial solar UV radiation—compared to the conditions on Earth—are found in the absorption characteristics of the DNA which is the crucial target for inactivation and mutation induction in this UV range. The UV action spectrum for inactivation of *B. subtilis* spores (Fig. 7.2) spans over seven orders of magnitude and declines exponentially with increasing wavelength at wavelengths longer than the peak wavelength around 260 nm.[3-7] Its spectral profile is similar to that showing generalized DNA damage.[8]

A decrease in total column ozone shifts the edge of the solar spectrum that reaches the surface of the Earth towards shorter wavelengths. It is this highly wavelength-specific absorption characteristic of ozone and the wavelength specificity of the biological action spectra in the UV-B range (Fig. 7.2) that give rise to the global concern about the impact of a depletion of the stratospheric ozone layer and thereby an increase in UV-B upon the biosphere. The shape of the action spectrum of the biological phenomenon under consideration determines whether

*The Effects of Ozone Depletion on Aquatic Ecosystems*, edited by Donat-P. Häder.
© 1997 R.G. Landes Company.

an incremental change in ozone results in significant changes in the biological effectiveness of solar UV radiation. This phenomenon stresses the need for a biological weighting of solar UV irradiance for assessing its effects on biological systems.

# QUANTIFICATION OF THE BIOLOGICAL EFFECTIVENESS OF SOLAR RADIATION

## WEIGHTED SPECTRORADIOMETRY

Three different approaches to quantify the biological effectiveness of environmen-

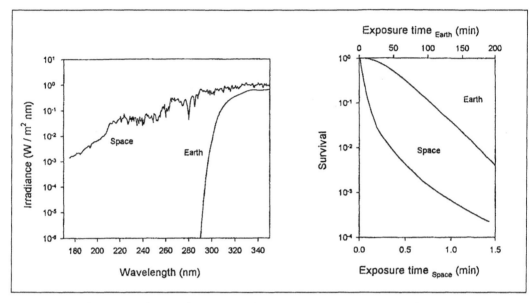

*Fig. 7.1. Survival of spores of B. subtilis Marburg after exposure to extraterrestrial solar radiation (>190 nm) during the Spacelab 1 mission (space) or to global UV radiation at the surface of the Earth around noon in San Francisco (37.5°N 122.2°W) on 10 August 1985 (Earth). Reprinted with permission from Horneck G, Brack A. In: Bonting SL ed. Advances in Space Biology and Medicine, Vol. 2 Greenwich, CT: JAI Press, 1992:229-262. © JAI Press Inc. 1992.*

*Fig. 7.2. Inactivation action spectra (50-400 nm) for spores of B. subtilis Marburg (UVR Marburg),[4,6] HA101 (UVR 101)[5] and its repair deficient mutant TKJ uvrA10 ssp-1 (UVS).[3,5,7]*

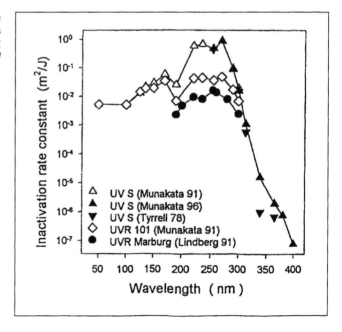

tal UV radiation are available: (1) weighted spectroradiometry; (2) the use of wavelength integrating chemical-based or physical dosimetric systems with spectral sensitivities similar to a biological response curve, and (3) biological dosimetry that directly weights the incident radiation in relation to the biological effectiveness of the different wavelengths and in relation to interactions between their effects. A discussion of the applicability of these three methods follows.

Spectroradiometry is the most fundamental radiometric technique. Biologically weighted radiometric quantities are derived from the spectral data by multiplication with an action spectrum of a relevant photobiological reaction, e.g. DNA damage, erythema formation, skin cancer, reduced productivity of terrestrial plants or UV sensitivity of aquatic ecosystems. The biologically effective irradiance $E_{eff}$ [W m$^{-2}$] is then determined as follows:[8]

$$E_{eff} = \int E_\lambda(\lambda) \bullet S_\lambda(\lambda)d\lambda \quad (1)$$

with $E_l(l)$ = solar spectral irradiance [W m$^{-2}$ nm$^{-1}$]

$S_l(l)$ = action spectrum [relative units]

$l$ = wavelength [nm]

Integration of the biologically effective irradiance $E_{eff}$ over time, e.g. a full day, gives the biologically effective dose $H_{eff}$ [J m$^{-2}$], e.g. daily dose.

Figure 7.3 shows an example where the inactivation action spectrum of UV-sensitive *B. subtilis* spores (data taken from Fig. 7.2) is utilized as a biological weighting function.[7] The solar biological effectiveness spectra, which are obtained by multiplication of the solar spectrum with the action spectrum, exhibit peaks in the range of 305-315 nm with a shift towards shorter wavelengths with decreasing solar zenith angle. Besides this strong

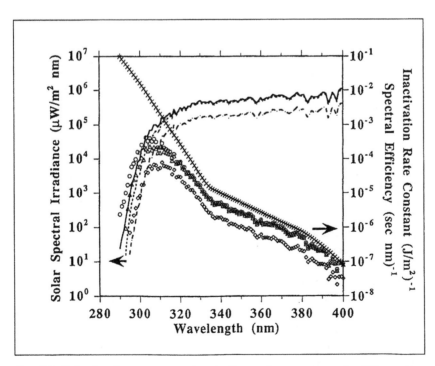

*Fig. 7.3. Solar irradiance spectra, inactivation action spectrum for UV sensitive B. subtilis spores and calculated solar biological effectiveness spectra for 28 July, 1993 noon (○), 14:00 (△), 16:00 (◊) at Tsukuba (36.1°N, 140.0°E), Japan. Reprinted with permission from Munakata N, Morohoshi F, Hieda K et al, Photochem Photobiol 1996; 63: 74-78. © 1996 American Society for Photobiology.*

dependence on the solar zenith angle, the weighted irradiance is influenced by clouds, aerosols and changes in the vertical ozone profile. With decreasing ozone concentrations the peaks of the biological effectiveness spectra are further shifted towards shorter wavelengths. Weighted spectroradiometry has provided experimental corroboration of the modeled relationship between ozone and UV-B. In the austral spring of 1993, at extremely low total ozone column amounts over Antarctica, DNA-weighted UV doses were measured at Palmer station (64.8°S, 64.0°W) that are equivalent to or exceeding those measured in mid summer in San Diego (32.4°N, 117.1°W).[9] Also in northern hemisphere mid latitudes, increases in weighted irradiance were measured due to reduced ozone amounts. Systematic spectroradiometric measurements in Southern Germany demonstrated that although the total irradiance as well as the UV-A irradiation were lower in summer 1993 than in summer 1992, the DNA-weighted irradiance in May 1993 was 30% higher than in May 1992.[10]

The advantages of weighted spectroradiometry lie in its high accuracy, the capability to identify influences by various parameters, like ozone concentration or cloudiness, to utilize a large variety of biological weighting functions and to use the data for the evaluation of model calculations. High demands are made on the instrument specifications, frequent calibrations and field intercomparisons with other spectroradiometers are indispensable (e.g. ref. 11). However, even when accepting highest accuracy of spectroradiometric data, the calculated $E_{eff}$ values can be erroneous due to insufficient weighting functions. Such errors may derive from action spectra of a too narrow waveband or too high measurement errors, especially in the tail of the action spectrum in the UV-A range where the biological sensitivity decreases by several orders of magnitude with increasing wavelength. Although ozone depletion affects only the UV-B edge of solar radiation, depending on the tail of the action

spectrum, the responses to UV-A and even to visible light may be important. Furthermore, Equation 1 assumes simple additivity of the effects of the various wavelength components. However, interactions between the effects of UV at different wavelengths have been reported for various biological effects (reviewed in ref. 12). Therefore, simple additivity might be an inadequate basis for biologically-weighted dosimetry and might not always provide the correct indicator for the biological responses to the solar radiation reaching the surface of the Earth.

## WAVELENGTH-INTEGRATING CHEMICAL AND PHYSICAL DOSIMETRY

For field measurements of environmental UV radiation, more simple devices than spectroradiometers are required. Widely used wavelength integrating UV detectors with spectral sensitivities similar to the standard erythema spectrum are the Robertson-Berger (RB) Meter[13] and photosensitive films, such as the polysulphone (PS) film.[14] Because their spectral sensitivity does not closely match with the mostly used action spectra,[15] the readings must be corrected considering the solar UV spectrum, the spectral response of the UV detector and the action spectrum of concern. The biologically effective UV dose $H_{eff}$ [J m$^{-2}$] is determined according to the following equation:

$$H_{eff} = \frac{\int E_\lambda(\lambda) \bullet S_\lambda(\lambda) d\lambda}{\int E_\lambda(\lambda) \bullet \upsilon_\lambda(\lambda) d\lambda} \bullet F \quad (2)$$

with $E_\lambda(\lambda)$ = solar spectral irradiance [W m$^{-2}$ nm$^{-1}$]

$S_\lambda(\lambda)$ = action spectrum [relative units]

$\upsilon_\lambda(\lambda)$ = response function of the sensor [relative units]

F  = equivalent dose of monochromatic radiation producing the same response of the detector [J m$^{-2}$]

$\lambda$  = wavelength [nm].

Because the spectral response characteristics of RB-type meters or PS-films are nearly wavelength-independent for wavelengths below 305 or 310 nm, respectively,[16] it is obvious that they are not sensitive to small variations of ozone that modulate the UV-B waveband.

## BIOLOGICAL UV DOSIMETRY

Biological dosimeters integrate the incident UV components of solar radiation thereby weighting them according to their biological effectiveness and to interactions between them. In contrast to the above mentioned chemical dosimeters, most biological dosimetry systems are characterized by an exponential increase in spectral sensitivity towards shorter wavelengths in the UV-B region (Figs. 7.2 and 7.4). Therefore they respond sensitively to small variations at the short-wavelengths edge of the solar spectrum that is modulated by ozone. Ideally, the spectral response of a biological dosimeter is identical to that of a photobiological effect of concern. In this case, the biologically effective dose $H_{eff}$ [J m$^{-2}$] is given by the following relation:

$$H_{eff} = F \qquad (3)$$

with $F$ = equivalent dose of monochromatic radiation producing the same response of the detector [J m$^{-2}$].

Because there is striking evidence that UV-induced DNA damage is the initiating event in a variety of critical photobiological reactions, such as immunosuppression, tumor promotion, virus induction and photocarcinogenesis in humans,[25] and plays also a central role in the adverse effects of UV-B radiation on terrestrial plants[26-27] and to a certain extent also in aquatic ecosystems,[22,28] biological dosimeters based on the DNA-damaging capacity of solar radiation have been widely used. The situation is more complex especially in the UV-A region, where additional endogenous chromophores and photodynamic reactions are involved.[26,29] In most cases simple test systems have been assigned to biological dosimetry (reviewed in refs. 15,30), such

as biomolecules, viruses, and bacteria. Their action spectra agree quite well with that for DNA damage (Fig. 7.4). Induction rates for lethality, mutagenesis and dimerization of pyrimidines have been quantified after insolation to directly determine the biologically effective dose of environmental UV radiation. In order to assess the implications of increased UV-B on ecosystems, more complex aquatic systems, such as flagellates[28] and *Daphnia*[22,31] or terrestrial plant systems, such as leaves or seedlings have been introduced.[23,32-33] Biological dosimeters that refer directly to a photobiochemical process can be classified as biomarkers for UV exposure, those that reflect a cellular or organism response as biomarkers for UV effects.

The advantage of a biological dosimeter heavily depends on the equivalence of its spectral response with that of the photobiological phenomenon of concern. Although nearly all action spectra for biologically adverse effects of solar radiation decrease with increasing wavelength in the UV-B and UV-A range, the rate of decline varies considerably (see, e.g. ref. 15). Even assuming the same endpoint, they may be different for different species or even strains (Fig. 7.2).[15,34-35] This phenomenon reflects the complex interaction of a variety of cellular mechanisms for processing UV-induced damage.

Biological dosimeters integrate over the exposure time, thereby providing an accurate record of the total biologically effective dose over an designated period regardless of changes in weather conditions. Even low levels of UV, as they occur in water at deeper layers, can be recorded with high sensitivity by prolonged exposure of biological dosimeters.

Biological dosimeters have to fulfill the same criteria as any other dosimetric system. This concerns the reproducibility of the system (standardized procedure, genetically defined homogenous material, robust against unspecific environmental stress parameters, long shelf life) and its dosimetric properties (calibration curve, linearity of response, dynamic range, no dose rate

Fig. 7.4. Relative action spectra of biological dosimeters: (A) uracil dimerization,[17] T7 bacteriophage inactivation,[18] spore inactivation,[7] biofilm response,[19] and (B) 7 DHC conversion to pre D3,[20] inhibition of motility in Euglena,[21] mortality of Daphnia,[22] inhibition of phytochrome-induced anthocyanin synthesis in Sinapis alba,[23] reduction of hypocotyl growth in Lepidium sativum seedlings.[24] For comparison, the action spectrum for DNA damage from ref. 8 is shown.

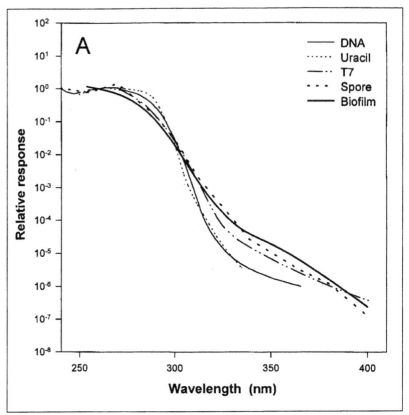

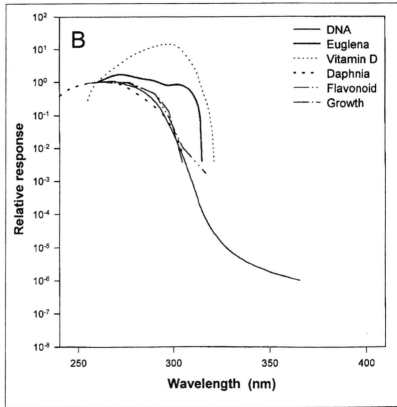

effect, action spectra using both, mono- and polychromatic radiation, dependence of response on temperature, relative humidity and angle of incidence). For assessing the UV-exposure of aquatic ecosystems with randomly oriented and actively moving organisms, a spherical target is more suitable than a horizontal surface. With respect to multicellular organisms, the spectral transmission of UV through surface layers before reaching the sensitive target must be considered or simulated. If precisely characterized, biological dosimeters are well suited to validate weighted spectroradiometry data.[7,36-37] For applications as field dosimeters the availability, suitability for routine measurements, easy handling, automatic registration—if possible—and low costs are important requirements.

## DOSAGE UNIT OF BIOLOGICALLY EFFECTIVE UV RADIATION

The terms biologically weighted UV irradiance, e.g. DNA damage weighted irradiance, in units of W m$^{-2}$ and biologically weighted UV dose, e.g. DNA weighted dose, with units of J m$^{-2}$ have been widely accepted in weighted spectroradiometry to quantify the biological effectiveness of environmental UV radiation.[9-10,38] In trend analyses on the implications of atmospheric ozone changes for biologically active UV radiation reaching the Earth's surface, relative action spectra representing generalized DNA damage,[8] generalized plant damage,[26] erythema induction[39] and skin cancer induction[40] are the mostly used biological weighting functions to convert the physical dose parameters into biologically meaningful dose parameters according to Equation 1.[38,41-42] However, it should be stressed that for trend analyses the action spectra are generally normalized to unity at 300 nm; therefore they represent only the relative biological responses at different wavelengths and not the absolute dimensions of the biological effects. Therefore, the biologically effective doses, calculated according to Equation 1 for two different photobiological processes cannot be compared numerically, only with respect to relative changes, e.g. of location, time or ozone concentration.

Biological UV dosimetry combines both—the wavelength-weighted responses of living organisms to environmental UV radiation including potential interactions between the effects of different wavebands, and the absolute sensitivity of a biological phenomenon to a certain incident radiation of concern. In order to compare the biological UV measurements with each other and with those obtained by weighted spectroradiometry and model calculations, it is desirable to express the biologically determined doses also in SI units, i.e. in J m$^{-2}$. A feasible approach has been proposed by Tyrrell[43] and is specified in Equation 3, where the biologically damaging exposure to environmental UV-radiation is equivalent to the incident dose of monochromatic radiation at a reference wavelength $\lambda_c$, which produces the same response as the actual radiation under consideration. The quantity $H_{eff}$ is the biologically effective dose given in J m$^{-2}$ (Fig. 7.5). Equal biologically effective doses $H_{eff}$ in J m$^{-2}$ would reflect equal biological responses regardless of the spectral composition of the incident radiation. From the biologically effective dose $H_{eff}$ the effective actinic irradiance $E_{eff}$ in the SI units W m$^{-2}$ can be obtained by division through the irradiation time. Knowing the action spectrum for the response of a certain biological dosimeter, the biologically measured data can be easily compared with data of weighted spectroradiometry,[7] obtained according to Equation 1 or those of model calculations.[37]

## CHARACTERISTICS OF BIOLOGICAL DOSIMETERS

### BIOMOLECULES AND VIRUSES

The characteristics of biomolecules and viruses that have been applied as biological dosimeters are listed in Table 7.1.

Fig. 7.5. Biologically effective dose $H_{eff}$, measured with the biofilm dosimeter from the calibration curve at the reference wavelength 254 nm according to Equation 3 using spores of B. subtilis UVR 101 (A) or UVS (B). These data were obtained in cooperation with W. Strauch and K. Lohmann.

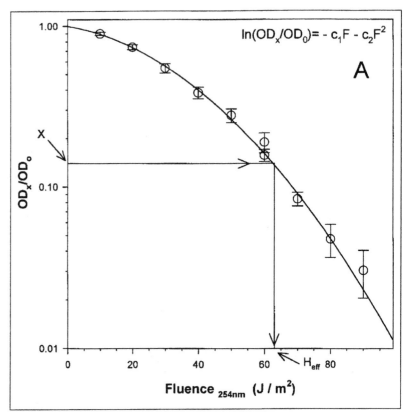

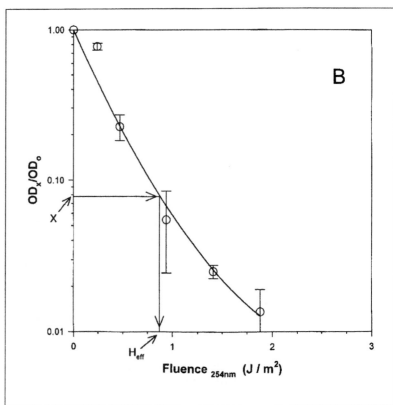

Table 7.1. Characteristics of biological dosimeters on the basis of biomolecules or viruses

| Biological dosimeter | Assay system | Biological endpoint | Dosimetric application | Dosage unit | Device | Application | Status | Ref. |
|---|---|---|---|---|---|---|---|---|
| Uracil dosimeter | crystalline uracil | dimer formation | comparison with T7 dosimeter | dose to reduce O.D. by $e^{-1}$ | thin layer on quartz plate | long-term monitoring | field measurements | 17 |
| Vitamin D dosimeter | 7 DHC | isomerization to pre-D3 | efficiency of vitamin D synthesis | % conversion of 7 DHC | quartz vessels with 7 DHC solution | seasonal and latitudinal exposures | field measurements | 20,45-46 |
| DNA dosimeter | $^{14}C$ labelled DNA of human fibroblasts | dimer formation | vertical attenuation coefficient of $H_{eff}$ in water | effective dimer producing daily dose, equivalent to 254 nm [J m$^{-2}$] | quartz tubes with DNA solution | clear tropical marine water (0-3 m depth) | demonstration experiment | 47-48 |
|  | DNA of φX174 bacteriophage | inactivation of plaque formers | vertical attenuation coefficient of $H_{eff}$ in water | effective DNA inactivating dose, equivalent to 254 nm [J m$^{-2}$] | quartz tubes with DNA solution | clear tropical marine water (0-3 m depth) | field measurement | 47 |
| Bacteriophage dosimeter | bacteriophage T1 | inactivation of plaque formers | annual profile of $H_{eff}$ sensitivity to ozone | inactivation rate constant (1/h) | quartz cuvette with phage suspension | at noon during one year in Japan | field measurement | 49-50 |
|  | bacteriophage T2 | inactivation of plaque formers | vertical attenuation coefficient of $H_{eff}$ in water | % loss of infectivity | quartz tubes with phage suspension w/ or w/o Mylar | clear lake (0-10 m depth) | intercomparison campaign | 36 |
|  | bacteriophage T4vx | inactivation of plaque formers | comparison with other biological dosimeters | inactivation rate constant [m$^2$ J$^{-1}$] | phage suspension | action spectroscopy | laboratory experiment | 3 |
|  | bacteriophage T7 | inactivation of plaque formers | global and direct $H_{eff}$ diurnal and annual profile, vertical attenuation coefficient of $H_{eff}$ in water | average number of hits in the population [\|ln(N/N$_0$)\|] | quartz cuvette with phage suspension | long-term/continuous monitoring at different sites; measurements in lakes, rivers, ocean | in operation for long-term and continuous dosimetry | 17-18,30 |

Uracil, one of the pyrimidine bases of nucleic acids, has been used as biomarker for UV exposure to measure the UV-induced pyrimidine dimerization.[17,51] Because pyrimidine dimers are among the predominant photochemical injuries of DNA in cells exposed to UV, especially at shorter wavelengths, the uracil dosimeter provides a measure of the DNA damaging efficiency of solar UV radiation. The action spectrum for uracil dimerization is shown in Figure 7.4.[17] The dimerization of uracil leads to a decrease in the absorption characteristics at 280-290 nm, which is determined spectrophotometrically. The biologically effective dose equal to unity is defined as the exposure that decreases the optical density (O.D.) of the uracil layer to the e-th part of the difference between the original (unirradiated) and the saturated values. The uracil dosimeter is about 3000 times less sensitive to solar UV than the bacteriophage T7 dosimeter (Table 7.1).[17] It is stable in response, easy to handle and has been utilized at several sites for long-term UV monitoring.

DNA in solution has been utilized as both, biomarker for UV exposure, indicating the formation of cyclobutane pyrimidine dimers, and for UV effects, indicating the inactivation of plaque forming activity of $\phi$X174 DNA.[47-48] The biologically effective dose is given in J m$^{-2}$, equivalent to a dose at 254 nm producing the same effect. The dosimeter was tested in marine waters at different depths.

7-Dehydrocholesterol (7 DHC), the precursor of vitamin D3 in methanol, has been used as biomarker for UV exposure, indicating the photoisomerization of provitamin D3 (7 DHC) to previtamin D3 which is the initiating step of cutaneous vitamin D3 synthesis.[20,45-46] Only low doses of UV are required at wavelengths below 315 nm (Fig. 7.4), e.g. in mid latitudes summer, after 1 h insolation around noon, 9% of 7 DHC are converted to previtamin D3.[45] The photoconversion of 7 DHC to pre D3 is determined spectrophotometrically[46] or by HPLC.[45] The vitamin D

dosimeter was utilized to determine the seasonal and latitudinal changes on the potential of solar radiation to initiate cutaneous production of vitamin D3 which is one of the beneficial effects of solar UV.[45]

Bacteriophages of the T series in suspension have been applied as biomarkers for UV effects indicating survival after UV exposure, which implies infection and processing by the host bacterium.[3,17-18,30,36,50,52] The UV effect is determined either by the plaque test[3,17-18,30,36,49-50,52] or in an on-line system photometrically from the lysis of the host cells.[30] From the inactivation action spectrum (Fig. 7.4) it is obvious that DNA damage is the crucial event. The biologically effective dose is given in terms of inactivation rate constant,[49-50] percent loss of infectivity[36] or as the absolute value of the term $\ln(N/N_o)$ with N = number of infectious individuals after irradiation and $N_o$ = number of viable individuals without irradiation.[18,30] Bacteriophage dosimeters have been utilized at different terrestrial sites and in different aquatic environments.

## CELLULAR SYSTEMS

The characteristics of cellular systems that have been applied as biological dosimeters are listed in Table 7.2.

Bacterial systems have been frequently used as biodosimeters, indicating lethality as the prime cellular UV effect that essentially reflects unrepaired DNA damage.[8] They include cells of E. coli in suspension[3,54-55,57] or spores of B. subtilis in suspension,[3,43,69-70] as dry layers[58-60,62,64] or immobilized in a biofilm.[15,19,37,44] Due to their exponential inactivation kinetics repair-deficient strains are especially suited for biological dosimetry. They respond with high sensitivity to low doses of UV-light, because they accumulate the DNA photoproducts and consequently are killed at relatively low numbers of lesions in their DNA. The E. coli CSR 603 *uvrA6 recA1 phr-1* strain which is deficient in all dark repair as well as photorepair pathways could be an ideal "worst case" indicator, because it quantifies the incidence of primary lethal damage by solar radiation.[54] Only 0.1% of the population survives an exposure to solar radiation of 30 s.[54] This extremely high degree in sensitivity makes it too difficult to handle these cells as field dosimeter. Its *rec*[+] variant *E. coli* CSR 06 *uvrA6 phr-1*[57] and spores of *B. subtilis uvrA10 ssp-1*[3,7,58-62] or *B. subtilis uvrA10 ssp-1 polA151*[64] are the most frequently used bacterial systems in biological dosimetry. The biologically effective dose is determined either from the inactivation rate constant[3,6,58-62,64] as number of lethal hits[7,58-61] or as percent lethality.[36,57] Because of their spectral response (Figs. 7.2 and 7.4), their long shelf-life, their resistance against other environmental parameters such as heat and moisture, the availability of repair-deficient mutants with monoexponential inactivation curves and the lack of a photoreactivation system, spores of *B. subtilis* especially meet many of the criteria for reliable biological dosimetry of solar radiation. The spore dosimeter has been utilized at different latitudinal and seasonal conditions, in South America,[3] in Japan[7,58-61] and in Antarctica[64] as well as in quality control of water disinfection by UV (254 nm),[69-70] whereas the *E. coli* cell dosimeter has mainly been used in aquatic environments in Antarctica and in freshwater lakes in the U.S.[36,57]

Based on dry spores of *B. subtilis* immobilized on transparent polyester sheets, a biofilm has been developed which combines both, the direct biological weighting of a biological dosimeter and the robustness and simplicity of a chemical film dosimeter.[15,19,37,44] After exposure to solar radiation and calibration at 254 nm, the biofilm is incubated in nutrient broth medium and the proteins synthesized after spore germination and cell growth are stained (Fig. 7.6). The UV effect measured in this system is inhibition of biological activity, determined as relative O.D. either photometrically[19,44] or by image analysis from the protein synthesized after incubation (Fig. 7.5).[37] Its action spectrum (Fig. 7.4) deviates from the standard erythema action spectrum[39] by less than a factor of 2 over a wide range from UV-B to UV-A.[15] The biologically effective dose is given as equivalent dose to that at 254 nm [J m$^{-2}$], which produces the same effect as the environmental radiation according to Equation 3 using the calibration curve (Fig. 7.5).[37] The response of the biofilm to UV follows the reciprocity law in the range of fluence rates that are characteristic for the terrestrial UV radiation climate (Fig. 7.7).[44] The response is independent of temperature between -20°C and 70°C and humidity in the range of 20% to 80% relative humidity. The biofilm can be stored at room temperature for up to 9 months without detectable changes in viability of the spores.[44] The biofilm has been used as a UV dosimeter in space and at different land sites in Europe and in Antarctica.[15,19,37,44]

In order to assess the potential hazards of enhanced solar UV-B radiation on phytoplankton productivity, sensitive test systems for freshwater (*Euglena*)[65-67] or marine habitats (dinoflagellates)[68] have been used as biomarkers for UV effects. The test parameters include impairment of motility, photo-orientation and growth. The action spectrum of inhibition of motility in *Euglena* (Fig. 7.4) suggests that in addition to DNA, other cellular components constitute the sensitive target for solar UV; proteins, above all the photoreceptor proteins in the paraflagellar body may also be involved.[71] The latter supposition is supported by the fast response within 10 min after insolation and by the absence of any restoration by DNA repair processes.[21,66-67] The speed, percentage of motile cells in the population and the phototactic orientation of individual cells is determined using a microscope with an IR transmission cut-off filter and an automatic computer-controlled video tracking system. The biologically effective UV radiation is quantified in terms of decrease in growth rate,[68] in percentage of motile cells, in speed (μm/s) or in degree of phototactic orientation.[65-68] A population of *Euglena* cells is

Table 7.2. Characteristics of biological dosimeters on the basis of cellular systems

| Biological dosimeter | Assay System | Biological endpoint | Dosimetric application | Dosage unit | Device | Application | Status | Ref. |
|---|---|---|---|---|---|---|---|---|
| Bacterial cell dosimeter | E. coli CSR603 uvrA recA phr | inactivation | polychromatic action spectrum of solar UV-B | inactivation rate constant | quartz covered watch glass with bacteria in suspension on ice | noon in Dallas, Texas | field measurements | 53-54 |
| | E. coli AB2480 uvrA recA; AB2463 recA; AB1886 uvrA; AB1157 | inactivation; role of repair processes; interactions | comparison with spore dosimeter | equivalent dose at 254 nm using spore dosimeter | quartz covered watch glass with bacteria in suspension at ≤6°C | clear cloudless days at noon in Rio de Janeiro, Brazil | field measurements | 3, 55 |
| | E. coli B/r uvrA; E. coli B/r | mutagenesis; interactions | comparison with spore dosimeter | equivalent dose at 254 nm using spore dosimeter | quartz cuvette with bacterial in suspension on ice | Rio de Janeiro, Brazil | field measurements | 56 |
| | E. coli CSR06 uvrA6 phr-1 | inactivation | sensitivity of $H_{eff}$ to ozone; vertical attenuation coefficient of $H_{eff}$ in water | % survival; % enhancement of survival by removal of UV-B | Whirlpak bags with bacteria in suspension w or w/o Mylar | Antarctic water (0-30 m depth); clear lake (0-10 m depth) | field measurements | 36, 57 |

| | | | | | | | | |
|---|---|---|---|---|---|---|---|---|
| Spore dosimeter | Spores of *B. subtilis* UVSSP uvrA42 ssp-1; (SSP-1 ssp-1) | inactivation; induction of spore photoproduct | diurnal, seasonal variation of $H_{eff}$ | equivalent incident dose at 254 nm | quartz cuvette with spores in suspension | Rio de Janeiro, Brazil | long-term dosimetry | 3, 43 |
| | Spores of *B. subtilis* UVS uvrA10 ssp-1; UVR HA101 | inactivation; mutagenesis | global $H_{eff}$ comparison with weighted spectroradiometry | inactivation rate constant; mutation doubling constant | membrane filters with dry spore layer | Tokyo and Tsukuba, Japan | long-term dosimetry | 7, 58-61 |
| | Spores of *B. subtilis* UVS uvrA10 ssp-1 plus UVR | relative inactivation of UVS to UVR spores | global $H_{eff}$ | equivalent dose at 313 nm | glass cover slip with dry spore layer | Dallas, Texas | demonstration experiment | 62 |
| | spores of *B. subtilis* uvrA10 ssp-1 polA151 | inactivation | sensitivity of $H_{eff}$ to ozone | ratio of inactivation rate constant w/ UV-B to that w/o UV-B | multifilter device with quartz discs with dry spore layers | space; Antarctica | space experiment; field measurements | 6, 63-64 |
| Biofilm dosimeter | spores of *B. subtilis* Marburg | loss of biological activity | sensitivity of $H_{eff}$ to ozone | $H_{eff}$ equivalent to 254 nm | monolayer of dried spores in agarose | Antarctica; space | long-term field monitoring; space experiment | 15,19, 37, 44 |
| Phytoplankton test | *Euglena gracilis* Z | inhibition of motility and photoorientation | sensitivity of UV effect to ozone | speed; % of motile cells; degree of photo-tactic orientation | cell culture below Plexiglas cuvette w/ or w/o ozone or cut-off filters | Lisbon, Portugal | field measurements | 65-67 |
| | Dinoflagellates | inhibition of motility and growth rate | sensitivity of UV effect to ozone | speed, % motile cells; growth rate | Plexiglas cuvette or quartz bottle with cultures | marine water (0-1.2 m depth) | field measurement | 68 |

Fig. 7.6. Biofilm with calibration fields (squares) and measurement fields (circles) after exposure, processing and staining and micrographs of the cells grown from spores during incubation. The biological activity is determined as quotient of optical density $OD_x/OD_o$ with $OD_x$ = optical density after irradiation and $OD_o$ = optical density without irradiation.

Fig. 7.7. Biologically effective dose $H_{eff254}$ measured with the biofilm technique on 6 August 1993 in Cologne, either for 2.5 h under different neutral density filters (squares) or for different time intervals without any entrance optics (circles). Exposure times are equally distributed before and after noon. These data were obtained in cooperation with P. Rettberg and E. Rabbow.

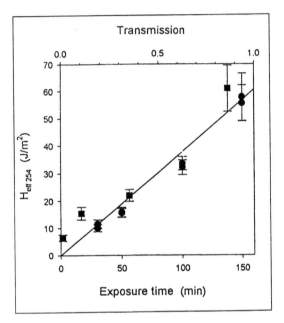

significantly affected by unfiltered solar radiation after about 1-2 h of exposure. Phytoplankton test systems have been utilized in different field measurements in Germany and Portugal[66-67] and in marine waters.[68]

## TISSUES, ORGANISMS AND COMMUNITIES

Plants as sessile organisms are especially affected by enhanced environmental UV-B radiation (reviewed in refs. 72-74). Leaf tissues,[23,33] seedlings,[24,32] whole cultivars[75] and natural communities[76-77] have been used as biomarkers for UV exposure, such as pyrimidine dimer production, photoreactivation and flavonoid metabolism or for UV effects, such as impairment of growth and productivity, of ecological balance and biodiversity (Table 7.3). Action spectra of most basic modes of UV damage to plant tissues (Fig. 7.4) have the common property of an increasing effectiveness with decreasing wavelengths, making them especially sensitive to ozone.[81]

*Daphnia pulex* has been used to quantify the implications of UV-B radiation on aquatic communities (Table 7.3).[22,31] The action spectrum for mortality (Fig. 7.4) suggests DNA damage as a key process in this response.

Human volunteers have served to quantify the minimal erythemal dose (MED)[39,78] of environmental UV radiation as well as the formation of pyrimidine dimers in the skin.[79] Furthermore, the photosynthesis of vitamin D was quantified

**Table 7.3. Characteristics of biological dosimeters on the basis of tissues, organisms or ecosystems**

| Biological dosimeter | Assay system | Biological endpoint | Dosimetric application | Dosage unit | Device | Application | Status | Ref. |
|---|---|---|---|---|---|---|---|---|
| Leaf epidermis dosimeter | Brassica oleracea, Sinapis alba | dimer formation and photoreactivation, flavonoid metabolism | global $H_{eff}$ | production rate | leaf discs covered w/ or w/o cut off filters | Freiburg, Germany | field measurements | 23, 33 |
| Seedling dosimeter | Lepidium sativum, Petroselinum hortense, Anethum graveolens, Daucus carota | growth rate of hypocotyl and root | global $H_{eff}$ | growth rate | seedlings w/ or w/o cut off filters | Freiburg, Germany | field measurements | 24, 32 |
| Daphnia test | Daphnia pulex | mortality, decrease in productivity, UV tolerance | vertical attenuation of $H_{eff}$ in water | % survival | population in open plastic container w/ or w/o cut off foil | alpine and arctic lakes | field measurements | 22, 31 |
| Human skin dosimeter | humans of Caucasian skin type | erythema, pyrimidine dimer formation | erythema inducing dose | minimal erythema dose (MED) | untanned skin, biopsies for dimer analysis | photodermatology, action spectroscopy | laboratory measurements | 39, 78-79 |
| Vitamin D test | human serum | 25(OH)D concentration | seasonal changes in vitamin D synthesis | percent volunteers with vitamin D deficiency | human population | Caucasian elderly population in a long-term care facility or free-living | medical study | 80 |
| Agricultural cultivar test | Helianthus annuus, Zea mays | influence on growth and photosynthetic activity | role of enhanced UV-B | % change in height, in leaf area, in net photosynthetic rate | climate controlled growth chamber with solar radiation and ozone filters | Lisbon, Portugal | long-term measurements | 75 |
| Ecosystem test | natural mountain heath vegetation | change in species composition | Role of enhanced UV-B | % change | natural habitat exposed to enhanced UV-B | Abisko Research Station, Sweden | long-term study | 76-77 |

from the 25(OH)D concentration in the serum of test persons exposed for different periods to solar radiation (Table 7.3).[80] Its action spectrum peaks around 298 nm (Fig. 7.4).

## SCOPE OF APPLICATION OF BIOLOGICAL DOSIMETERS

### SENSITIVITY OF THE BIOLOGICALLY EFFECTIVE SOLAR IRRADIANCE TO OZONE

Biological dosimetry has provided experimental proof of the high sensitivity of the biologically efficient UV-B doses to changes in total overhead atmospheric column of ozone and has thereby confirmed the predictions from model calculations[38] and from weighted spectroradiometry.[9] In Antarctica, during the spring depletion of stratospheric ozone, a significant increase in biologically effective UV-B was measured, using a bacterial dosimeter on the base of *E. coli*,[57] the spore dosimeter[64] and the biofilm dosimeter.[19] Cut-off filters were used, to separate the effects at different UV ranges. Figure 7.8 shows the annual profile of daily $H_{eff254}$ values separately for the UV-B and UV-A range, that were recorded at the Georg von Neumayer Station (70.6°S 8.4°W) using the biofilm dosimeter.[19] After the polar night, when stratospheric ozone destruction led to values down to

150 Dobson Units (DU), $H_{eff}$ for UV-B was elevated to values which were measured later on in mid summer.

For more complex biological dosimeters, such as phytoplankton species[66-67] or plant cultivars[75] increased ozone values were simulated in field experiments using two growth chambers, of which one was covered by an ozone floated cuvette.[82] Measurements in these chambers near Lisbon (38.4°N 9.1°W) where ambient levels of solar UV-B radiation are naturally high, showed that the inhibition of photo-orientation and motility of *Euglena* cells is largely attributed to the UV-B component of solar radiation.[66-67] Likewise, sensitive cultivars of higher plants, e.g. react to elevated levels of UV-B radiation with reduced growth rate and impaired photosynthetic activity (reviewed in ref. 73). In an alternative approach to simulate various ozone depletion scenarios UV-B was supplemented by filtered fluorescent sun lamps (reviewed in ref. 77). A six-year study revealed a significant reduction in crop yield in UV sensitive soybean cultivars.[83]

In a space experiment, extraterrestrial solar radiation was used as natural radiation source, filtered through a set of different cut-off filters to simulate the terrestrial UV radiation climate at different ozone concentrations down to very low values.[37] Biologically effective irradiances

*Fig. 7.8. Ozone data and biologically effective dose H$_{eff254}$ of solar UV-B and UV-A radiation recorded with the biofilm dosimeter at the Georg von Neumayer Station in Antarctica (modified from ref. 19).*

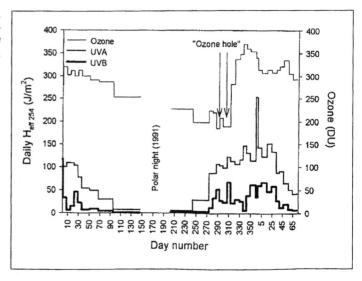

$E_{eff254}$ as a function of simulated ozone column thickness were directly measured with the biofilm technique and compared with expected irradiances (Fig. 7.9). A dramatic increase in $E_{eff254}$ was demonstrated with decreasing (simulated) ozone concentrations. The full spectrum of extraterrestrial solar radiation leads to an increment of $E_{eff}$ by nearly three orders of magnitude compared with the solar spectrum at the surface of the Earth for average total ozone columns (Fig. 7.9).

## VERTICAL ATTENUATION COEFFICIENT OF BIOLOGICALLY EFFECTIVE SOLAR IRRADIANCE IN NATURAL WATERS

The spectral transmission of UV radiation in natural waters may vary substantially from one water body to another in response to the concentration of humic material. To assess the risks from enhanced UV-B radiation for aquatic ecosystems, it is important to determine the vertical profile of the biologically effective UV radiation within the water column, separately for each habitat of concern. Vertical attenuation coefficients of biologically effective UV irradiance have been determined for various freshwater and marine environments using DNA, bacteriophage or bacterial dosimeters (Table 7.4). The biological attenuation coefficients, integrated over the whole spectral range of solar radiation,

correspond to the spectroradiometrically determined attenuation coefficients in the range of 320-340 nm.[36] This suggests a high contribution of UV-A to the harmful effect of solar radiation. During austral spring 1990, when stratospheric ozone concentrations fluctuated between 170 and 380 DU, the potential biological effect of in-water UV-B down to 29 m was closely correlated to stratospheric ozone levels.[84] Biological dosimetry bears several advantages for UV measurements in aquatic environments, such as: (1) direct biological weighting; (2) availability of $4\pi$ detectors; (3) measurement of total biologically effective dose over a designated period; (4) possibility of simultaneous monitoring at different depths; and (5) frequent monitoring at different sites.

## UV TOLERANCE AND PROTECTION MECHANISMS

Organisms, naturally exposed to solar radiation, have developed various strategies for dealing with prevention and repair of damage produced by solar UV-B radiation. The presence of UV-absorbing molecules in cells and tissues is an efficient means for protection against the harmful effects of UV. Biological dosimeters have been utilized to determine the UV protection by the outer body covering of adult marine invertebrates of Antarctica.[85] Best protection

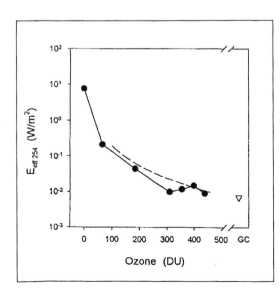

Fig. 7.9. Biologically effective irradiance $E_{eff254}$ for different ozone column thicknesses (•) determined by the biofilm technique in a space experiment and the $E_{eff254}$ value measured in Cologne on 6 August 1993 (▽). The dashed line shows the corresponding curve modeled for DNA damage (from ref. 38). Reprinted from J Photochem Photobiol B:Biol 1996; 32:189-196, Horneck G, Rettberg P, Rabbow E et al, Biological dosimetry of solar radiation for different simulated ozone column thickness, ©1996, with kind permission from Elsevier Science SA, Lausanne, Switzerland.

Table 7.4. Attenuation coefficient of different aquatic habitats, determined by biological dosimetry

| Aquatic habitat | Location | Depth | Biological dosimeter | Attenuation Coefficient (m⁻¹) from biodosimetry measurement | from physical | Date | Ref. |
|---|---|---|---|---|---|---|---|
| marine water | Palmer Station, Antarctica 64.8°S, 64.0°W | 0-30 m | bacteria: E. coli CSR06 | not determined | not determined | Oct./Nov. 1988 | 57 |
| clear, tropical marine water | Exuma Cays, Bahamas 23.7°N, 76.1°W | 0-3 m | DNA | 0.48 | 0.12 (305 nm) | July 1990 | 47-48 |
| lake | Lake Giles, Pennsylvania 41.4°N, 75.3°W | 0-10 m | bacteria: E. coli CSR06 | 0.73 full sun 0.65 UV-A | 1.2 (305 nm) 0.7 (320-340 nm) | Sept. 1993 | 36 |
| lake | Lake Giles, Pennsylvania 41.4°N, 75.3°W | 0-10 m | bacteriophage: T2 | 0.71 full sun 0.67 UV-A | 1.2 (305 nm) 0.7 (320-340 nm) | Sept. 1993 | 36 |
| river | Danube near Budapest 45.5°N, 19.0°E | 0-1.5 m | bacteriophage: T7 | 7.27 | not determined | June 1992 | 30 |
| lake | Lake Balaton, Hungary 47.0°N,18.0°E | 0-1.5 m | bacteriophage: T7 | 3.69 | not determined | July 1992 | 30 |
| lake | Lake Szalajka, Hungary 48.0°N, 20.5°E | 0-1.5 m | bacteriophage: T7 | 2.13 | not determined | July 1992 | 30 |
| marine water | Adriatic Sea 45.0°N, 15.0°E | 0-3 m | bacteriophage: T7 | 0.36 | not determined | Sept. 1992 | 30 |

was provided by the exteriors of limpet (*Nacella concinna*) or sea star (*Odontaster validus*), which attenuate the incident biologically effective UV radiation by more than 97%. However, after an exposure of 4 h to solar radiation during mid-day in Antarctica, enough solar radiation penetrates to kill a substantial fraction of cells. *Euglena gracilis*, if dark bleached, is more affected by solar radiation than the green variant.[66] This suggests that a protective UV screen is exerted by the pigments of the green cells. Flavonoids which are the major UV screening components in plant leaves are induced by UV in a great variety of species (Fig. 7.4).[23] Using a leaf epidermis dosimeter, it was shown that pyrimidine dimers, the predominant photoproduct induced by UV-B in the DNA, were only produced in flavonoid-free leaves of *Brassica oleracea*.[33] In leaves that had acclimated to UV by flavonoid accumulation induced by nondamaging UV, dimers were not detected after insolation. In *Daphnia*, melanin pigmentation provides an efficient photoprotection, as has been shown by comparison of the responses to solar UV radiation of transparent and pigmented animals of alpine and arctic origin.[31] UV-absorbing mycosporine-like amino acids have been found in phytoplankton after exposure to solar radiation.[29]

## RATIO OF BIOLOGICALLY EFFECTIVE DOSE OF UV-B TO THAT OF UV-A

In order to monitor the biological effectiveness of distinct wavebands of the solar spectrum, cut-off filters that largely remove the UV-B range of the solar spectrum have been used in biological dosimetry. The most frequently used filters are the polyester Mylar,[31,36,57] quartz filters, e.g. Schott WG 320,[19,23-24,32-33,37,64,66-67] or cuvettes floated with ozone.[66-67,75] Using the *E. coli* biodosimeter, Karentz and Lutze for the first time provided experimental proof of an enhanced UV-B/UV-A ratio during periods of reduced stratospheric ozone concentrations over Antarctica in

1988.[57] During periods of depleted stratospheric ozone the ratio UV-B/UV-A increased up to a factor of 10, from 0.3-0.4 to values between 2.5-3.4. This observation was subsequently confirmed with the spore dosimeter for the ozone depletion period in 1990[64] and with the biofilm for the following period in 1991 (Fig. 7.8).[19] In the latter case, the ratio UV-B/UV-A was raised by a factor of 2 during the ozone depletion period, from 0.1-0.3 to 0.5-0.6. It is interesting to note the high contribution of UV-A to the total biologically effective dose recorded by bacterial dosimeters,[19,57] *Euglena*,[66-67] and *Daphnia*.[31] Mylar cut-off filters have also been used to determine the biological attenuation coefficient in aquatic habitats separately for full solar radiation and for UV-A.[36] There is only a slight decrease in the coefficient if UV-B is removed (Table 7.4).

## COMPARISON OF MEASURED AND CALCULATED DATA OF THE BIOLOGICALLY EFFECTIVE DOSE

If biological dosimetry is applied concurrently with spectroradiometry, the experimentally determined biologically effective dose $H_{eff}$ can be compared with the calculated one, which is obtained by multiplication of the spectroradiometric data with the action spectrum of the biodosimeter according to Equation 1 (Fig. 7.3). This intercomparison allows one to assess, whether the component wavelengths contribute independently to the biological effect or whether interactions occur. Using the spore dosimeter, Munakata et al demonstrated that the biologically determined UV doses are slightly higher than the calculated ones.[7] The ratio of $H_{eff(observed)}$ to $H_{eff(calculated)}$ ranged between 1.03 and 1.60, depending on day and hour of exposure. Likewise, the biofilm dosimeter yielded ratios of experimental values of $H_{eff}$ to calculated data which are always larger than one.[37] Further intercomparisons are required to assess the complementary nature of biological dosimetry and weighted spectroradiometry.

## LONG-TERM TRENDS OF BIOLOGICALLY EFFECTIVE SOLAR RADIATION

Daily and annual profiles of biologically effective environmental UV radiation have been obtained using the spore dosimeter in Brazil[3,86] and in Japan,[60-61] the biofilm dosimeter in Antarctica (Fig. 7.8)[19] and the bacteriophage dosimeter in Hungary.[17,30] In the latter case, a continuous measuring device has been constructed with programmed pumping of a defined volume of bacteriophage suspension into a quartz cuvette for measurements and subsequent transfer into a microbial assay device for infection of *E. coli* host cells and incubation up to lysis of the cells. Using the spore dosimeter, Munakata showed in a 14-year study a trend of increase of the biologically effective dose of solar UV radiation in Tokyo from 1980 to 1993.[61] Using the biofilm dosimeter in a 1+-year UV-monitoring campaign in Antarctica, Quintern et al provided experimental proof of an enhanced level of biologically effective UV radiation during periods of stratospheric ozone depletion.[19] This enhancement is unequivocally attributed to the UV-B range of the spectrum (Fig. 7.8). These examples demonstrate the potential of biological dosimetry to assess long-term trends in changes of the UV-radiation climate on Earth. Biological dosimetry will contribute to the concerted efforts to monitor biologically efficient UV radiation in order to assess the implications of enhanced environmental UV radiation on human health, crop productivity and ecosystem balance.

## ACKNOWLEDGMENTS

This study has been supported by two contracts of the European Commission (No. EV-CT93-0342, No. ENV4CT950044). I acknowledge the help by many colleagues who provided me with relevant literature in their field of research, above all A. Cabaj, B. Diffey, E.C. deFabo, D.-P. Häder, D. Karentz, P. Knuschke, N. Munakata, W. Rau, J. Regan, G. Rontó, G. Seckmeyer, O. Siebeck, A. Teramura, I. Terenetskaya, M. Tevini, R. Tyrrell, T. Wang, A. Webb, E. Wellmann and I. Zellmer. I especially appreciate the valuable comments by P. Rettberg during the preparation of the manuscript.

## REFERENCES

1. Frederick JE, Snell HE, Haywood EK. Solar ultraviolet radiation at the earth's surface. Photochem Photobiol 1989; 51: 443-450.
2. Horneck G, Brack A. Study of the origin, evolution and distribution of life with emphasis on exobiology experiments in Earth orbit. In: Bonting SL, ed. Advances in Space Biology and Medicine, Vol. 2 Greenwich, CT: JAI Press, 1992:229-262.
3. Tyrrell R. Solar dosimetry with repair deficient bacterial spores: action spectra, photoproduct measurements and a comparison with other biological systems. Photochem Photobiol 1978; 27:571-579.
4. Lindberg C, Horneck G. Action spectra for survival and spore photoproduct formation of *Bacillus subtilis* irradiated with short-wavelength (200-300 nm) UV at atmospheric pressure and *in vacuo*. J Photochem Photobiol B:Biol 1991; 11:69-80.
5. Munakata N, Saito M, Hieda K. Inactivation action spectra of *Bacillus subtilis* spores in extended ultraviolet wavelengths (50-300 nm) obtained with synchrotron radiation. Photochem Photobiol 1991; 54: 761-768.
6. Horneck G. Responses of *Bacillus subtilis* spores to space environment: results from experiments in space. Origins of Life 1993; 23:37-52.
7. Munakata N, Morohoshi F, Hieda K et al. Experimental correspondence between spore dosimetry and spectral photometry of solar ultraviolet radiation. Photochem Photobiol 1996; 63:74-78.
8. Setlow RB. The wavelengths in solar radiation effective in producing skin cancer: a theoretical analysis. Proc Nat Acad Sci USA 1974; 71:3363-3366.
9. McKenzie RL, Blumthaler M, Booth CR et al. Surface ultraviolet radiation. In: UNEP/

WMO Scientific Assessment of Ozone Depletion, WMOCH9.DOC, 22 August 94, 1994:9.1-9.18.

10. Seckmeyer G, Mayer B, Erb R et al. UV-B in Germany higher in 1993 than in 1992. Geophys Res Lett 1994; 21:577-580.

11. Seckmeyer G, Mayer B, Bernhard G et al. Geographical differences in the UV measured by intercompared spectroradiometers. Geophys Res Lett 1995; 22:1889-1892.

12. Jagger J. Solar Actions on Living Cells. New York: Praeger, 1985.

13. Berger DS. The sunburning ultraviolet meter: design and performance. Photochem Photobiol 1976 24:587-593.

14. Davis A, Deane GHW, Diffey BL. Possible dosimeter for ultraviolet radiation. Nature 1976; 261:169-170.

15. Horneck G. Quantification of the biological effectiveness of environmental UV radiation. J Photochem Photobiol B:Biol 1995; 31:43-49.

16. Kollias N, Bager AH, Sadiq I. Measurements of the solar middle ultraviolet radiation in a desert environment. Photochem Photobiol 1988; 47:565-569

17. Rontó G, Gróf P, Gáspár S. Biological UV dosimetry—a comprehensive problem. J Photochem Photobiol B:Biol 1995; 31: 51-56.

18. Rontó G, Gáspár S, Bérces A. Phages T7 in biological UV dose measurements. J Photochem Photobiol B:Biol 1992; 12: 285-294.

19. Quintern LE, Puskeppeleit M, Rainer P et al. Continuous dosimetry of the biologically harmful UV-radiation in Antarctica with the biofilm technique. J Photochem Photobiol B: Biol 1994; 22:59-66.

20. Webb AR. Vitamin D synthesis under changing UV spectra. In: Young AR, Björn LO, Moan J et al, eds. Environmental UV Photobiology. New York: Plenum, 1993: 185-202.

21. Häder D-P. Effects of enhanced solar ultraviolet radiation on aquatic ecosystems. In: Tevini M, ed. UV-B Radiation and Ozone Depletion. Boca Raton: Lewis, 1993: 155-192.

22. Siebeck O. Risiken erhöhter UV-B-Strahlung für das Zooplankton. In: Rund-

gespräche der Kommission für Ökologie, Bd. 8 "Klimaforschung in Bayern". München: Pfeil, 1994:175-185.

23. Beggs CJ, Wellmann E. Photocontrol of flavonoid biosynthesis. In: Kendrick RE, Kronenberg GHM, eds. Photomorphogenesis in Plants. 2nd ed. Dordrecht: Kluwer Academic, 1994:733-751.

24. Steinmetz V, Wellmann E. The role of solar UV-B in growth regulation of cress (*Lepidium sativum* L.) seedlings. Photochem Photobiol 1986; 43:189-193.

25. Yarosh DB. The role of DNA damage and UV-induced cytokines in skin cancer. J Photochem Photobiol B: Biol 1992; 16:91-94.

26. Coohill TP. Ultraviolet action spectra (280 to 380 nm) and solar effectiveness spectra for higher plants. Photochem Photobiol 1989; 50:451-457.

27. Buchholz G, Ehmannn B, Wellmann E. Ultraviolet light inhibition of phytochrome-induced flavonoid biosynthesis and DNA photolyase formation in mustard cotyledons (*Sinapis alba* L.). Plant Physiol 1995; 108:227-234.

28. Häder D-P, Worrest RC. Effects of enhanced solar ultraviolet radiation on aquatic ecosystems. Photochem Photobiol 1991; 53:717-725.

29. Häder D-P. Risk of enhanced solar ultraviolet radiation for aquatic ecosystems. In: Round FE, Chapman DJ, eds. Progress in Phycological Research. Vol. 9. Biopress Ltd. 1993:1-45.

30. Rontó G, Gáspár S, Gróf P et al. Ultraviolet dosimetry in outdoor measurements based on bacteriophage T7 as a biosensor. Photochem Photobiol 1994; 59:209-214.

31. Zellmer ID. UV-B-tolerance of alpine and arctic *Daphnia*. Hydrobiologia 1995; 307:153-159.

32. Wellmann E. UV radiation in photomorphogenesis. In: Lange OL, Nobel PS, Osmond CB et al, eds. Encyclopedia of Plant Physiology, New Series, Vol 16. Berlin: Springer, 1983:745-756.

33. Buchholz G, Riegger L, Wellmann E. DNA damage and flavonoid formation in response

to solar UV in leaves of white cabbage (Brassica oleracea L.), Plant Physiol (submitted).

34. Calkins J, Barcelo JA. Some further considerations on the use of repair-defective organisms as biological dosimeters for broad-band radiation sources, Photochem Photobiol 1979; 30:733-737.

35. Sutherland BM. Action spectroscopy in complex organisms: potentials and pitfalls in predicting the impact of increased environmental UV-B. J. Photochem Photobiol B:Biol 1995; 31:29-34.

36. Kirk JTO, Hargreaves BR, Morris DP et al. Measurements of UV-B radiation in two freshwater lakes: an instrument intercomparison. Arch Hydrobiol Beih 1994; 43:71-88.

37. Horneck G, Rettberg P, Rabbow E et al. Biological dosimetry of solar radiation for different simulated ozone column thickness. J Photochem Photobiol B:Biol 1996; 32:189-196

38. Madronich S. The atmosphere and the UV-B radiation at ground level. In: Young AR, Björn LO, Moan J et al, eds. Environmental UV Photobiology. New York: Plenum, 1993:1-39.

39. McKinley AF, Diffey BL. A reference action spectrum for ultraviolet induced erythema in human skin. CIE J 1987; 6:17-22.

40. de Gruijl FR, Sterenborg HJCM, Forbes PD et al. Wavelength dependence of skin cancer induction by ultraviolet radiation of albino hairless mice. Cancer Res 1993; 53:53-60.

41. Madronich S. Implications of recent total atmospheric ozone measurements for biologically active ultraviolet radiation reaching the Earth's surface. Geophys Res Lett 1992; 19:37-40.

42. Madronich S, de Gruijl FR. Stratospheric ozone depletion between 1979 and 1992: implications for biologically active ultraviolet-B radiation and non-melanoma skin cancer incidence. Photochem Photobiol 1994; 59:541-546.

43. Tyrrell RM. Solar dosimetry and weighting factors. Photochem Photobiol 1980; 31: 421-422.

44. Quintern LE, Horneck G, Eschweiler U et al. A biofilm used as ultraviolet dosimeter. Photochem Photobiol 1992; 55:389-395.

45. Webb AR, Kline L, Hollick MF. Influence of season and latitude on the cutaneous synthesis of vitamin $D_3$: exposure to winter sunlight in Boston and Edmonton will not promote Vitamin $D_3$ synthesis in human skin. J Clin Endocrin Metabol 1988; 67:373-378.

46. Terenetskaya IP. Provitamin D photoisomerisation as possible UV-B monitor: kinetic study using a tuneable dye laser. SPIE Proc Int Conf Biomedical optics, Vol 2134 B, 1994:135.

47. Regan JD, Carrier WL, Gucinski H et al. DNA as a solar dosimeter in the ocean. Photochem Photobiol 1992; 56:35-42.

48. Regan JD, Yoshida H. DNA UV-B dosimeters. J Photochem Photobiol B:Biol 1995; 31:57-61.

49. Furusawa Y, Suzuki K, Sasaki M. Biological and physical dosimeters for monitoring solar UV-light. J Radiat Res 1990; 31:189-206.

50. Sasaki M, Takeshita S, Sugimura M et al. Ground-based observation of biologically active solar ultraviolet B irradiance at 35°N latitude in Japan. J Geomag Geoelectr 1993; 45:473-485.

51. Tsyganenko NM, Kiseleva MN, Alekseyev AB et al. Photodimerisation of uracile as films and its possible application for dosimetry of genetically active ultraviolet radiation. Biofizika (USSR) 1987; 32:7-11.

52. Gáspár S, Bérces A, Rontó G et al. Biological effectiveness of environmental radiation in aquatic systems, measurements by T7-phage sensor. J Photochem Photobiol B: Biol 1996; 32:183-187.

53. Harm W. Biological determination of the germicidal activity of sunlight. Rad Res 1969; 40:63-69.

54. Harm W. Relative effectiveness of the 300-320 nm spectral region of sunlight for the production of primary lethal damage in *E. coli* cells. Mut Res 1979; 60:263-270.

55. Tyrrell RM, Souza-Neto A. Lethal effects of natural solar ultraviolet radiation in repair proficient and repair deficient strains of *Escherichia coli*: actions and interactions. Photochem Photobiol 1981; 34:331-337.

56. Tyrrell RM. Mutation induction and mutation suppression by natural sunlight. Biochem Biophys Res Comm 1979; 91: 1406-1414.

57. Karentz D, Lutze LH. Evaluation of biologically harmful ultraviolet radiation in Antarctica with a biological dosimeter designed for aquatic environments. Limnol Oceanogr 1990; 35:549-561.

58. Munakata N. Killing and mutagenic action of solar radiation upon *Bacillus subtilis* spores: a dosimetric system. Mut Res 1981; 82:263-268.

59. Munakata N. Genotoxic action of sunlight upon *Bacillus subtilis* spores: monitoring studies at Tokyo, Japan. J Radiat Res 1989; 30:338-351.

60. Munakata N. Biologically effective dose of solar ultraviolet radiation estimated by spore dosimetry in Tokyo since 1980. Photochem Photobiol 1993; 58:386-392.

61. Munakata N. Continual increase in biologically effective dose of solar UV radiation determined by spore dosimetry from 1980 to 1993 in Tokyo. J Photochem Photobiol B:Biol 1995; 31:63-68.

62. Wang TV. A simple convenient biological dosimeter for monitoring solar UV-B radiation. Biochem Biophys Res Comm 1991; 177:48-53.

63. Horneck G, Bücker H, Reitz G, Requardt H, Dose K, Martens KD, Mennigmann HD, Weber P. Microorganisms in the space environment. Science 1984; 225:226-228.

64. Puskeppeleit M, Quintern LE, ElNaggar S et al. Long-term dosimetry of solar UV radiation in Antarctica with spores of *Bacillus subtilis*. Appl Envirnm Microbiol 1992; 58:2355-2359.

65. Häder D-P. Effects of solar and artificial UV irradiation on motility and phototaxis in the flagellate, *Euglena gracilis*. Photochem Photobiol 1986; 44:651-656.

66. Häder D-P, Häder MA. Ultraviolet-B inhibition of motility in green and dark bleached *Euglena gracilis*. Curr Microbio 1988; 17:215-220.

67. Häder D-P, Häder MA. Inhibition of motility and phototaxis in the green flagellate, *Euglena gracilis*, by UV-B radiation. Arch Microbiol 1988; 150:20-25.

68. Nielsen T, Björn LO, Ekelund NGA. Impact of natural and artificial UV-B radiation on motility and growth rate of marine dinoflagellates. J Photochem Photobiol B: Biol 1995; 27:73-79

69. Sommer R, Weber G, Cabaj A et al. UV inactivation of microorganisms in water. Zbl Hyg 1989; 189:214-224.

70. Sommer R, Cabaj A. Evaluation of the efficiency of a UV plant for drinking water disinfection. Wat Sci Tech 1993; 27: 357-362.

71. Brodhun B, Häder D-P. UV-induced damage of photoreceptor proteins in the paraflagellar body of *Euglena gracilis*. Photochem Photobiol 1993; 58:270-274.

72. Bornman JF, Teramura AH. Effects of ultraviolet-B radiation on terrestrial plants. In: Young AR, Björn LO, Moan J et al, eds. Environmental UV Photobiology. New York: Plenum, 1993:427-471.

73. Tevini M. Effects of enhanced UV-B radiation on terrestrial plants. In: Tevini M, ed. UV-B Radiation and Ozone Depletion. Boca Raton: Lewis, 1993:125-153.

74. Tevini M. UV-B effects on terrestrial plants and aquatic organisms. Progr Botan 1994; 55:174-189.

75. Mark U, Tevini M. Combination effects of UV-B radiation and temperature on sunflower (*Helianthus annuus* L. cv. Polstar) and Maize (*Zea mays* L. cv. Zenit 2000) seedlings. J Plant Physiol 1996 (in press).

76. Johanson U, Gehrke C, Björn LO et al. The effects of enhanced UV-B radiation on a subarctic heath system. Ambio 1995; 24:106-111.

77. Caldwell MM, Teramura AH, Tevini M et al. Effects of increased solar ultraviolet radiation on terrestrial plants. Ambio 1995; 24:166-173.

78. Parrish JA, Jaenicke KF, Anderson RR. Erythema and melanogenesis action spectra of normal human skin. Photochem Photobiol 1982; 36:187-191.

79. Freeman SE, Hacham H, Gangue RW et al. Wavelength dependence of pyrimidine dimer formation in DNA of human skin irradiated *in situ* with ultraviolet light. Proc Natl Acad Sci USA 1989; 86:5605-5609.

80. Webb AR, Pilbeam C, Hanafin N et al. An evaluation of the relative contributions of exposure to sunlight and of diet to the circulating concentrations of 25-hydroxyvitamin D in elderly nursing home population in Boston. Am J Clin Nutr 1990; 51: 1075-1081.

81. Caldwell MM, Camp LB, Warner CW et al. Action spectra and their key role in assessing biological consequences of solar UV-B radiation change. In: Worrest RC, Caldwell MM, eds. Stratospheric Ozone Reduction, Solar Ultraviolet Radiation and Plant Life. Berlin: Springer, 1986:87-111.

82. Tevini M, Mark U, Saile M. Plant experiments in growth chambers illuminated with natural sunlight. In: Payer HD, Pfirrmann T, Mathy P, eds. Environmental Research with Plants in Closed Chambers, Air Pollution Research Report 26. Brussels: Commission of the European Communities, 1990:240-251.

83. Teramura AH, Sullivan JH, Lydon J. Effects of UV-B radiation on soybean yield and seed quality: a six year study. Physiol Plant 1990; 80:5-11.

84. Karentz D, Spero HJ. Response of natural *Phaecocystis* population to ambient fluctuations of UV-B radiation caused by Antarctic ozone depletion. J Plankton Res 1995; 17:1771-1789.

85. Karentz D. Prevention of ultraviolet radiation damage in Antarctic marine invertebrates. In: Biggs RH, Joyner MEB, eds. Stratospheric Ozone Depletion/UV-B Radiation in the Biosphere. Berlin: Springer, 1994:175-180.

86. Tyrrell RM. Biological dosimetry and action spectra. J Photochem Photobiol B:Biol 1995; 31:35-41.

# CHAPTER 8

# ROLE OF ULTRAVIOLET RADIATION ON BACTERIOPLANKTON ACTIVITY

Gerhard J. Herndl

## INTRODUCTION

Over the last two decades our view of the role of bacterioplankton in aquatic systems changed substantially. While traditionally heterotrophic bacterioplankton have been seen as the decomposer compartment of food webs, this view has been altered with the notion that bacterioplankton are about three orders of magnitude more abundant if enumerated under the epifluorescence microscope instead of using plate counting techniques.[1,2] Moreover, bacterioplankton production has been shown to be within the range of phytoplankton primary production; under distinct ecological conditions even exceeding primary production.[3,4] Their short turnover time which ranges from 3 to 0.1 $d^{-1}$ indicates high predatory losses due to heterotrophic protists and viruses.[5] All these findings lead to the revision of the traditional concept of aquatic food web structures where bacterioplankton played only a terminal role as decomposers of organic matter; the "microbial loop hypothesis" formulated more than a decade ago placed the bacterioplankton in the center of a microbial food web with phytoplankton and protists as the other living compartments and a dissolved organic matter (DOM) pool and an inorganic nutrient pool as the nonliving compartments (Fig. 8.1).[1,6,7] It was hypothesized that carbon and energy are efficiently retained within the microbial loop and that only a comparatively small portion of the carbon and energy cycled through this microbial loop becomes available to higher trophic levels.[8]

It is now well-established that bacterioplankton play an important role in the carbon flux through aquatic ecosystems, dominating, in terms

*The Effects of Ozone Depletion on Aquatic Ecosystems,* edited by Donat-P. Häder.
© 1997 R.G. Landes Company.

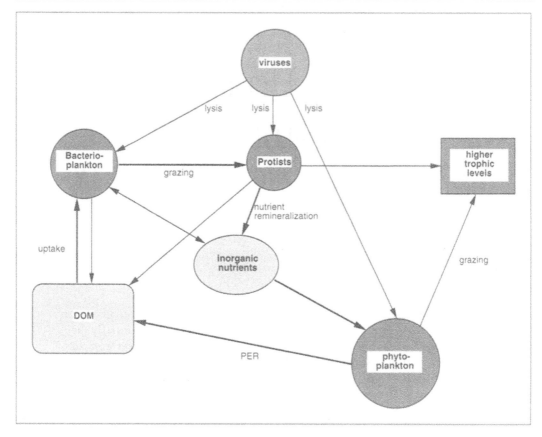

*Fig. 8.1. The main material pathways (indicated by bold arrows) through the microbial loop: phytoplankton releases photosynthetic extracellular release (PER) and fuels the dissolved organic matter (DOM) pool. The principal consumers of the DOM pool are the bacterioplankton which are, in turn, preyed upon by flagellates. Viruses are responsible for bacterial lysis and consume roughly the remaining half of the bacterial production. Protists are the principal remineralizers of nutrients which are used for phytoplankton production.*

of biomass, even over phytoplankton under oligotrophic conditions in the euphotic zone.[8-12] Within the euphotic zone a variable portion of the water column is subjected to mixing processes mainly mediated by wind and tides. Planktonic organisms within this so-called "upper mixed layer" are subjected to rapidly fluctuating physical conditions; physical parameters varying most rapidly in the upper mixed layer are turbulence and radiation.

In light of recent findings that, due to ozone depletion, UV-B radiation increases not only over polar regions but also in temperate latitudes,[13] research on the effects of UV on aquatic organisms and on the system's metabolism in general has become one of the hot issues in aquatic ecology. Although a number of excellent

reviews have been published recently addressing the role of UV on aquatic microorganisms,[14-17] in this contribution I will specifically focus on the interaction between DOM and heterotrophic bacterioplankton. The uptake of DOM by bacterioplankton is the major route of transformation and remineralization of carbon from the primary producers to the heterotrophic compartment and therefore of ultimate importance for the overall ecosystem's productivity.

## ROLE OF ULTRAVIOLET RADIATION ON BACTERIOPLANKTON CELLS

The role of UV radiation on aquatic microorganisms is dependent on: (1) the attenuation of the different UV wavelength

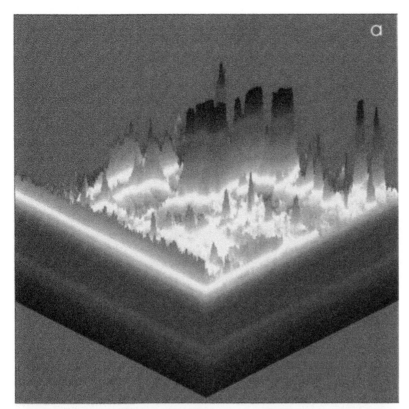

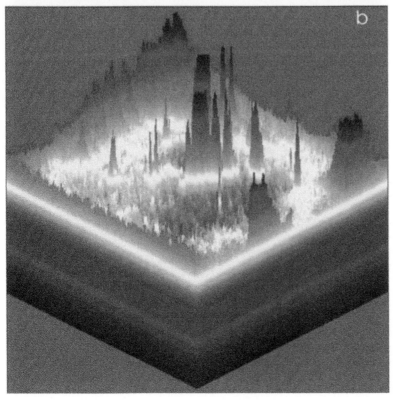

Fig. 3.5. 3D representation of 2D gels of the proteins isolated from the paraflagellar body, the putative photoreceptor of the flagellate Euglena gracilis, before (a) and after (b) UV-B irradiation.[41]

ranges, and (2) on the turbulence which, in turn, determines the exposure time to high levels of solar radiation close to the surface and the time an organism or a parcel of water spends in the more radiation-protected, deeper horizons of the upper mixed layer. The attenuation of the UV radiation is dependent on the quantity and quality of the dissolved organic matter (DOM) present in the water column.[18-20] In open oligotrophic oceans, the 1% depth of surface radiation for 305 nm wavelength is in 10-20 m, while in coastal zones the

corresponding depth is only 2-8 m and in humic-rich lakes the 1% depth for 305 nm is in only 0.01-0.05 m (Fig. 8.2). Thus, turbulent mixing causes planktonic organisms to live under permanently changing radiation regimes. These rapidly changing radiation conditions have to be taken into consideration if we want to understand the ecological role of (elevated) UV radiation in aquatic ecosystems.

As far as is currently known, natural bacterioplankton species obviously lack UV-protective pigmentation. Other than

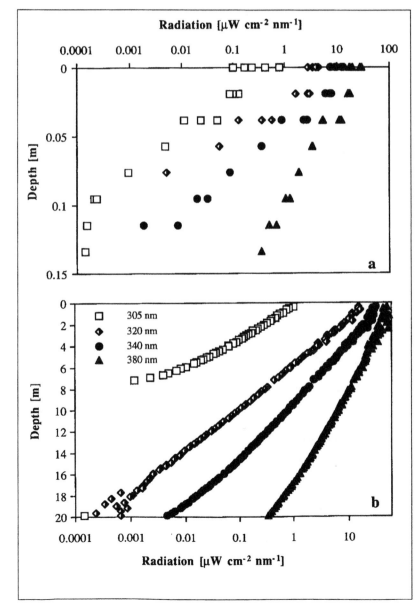

Fig. 8.2. Attenuation of UV radiation in the water column of (a) the humic substances-rich Lake Neusiedl (Austria) and (b) the northern Adriatic Sea on a cloudless day in August 1995. While the radiation at the air-water interface is similar for both sites, the radiation in the 5 cm depth horizon in Lake Neusiedl is similar to the radiation level at 5 m depth in the northern Adriatic.

phytoplankton, which produces UV absorbing compounds such as mycosporine-like amino acids, no such pigmentation has been detected in bacterioplankton.[15] In accordance with these findings, recent theoretical considerations have predicted that bacterioplankton are too small to develop protective pigments against the detrimental effects of UV radiation.[21] Thus one would suggest that bacteria are more efficiently damaged upon UV exposure than other microorganisms such as phytoplankton or heterotrophic protists. Indeed, there is evidence that the smallest organisms of the microbial loop, the bacterioplankton, receive more DNA-damage than the larger organisms. Measuring the cyclobutane dimer formation in size-fractionated samples in the Gulf of Mexico over a diel cycle, it has been found that the bacterioplankton fraction produced about twice as much cyclobutane dimers than the prokaryotic fraction.[22] This dimer formation exhibited a pronounced diel cycle with highest dimer formation around noon.[22] However, not only the DNA receives damage (Fig. 8.3), leading to cyclobutane and pyrimidine pyrimidone dimer formation, but also other macromolecules synthesized by the bacterial

cell are affected. Bacterial ectoenzymatic activity is efficiently retarded if exposed to UV radiation.[23,24] These ectoenzymes are responsible for cleavage of molecules in the intimate vicinity of the outer cell membrane; the cleaved molecules are subsequently taken up through the pores of the cell membrane of Gram-negative bacteria which represent the vast majority of the bacterial species present in the water column.[25] The action of UV on the dominant cellular macromolecules, the DNA and the proteins, obviously leads to an overall reduction of the bacterial activity and metabolism.

It has been shown that bacterioplankton metabolism is reduced if directly exposed to surface solar radiation levels (Fig. 8.4). Recovery from the previous UV stress is highly dependent on the radiation range to which the bacterioplankton are exposed (Fig. 8.4). While the recovery of bacterial activity is fastest under UV-A exposure, the recovery is significantly slower under photosynthetic active radiation (PAR, 400-700 nm) and almost no recovery is detectable if the bacterioplankton are held in the dark (Fig. 8.4). This strongly indicates that the DNA-dam-

*Fig. 8.3. Scheme of DNA-damage and possible repair mechanisms. UV-B radiation induces the production of photoproducts which can be potentially removed by three different repair mechanisms. As shown in Fig. 8.4, the photoenzymatic repair seems to be the most important repair mechanism in bacterioplankton.*

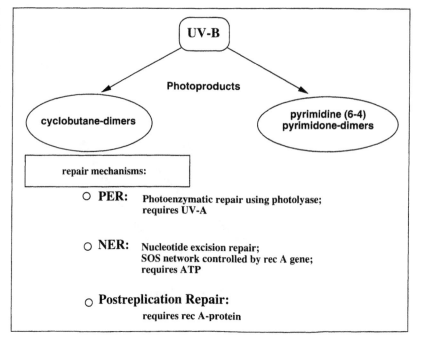

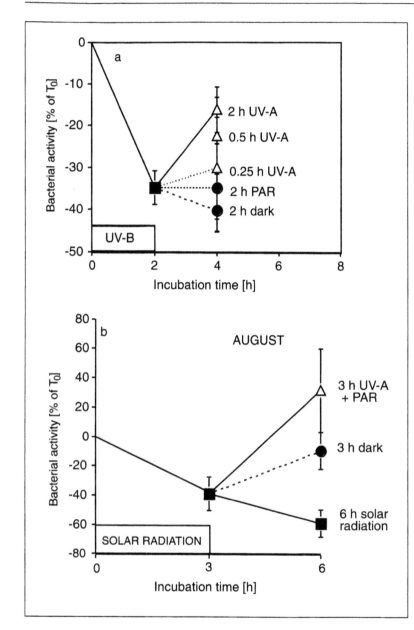

Fig. 8.4. Development of bacterial activity (a) after exposure to UV-B and subsequent exposure to the following radiation regimes: UV-A, photosynthetic active radiation (PAR, 400-700 nm), and darkness and (b) in 0.8 μm filtered seawater (to reduce possible bacterivory by flagellates and the influence of phytoplankton production) under different solar wavelength ranges as compared to the activity prior to starting the experiment (t₀). Bacterial activity was measured by leucine incorporation.[44]

age is repaired primarily by the photoenzymatic repair mechanism (Fig. 8.3). This repair mechanism does not require ATP for activating the enzyme photolyase, but the wavelength range between 360-430 nm is used for inducing photolyase expression.[26] This repair mechanism leads not only to a rapid recovery of the bacterial metabolism to the initial rate but to elevated bacterial activity (Fig. 8.4b). The possible reasons for the elevated bacterial activity after previous exposure to high levels of UV-B

might be caused by the concomitant photolysis of DOM as discussed in detail below.

Obviously, bacterioplankton are more effectively damaged upon UV-B exposure than other planktonic organisms, but the recovery from this UV-stress is rapid as well, using predominately the photoenzymatic repair. UV exposure, however, could influence the species composition of bacterioplankton if interspecific differences exist in the sensitivity against UV and/or in the efficiency in repairing the DNA-damage. In a set of experiments using denaturing gradient gel electrophoresis (DGGE) to characterize the bacterial community, no evidence was found of UV mediated shifts in the species composition, even if exposed to surface solar radiation levels for 2 consecutive days.[27]

Bacteria are efficiently controlled by viruses and bacterivorous flagellates.[5,28] The destruction of viral infectivity in natural waters is believed to be primarily biologically mediated, however, recently it was shown that solar radiation is probably the main cause for viral decay in surface waters.[29] The losses of infectivity are correlated to the formation of cyclobutane pyrimidine dimers (CPDs) in the bacterial DNA (Weinbauer, personal communication). (6-4) photoproducts comprise less than 10% of CPD concentrations in the DNA of natural viral communities collected along a transect from neritic to oceanic waters in the Gulf of Mexico (Weinbauer et al, in preparation). Considering the high sensitivity of viruses against UV radiation the high concentrations of infective viruses found in the presence of solar radiation pose a paradox.

The role viruses play in microbial food webs depends on the means of viral replication. Whereas lytic phage production depends on the encounter rates between viruses and host, lysogenic phage production depends on the presence of an inducing agent. In the Gulf of Mexico, usually less than 10% of the bacterial community is lysogenized (Weinbauer and Suttle, submitted). Solar radiation and hydrogen peroxide induce bacteriophage production in natural bacterial communities; however, this viral production is unlikely to contribute significantly to total (lytic + lysogenic) viral production (Weinbauer and Suttle, in preparation). Thus it is also unlikely that induction of lysogenized bacteria due to solar radiation can explain the paradox of high concentrations of viruses in the presence of solar radiation. Therefore, the impact bacteriophages can have on bacterial mortality seems to depend predominantly on lytic infection (Weinbauer, personal communication). Another possible explanation for the parodox could be that damaged DNA is repaired after its injection into the host cell. Preliminary data show that light-dependent repair mechanisms, probably photoreactivation, could restore infectivity to 21-52% of UV-damaged natural viral communities (Weinbauer, personal communication). Although our knowledge on the interaction between viruses and bacterioplankton is still rather limited, it appears that UV radiation considerably influences this interaction.

In an evolutionary context, the strategy of bacterioplankton to combat UV radiation is remarkably different from other organisms: (1) because of the obvious lack of protective pigmentation but (2) the high efficiency in repairing UV-induced damage using the long UV-A range and the shorter PAR wavelength range for activating repair at energetically low costs with (3) no interspecific differences detectable in either the sensitivity against UV or the repair efficiency. This delicate balance between UV-mediated damage and repair is only possible if mixing processes in the upper layers take place and transport the bacterioplankton community from the near-surface layers to deeper layers where UV-B is already largely attenuated but sufficient UV-A and PAR is available for inducing repair. Thus, a delicate interplay between turbulence and radiation determines bacterioplankton metabolism in the upper layers of the water column.

## ROLE OF ULTRAVIOLET RADIATION ON DISSOLVED ORGANIC MATTER (DOM)

The elevated bacterial metabolism following UV-B exposure (Fig. 8.4) indicates not only efficient recovery from the UV-stress but also the more efficient utilization of the DOM previously exposed to UV. Photolytic activity on DOM is dependent on the organic compounds subjected to cleavage. Humic substances, for example, known to be resistant against bacterial degradation are more intensively cleaved by radiation than nonhumic material (Fig. 8.5). Only about one-third of this photolytic activity is caused by UV-B (Fig. 8.5). The products of this photolytic activity are predominantly low molecular carbonyl compounds (formaldehyde, acetaldehyde, glyoxylate, and pyruvate) which might be taken up rapidly by bacterioplankton.[30] These low molecular compounds are present in the nanomolar range in the upper mixed water column exhibit-ing pronounced diel fluctuations in their concentrations and low turnover rates.[31-33] Photoproduction rates of these carbonyl compounds vary from ≈500 nM C h$^{-1}$ in the Everglades to 2-4 nM C h$^{-1}$ in the Sargasso Sea.[33-35] Despite the formation of low molecular, labile compounds upon photolysis, refractory material might resist photolysis at least partially and thereby become even more refractory than prior to the exposure

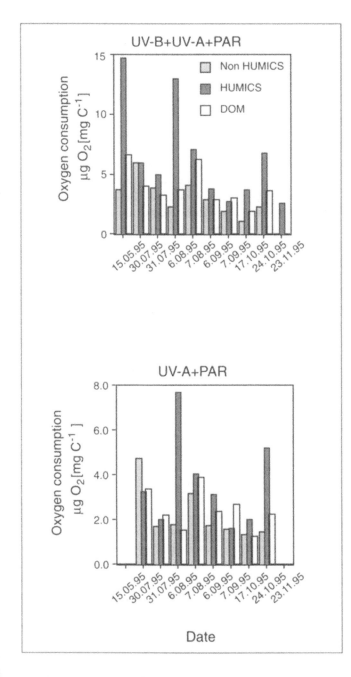

Fig. 8.5. Photolytic activity (expressed as oxygen consumed per mg C available) of surface solar radiation on the humic and the nonhumic fraction and on the unfractionated DOM at Lake Neusiedl. The photolytic activity is shown for the full solar spectrum (upper panel) and for the solar spectrum where UV-B was excluded by Mylar D foil (lower panel).

to radiation. The inorganic carbon produced from photolysis is mainly carbon dioxide, and to a lesser extent, also carbon monoxide.[33] Thus photolysis is a mechanism which makes a fraction of the dissolved organic carbon unavailable for bacteria by converting it into inorganic carbon.

Another important role of photolysis of DOM is the formation of photosensitizers which catalyze reactions of substances which do not absorb UV and are therefore not subjected to UV-photolysis directly. An example for such a reaction is dimethyl sulfide (DMS) which is microbially cleaved from dimethyl sulfoniopropionate (DMSP), most likely an osmoregulator in phytoplankton.[36-39] In the presence of DOM, DMS is removed from the water column; the removal rate of DMS is directly related to the concentration of DOM.[39] As far as we know now, upon exposure to UV only 10-15% of the DMS is transformed to dimethyl sulfoxide (DMSO), the compound(s) formed are not yet known.[40]

The oxidation of DOM by photolysis is of the same order of magnitude or even higher than that mediated by bacterial respiration in the upper layers of the water column.[41-43] Thus photolysis of DOM might be far more important than previously assumed. More information, however, is needed to elucidate the role of different fractions and compounds of the DOM pool on the bulk photolytic activity of the DOM.

## INTERACTION BETWEEN BACTERIOPLANKTON ACTIVITY AND DOM IN THE UPPER MIXED WATER COLUMN

As shown above, the radiation regime in the water column changes rapidly with depth (Fig. 8.2). This indicates that bacterioplankton and DOM are subjected to rapidly changing radiation in both quantity and quality. In the layers close to the air-water interface bacterioplankton activity is severely inhibited (Fig. 8.4) while the DOM is subjected to photolysis, making

labile cleavage products potentially available for bacterial uptake (Fig. 8.5). This bacterial uptake is inhibited, however, in the surface layer due to the UV-mediated destruction of the ectoenzymes and most probably also the uptake systems in addition to the DNA-damage. Turbulent mixing transports bacterioplankton and DOM from the surface into the deeper layers of the mixed water column where UV-B is largely attenuated and the long-wavelength UV-A and the short-wavelength PAR are used to repair the DNA-damage cost-effectively by saving ATP. Moreover, the labile components of the photocleavage of DOM derived from the surface waters are efficiently utilized in the deeper layers, leading to elevated bacterial production (Fig. 8.6). Thus, UV radiation in the water column leads to a decoupling of cleavage of DOM from its bacterial utilization on a temporal as well as spatial scale (Fig. 8.7). Close to the surface, cleavage prevails which should result in an accumulation of easily metabolizable cleavage products, while in the deeper layers where UV-B is almost completely attenuated, uptake of cleavage products prevail over photolysis of DOM. The intensity and depth of mixing determines therefore the time a planktonic organism spends in a specific radiation regime and ultimately its metabolic activity.

## FUTURE RESEARCH DIRECTIONS

One of the main purposes of this contribution was to demonstrate that the ecological role of UV on planktonic organisms can only be assessed adequately if the dynamic nature of the upper mixed water column is taken into account. Other than in laboratory experiments, in the real world bacterioplankton and DOM are probably never exposed to a constant radiation environment for more than a few seconds. Thus there is probably a delicate balance established between damage and repair in bacterioplankton. These short-term changes from damage of bacteria and photolysis of DOM to repair and uptake have not been

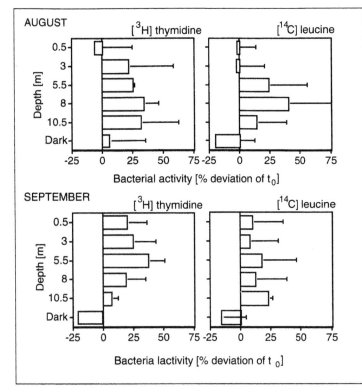

Fig. 8.6. Pattern of the bacterial activity after a 3 h exposure to surface solar radiation levels and subsequent incubation at different depth layers of the northern Adriatic Sea. Bacterial activity is given as the percent deviation from the activity measured before exposing the samples to surface solar radiation levels. For comparison, bacterial activity of a sample held in the dark after exposure to surface solar radiation is also given. Bacterial activity was measured by the dual labeling technique.[45] Bars represent the mean of 5 measurements in August and 6 measurements in September 1995; horizontal bars indicate standard error.

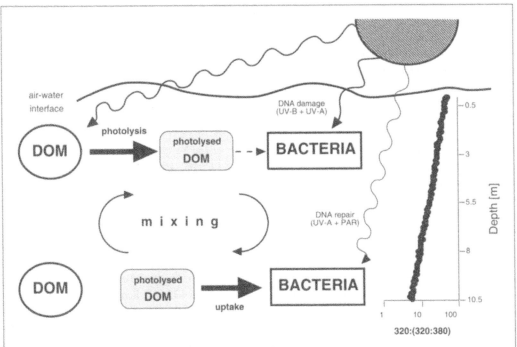

Fig. 8.7. Scheme on the major interactions between bacterioplankton and DOM as influenced by UV radiation. In the top surface layers of the water column DOM is photolytically cleaved and bacterioplankton activity retarded due to high levels of UV-B. If mixed into deeper layers, bacterioplankton repair damage primarily via photoenzymatic repair induced by UV-A and the photolyzed DOM is taken up. Thus UV causes a separation of cleavage of DOM and its bacterial uptake in both, time and space.

addressed thus far. Furthermore, focus should be put on the UV-mediated cleavage of DOM; it is likely that some compounds of the DOM are more susceptible to photolysis than others and that the photolysis products are utilized by bacterioplankton in different ways. A more detailed quantification of the DNA-damage in bacterioplankton and its repair is essential to elucidate the delicate interaction between bacteria and DOM in the upper layers of the water column.

The ecological consequences of increased levels of UV-B reaching the earth's surface is difficult to predict for the aquatic microbial community. With the increase in UV-B and the UV-A radiation levels remaining stable, the delicate balance between detrimental and repairing radiation levels is shifted towards the damaging radiation level. It is a major challenge for years to come to decipher the adaptation capacities of microorganisms to such changes.

## ACKNOWLEDGMENTS

I want to thank my working group at the Biocenter of the University of Vienna for providing data, particularly to A. Brugger, E. Kaiser, I. Obernosterer and B. Reitner, and M. Weinbauer for making unpublished data available. The hospitality of the staff of the Biological Station Neusiedler See at Illmitz (Austria) and at the Center for Marine Research at the Ruder Boskovic Institute at Rovinj (Croatia) is gratefully acknowledged. Funding support for work presented in this chapter was provided by grants from the Austrian Science Foundation (Fonds zur Förderung der wissenschaftlichen Forschung, FWF project #10023 to G.J.H.) and by the Environment and Climate Program of the European Union (Microbial community response to UV-B stress in European waters, project # EV5V-CT94-0512).

## REFERENCES

1. Pomeroy LR. The ocean's food web, a changing paradigm. BioScience 1974; 24: 499-504.

2. Zimmermann R, Meyer-Reil L-A. A new method for fluorescence staining of bacterial populations on membrane filters. Kieler Meeresforsch 1974; 30:24-27.

3. Sorokin YI. The heterotrophic phase of plankton succession in the Japan Sea. Mar Biol 1977; 41:107-117.

4. Scavia D, Laird GA. Bacterioplankton in Lake Michigan: dynamics, controls, and significance to carbon flux. Limnol Oceanogr 1987; 32:1017-1033.

5. Fuhrman JA, Noble RT. Viruses and protists cause similar bacterial mortality in coastal seawater. Limnol Oceanogr 1995; 40:1236-1242.

6. Azam F, Fenchel T, Field JG et al. The ecological role of water-column microbes in the sea. Mar Ecol Prog Ser 1983; 10: 257-263.

7. Pomeroy LR, Wiebe WJ. Energetics of microbial food webs. Hydrobiologia 1988; 159:7-18.

8. Cho BC, Azam F. Significance of bacterioplankton biomass in the epipelagic and mesopelagic zones in the Pacific Ocean. Eos 1987; 68:1729.

9. Cho BC, Azam F. Biogeochemical significance of bacterial biomass in the ocean's euphotic zone. Mar Ecol Prog Ser 1990; 63:253-259.

10. Cho BC, Choi J-K, Chung C-S et al. Uncoupling of bacteria and phytoplankton during a spring diatom bloom in the mouth of the Yellow Sea. Mar Ecol Prog Ser 1994; 115:181-190.

11. Fuhrman JA, Sleeter TD, Carlson CA et al. Dominance of bacterial biomass in the Sargasso Sea and its ecological implications. Mar Ecol Prog Ser 1989; 57:207-217.

12. Herndl GJ. Microbial biomass dynamics along a trophic gradient at the Atlantic Barrier Reef off Belize (Central America). P.S.Z.N.I: Mar Ecol 1991; 12:41-51.

13. Blumthaler M, Ambach W. Indication of increasing solar ultraviolet-B radiation flux in Alpine regions. Science 1990; 248: 206-208.

14. Zafiriou OC, Joussot-Dubien J, Zepp RG et al. Photochemistry of natural waters. Environ Sci Technol 1984; 18:358-371.

15. Karentz D, Bothwell ML, Coffin RB et al. Impact of UVB radiation on pelagic freshwater ecosystems: report of the working group on bacteria and phytoplankton. Arch Hydrobiologia 1994; 43:31-69.

16. Zepp RG, Callaghan TV, Erickson DJ. Effects of increased solar ultraviolet radiation on biogeochemical cycles. Ambio 1995; 24:181-187.

17. Häder D-P, Worrest RC, Kumar HD et al. Effects of increased solar ultraviolet radiation on aquatic ecosystems. Ambio 1995; 24:174-180.

18. Kirk JTO, Hargreaves BR, Morris DP et al. Measurements of UV-B radiation in two freshwater lakes: an instrument intercomparison. Arch Hydrobiol 1994; Beiheft 43:71-99.

19. Scully NM, Lean DRS. The attenuation of ultraviolet radiation in temperate lakes. Arch Hydrobiol 1994; Beiheft 43:135-144.

20. Morris DP, Zagarese H, Williamson CE et al. The attenuation of solar UV radiation in lakes and the role of dissolved organic carbon. Limnol Oceanogr 1995; 40:1381-1391.

21. Garcia-Pichel F. A model for the internal self-shading in planktonic organisms and its implications for the usefulness of ultraviolet sunscreens. Limnol Oceanogr 1994; 39:1704-1717.

22. Jeffrey WH, Pledger RJ, Aas P et al. Diel and depth profiles of DNA photodamage in bacterioplankton exposed to ambient solar ultraviolet radiation. Mar Ecol Prog Ser in press

23. Herndl GJ, Müller-Niklas G, Frick J. Major role of ultraviolet-B in controlling bacterioplankton growth in the surface layer of the ocean. Nature 1993; 361:717-719.

24. Müller-Niklas G, Heissenberger A, Puskaric S et al. Ultraviolet-B radiation and bacterial metabolism in coastal waters. Aquat Microb Ecol 1995; 9:111-116.

25. Gottschalk G. Bacterial Metabolism. New York: Springer-Verlag, 1986:359.

26. Friedberg EC. DNA Repair. New York: W.H. Freeman and Company, 1985:614.

27. Hager S, Buchholz B, Herndl GJ. Does ultraviolet radiation alter the bacterioplankton community structure in marine surface waters? Appl Environ Microbiol (in preparation).

28. Fuhrman JA, Suttle CA. Viruses in marine planktonic systems. Oceanography 1993; 6:51-63.

29. Suttle CA, Chen F. Mechanisms and rates of decay of marine viruses in seawater. Appl Environ Microbiol 1992; 58:3721-3729.

30. Wetzel RG, Hatcher PG, Bianchi TS. Natural photolysis by ultraviolet irradiance of recalcitrant dissolved organic matter to simple substrates for rapid bacterial metabolism. Limnol Oceanogr 1995; 40: 1369-1380.

31. Kieber DJ, McDaniel J, Mopper K. Photochemical source of biological substrates in sea water: implications for carbon cycling. Nature 1989; 341:637-639.

32. Kieber RJ, Zhou X, Mopper K. Formation of carbonyl compounds from UV-induced photodegradation of humic substances in natural waters: fate of riverine carbon in the sea. Limnol Oceanogr 1990; 35:1503-1515.

33. Mopper K, Zhou X, Kieber RJ et al. Photochemical degradation of dissolved organic carbon and its impact on the oceanic carbon cycle. Nature 1991; 353:60-62.

34. Mopper K, Kieber DJ. Distribution and biological turnover of dissolved organic compounds in the water column of the Black Sea. Deep Sea Res 1991; 38:1021-1047.

35. Mopper K, Stahovec WL. Sources and sinks of low molecular weight organic carbonyl compounds in seawater. Mar Chem 1986; 19:305-321.

36. Andreae MO, Barnard WR. The marine chemistry of dimethylsulfide. Mar Chem 1984; 14:267-279.

37. Dickson DMJ, Kirst GO. The role of β-dimethylsulphonoipropionate, glycine betaine and homarine in the osmoacclimation of *Platymonas subcordiformis*. Planta 1986; 167:536-543.

38. Kiene RP. Dimethyl sulfide production from dimethylsulfoniopropionate in coastal seawater samples and bacterial cultures. Appl Environ Microbiol. 1990; 56: 3292-3297.

39. Brugger A, Slezak DM, Herndl GJ. (in preparation).

40. Kieber DJ, Jiao J, Kiene RP et al. Impact of dimethylsulfide photochemistry on methyl sulfur cycling in the equatorial Pacific Ocean. J Geophysic Res 1996; 101:C2, 3715-3722.

41. Lindell MJ, Rai H. Photochemical oxygen consumption in humic waters. Arch Hydrobiol 1994; Beiheft 43:145-155.

42. Amon RMW, Benner R. Photochemical and microbial consumption of dissolved organic carbon and dissolved oxygen in the Amazon River system. Geochim Cosmochim Acta 1996; 60:1783-1792.

43. Reitner B, Herzig A, Herndl GJ. Role of ultraviolet-B radiation on photochemical and microbial oxygen consumption in a humic-rich shallow lake. Limnol Oceanogr, in press.

44. Simon M, Azam F. Protein content and protein synthesis rates of planktonic marine bacteria. Mar Ecol Prog Ser 1989; 51: 201-213.

45. Chin-Leo G, Kirchman DL. Estimating bacterial production in marine waters from the simultaneous incorporation of thymidine and leucine. Appl Environ Microbiol 1988; 54:1934-1939.

# CHAPTER 9

# OPTICAL PROPERTIES AND PHYTOPLANKTON COMPOSITION IN A FRESHWATER ECOSYSTEM (MAIN-DONAU-CANAL)

Maria A. Häder and Donat-P. Häder

## INTRODUCTION

More than 70% of the earth's surface is covered with water. Most of the primary productivity is restricted to the top layers of the water column.[1] The primary production of aquatic ecosystems contributes about half of the total primary production of the earth. Besides their fundamental position in the food web, aquatic producers play an essential role in the carbon dioxide and oxygen household of the earth.

The decrease in ozone concentration will result in an increase of UV-B at the earth's surface and affect man, animal, plant, and materials.[2-8] Solar UV-B radiation has been found to damage fish, shrimp, crab and other animals during critical phases of their development.[9] Furthermore, UV-B radiation is able to impair enzyme activity, photosynthesis and nitrogen fixation, to affect DNA and cellular pigments, and to inhibit motility and orientation.[10-14] Impacts on phytoplankton will cause a decrease in the main $CO_2$ sink, phytoplankton losses and diminish the biomass productivity, which affects the entire food chain.

Phytoplankton need solar energy for growth and metabolism. Close to the surface phytoplankton are affected by high levels of solar irradiation, especially UV-B.[15-18] Photosynthesis and penetration of solar radiation into the water depend on the optical properties of the water.[19-20]

*The Effects of Ozone Depletion on Aquatic Ecosystems,* edited by Donat-P. Häder.
© 1997 R.G. Landes Company.

The attenuation of solar radiation in the water column is wavelength dependent. Yellow humic substances are dissolved decomposition products of organic compounds.[20-21] They are responsible for a significant absorption, starting in the yellow spectral range and increasing with decreasing wavelengths. In addition, particulate substances in the water cause absorption and scattering. However, 1% of the surface irradiation has been found between 4 and 8.7 m in Lake Michigan at 310 nm[22] and about 4.5 m in Lake Giles, PA, U.S.[23] In a clear Argentinean Lake 305 nm radiation has been found as deep as 27 m.[24] Even in less transparent lakes phytoplankton will be exposed to high UV-B irradiation since water circulation moves the organisms close to the surface.

Life in both freshwater and marine ecosystems can be divided in two large groups: autotrophic organisms which synthesize organic substances from anorganic material and heterotrophs which need organic substances for their metabolism.[25] The heterotrophic organisms can be further divided: the destruents consume dead organic matter; the primary consumers prey directly on photosynthetic organisms as first links in the food web and, in turn, feed the secondary consumers. The consumers are negatively affected by decreased production of primary producers.[13,15,26]

In climatic zones with seasonal changes most of the plankton are confronted with bad or lethal environmental conditions.[27] In such periods it is important that enough individuals survive in order to serve as inoculum under improved environmental conditions.

Besides harmful effects as consequence of unfavorable environmental conditions, the phytoplankton composition can be changed by colonization. Ships can act as transport vectors since they carry ballast water. When impacts on phytoplankton or composition changes occur, the biological equilibrium is disturbed.

The Main-Donau-Canal is basically a stagnant freshwater system which is continuously mixed by regular ship traffic.

It extends from Bamberg/Germany to Kehlheim/Germany and has a length of 171 kilometers and a depth of 3.5 m. The canal was completed in 1992, connects the North Sea and the Black Sea over a distance of 3500 km. It provides an excellent model system for the study of a temperate zone freshwater ecosystem influenced by ship traffic.

## MATERIALS AND METHODS

### WATER SAMPLES

The Main-Donau-Canal was studied in 1994 and 1995 at eight locations between Nürnberg and Bamberg (Bavaria/Germany), covering a section of 73 km: Nürnberg (harbor); Nürnberg (lock); Erlangen (harbor); Erlangen (lock); Möhrendorf; Seebach inlet; Hausen (below Regnitz inlet); Bamberg (harbor area). Water samples were taken from a depth of 20-50 cm, unless mentioned differently; and analyzed immediately.

The transparency of water was measured with the Secchi disc, a standardized white disc. The Secchi depth is the depth at which the disc moves out of sight.

### SPECTROSCOPY

Absorption spectra from 250-750 nm were determined with a turbidity cuvette (size: 9.7 cm x 1 cm x 4.5 cm) in a spectrophotometer (DU 70 Beckman, Palo Alto, CA, U.S.). Fluorescence spectra were measured in a Shimadzu fluorimeter (RF5000, Japan); the size of the optical cuvette was 1 cm x 1 cm x 4.5 cm. Experiments were carried out with turbid samples, with filtered samples which were obtained by using 200 nm micropore filters and with concentrated samples. Water samples were centrifuged for 10 min at 1000 rpm (120 g) in order to concentrate particles (GP Centrifuge, Beckman, Palo Alto, CA, U.S.).

### IRRADIATION MEASUREMENTS

The spectral distribution of solar radiation in the water column was measured with a double monochromator spectroradiometer. A hollow, internally frosted quartz sphere (60 mm in diameter) served as a $4\pi$ receiver and was connected to the spectroradiometer (Optronic, model 752, Orlando, FL, U.S.) by a 10 m quartz fiber bundle.[28] This device allows one to determine spectra in the water column in dependence of depth and solar zenith angle. Measurements of the downwelling radiation were enabled by covering the lower half of the quartz sphere with an opaque half sphere. Radiation loss from one depth level in the water column to the next one was determined by attenuation coefficients which were calculated by the following equation:

$$a = -1/z \ln I(z_1) / I(z_2)$$

where a is the attenuation coefficient, I the measured irradiance and z the depth.

## PHYTOPLANKTON QUANTIFICATION

For determining the cell number water samples were centrifuged for 5 min at 1000 rpm (120 g) (GP Centrifuge, Beckman, Palo Alto, CA, U.S.). The pellets were stained with acridine orange (Fluka, Neu-Ulm, Germany) at a concentration of 1 µg/ml sample and analyzed after an incubation of 10 min. Fluorescing organisms were counted under a 10 x objective (Plan Neofluar) using a fluorescence microscope (Axioplan, Zeiss, Jena, Germany) with the filter combination BP 436, FT 580, LP 590 and a computer image analysis program.[29] The excitation wavelength band for acridine orange bound to RNA in water is 426-458, the corresponding emission wavelength band 630-644 nm.[30-31] The fluorescence microscope was connected to an image intensifier video camera (Proxon, Proxitronic, Bensheim, Germany).

Plankton organisms were analyzed in Lugol-fixed samples in a counting chamber with a volume of 0.05 ml.[32-33] Lugol solution (IKI) was added to a 1-l water sample until a cognac color was reached. After 24 h the sediment of 100 ml was removed and allowed to settle for another 24 h. The cells were concentrated in a volume of 10 ml. The cell number of the predominant algae families and genera was determined as a percentage of the total cell number.

## PHOTOSYNTHESIS MEASUREMENTS

Photosynthetic measurements (oxygen production) were carried out by an apparatus described by Häder and Schäfer.[34-35] During a bloom of Chlorococcales in July 1994, water samples were taken at the location Möhrendorf and concentrated 64 x by tangential filtration (Filtron, Dreieich, Germany). Then, 20 ml of the concentrated sample was added to the photosynthesis chamber of the instrument. The oxygen concentration in the photosynthesis chamber is measured by a Clark electrode. A magnetic stirrer warrants even distribution of oxygen. The oxygen concentration and net exchange are analyzed on-line and stored in a computer. Irradiance and temperature values are registered in parallel. The location Erlangen harbor with a depth of 3.2 m was suitable for experiments with the photosynthetic unit. Absolute oxygen values were calculated after standard calibration.

## RESULTS

### OPTICAL PROPERTIES

The Secchi disc measurements allow easy quantification of the turbidity of waters. Measurements were carried out at the locations Seebach inlet and Erlangen lock at short time intervals (Fig. 9.1a). The Main-Donau-Canal showed a high degree of turbidity in the area of the Seebach inlet. A comparison of the Secchi depths at eight locations between Bamberg and Nürnberg is shown in Figure 9.1b. Possible fluctuations at one location and within the section are obvious. The more detailed analysis from Figure 9.1a includes the fluctuations occurring in the whole section and indicates short-term effects.

Since the absorption (O.D.) increases proportionally to the quantity of the absorbing components, absorption spectra reveal the amount of particulate and dissolved substances. The absorption of turbid and filtered water samples is shown in Figure 9.2a. The difference between turbid and filtered samples indicates the amount of particles >200 nm. The strong absorption in the UV range of the spectrum is caused by yellow substances. In order to compare absorption spectra throughout the year absorption values at 250 nm were identified (Fig. 9.2b). Figure 9.2c shows the absorption values in the whole section from Bamberg to Nürnberg. There is a

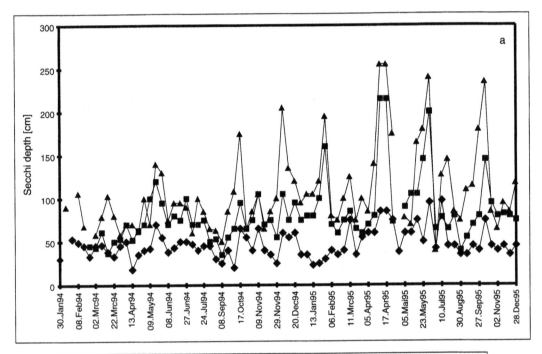

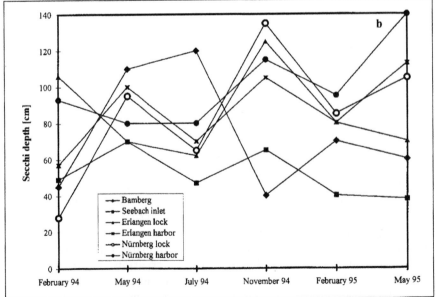

Fig. 9.1. (a) Secchi disc transparency at the locations Seebach inlet (diamonds); Erlangen lock, lower basin (squares); Erlangen lock, upper basin (triangles). (b) Comparison of Secchi disc transparency along the section Bamberg-Nürnberg.

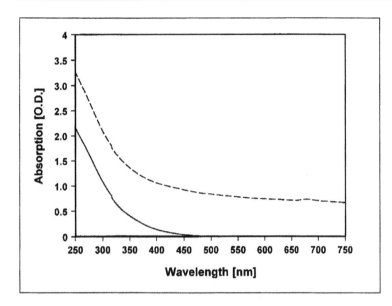

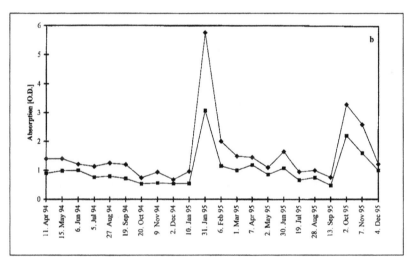

Fig. 9.2. (a) Absorption curves of canal water (turbid sample (broken line); filtered sample (continuous line)) at the location Möhrendorf, collected in October 1995. (b) Absorption (O.D.) at 250 nm in different months throughout 1994 and 1995 at the location Möhrendorf (turbid sample (diatoms); filtered sample (squares)). (c) Absorption (O.D.) of turbid water samples at 250 nm along the section Bamberg-Nürnberg.

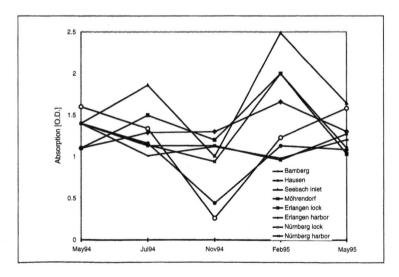

predominant turbidity peak at the location Seebach inlet, confirming results found with the Secchi disk. Absorption values in May 1994 and 1995 showed similar results and reveal another small peak at the location Nürnberg lock.

The spectral distribution of radiation in the water column is an important parameter with respect to the photosynthesis of phytoplankton organisms.[36-37] Since solar UV radiation is known to have inhibitory effects on phytoplankton organisms, the penetration of UV radiation was analyzed in summer, close to the solar solstice at the location Erlangen harbor. Figure 9.3a shows a depth profile of downwelling radiation, i.e. exclusively radiation penetrating from above into the water column, but not reflected or refracted radiation coming from below. In the UV range the transmission decreased strongly. The radiation loss from one depth level of the water column to the next one was determined by attenuation coefficients (Fig. 9.3b). Curve 5 represents the mean of the attenuation coefficients integrating over curves 1 to curve 3 corresponding to equal depth intervals. The Secchi disc value, measured in parallel, was 78 cm. The water depth to which 0.1% of the surface radiation penetrates is shown in Figure 9.3c. The 0.1% values can be derived from the attenuation coefficients for the different wavelengths. The euphotic zone is defined as the water depth with active photosynthesis and corresponds to 0.1% of the surface radiation.[1]

## Spectral Analysis

Fluorescence and absorption spectra reveal the pigment content in the water. Chlorophyll, phycocyanin and yellow substances were analyzed by excitation spectra. Synchronous spectra with an interval of 15 nm between excitation and emission wavelengths should answer the question whether additional pigments are present. Absorption spectra of concentrated samples were analyzed between 250 and 750 nm.

Chlorophyll emission spectra were registered with an excitation at 435 nm and an emission from 450-750 nm. At the beginning of 1994, a weak chlorophyll fluorescence at about 680 nm occurred in turbid water samples. In April 1994 the emission peak increased in parallel to increasing phytoplankton populations. The signal improved when the sample was concentrated 80-fold (Fig. 9.4a). Samples of June and July showed a very significant chlorophyll fluorescence in concentrated samples (Fig. 9.4b). The strong chlorophyll signal occurring in June 1994 could also be identified in the absorption spectra when samples were concentrated 25-fold and 33-fold, respectively (Fig. 9.5). Strong absorption peaks at 675 nm could be detected. Since the centrifuged samples gave improved results, fluorescence experiments in 1995 were carried out in concentrated samples (80 x). The results of the previous year could be verified.

Phycocyanin spectra were measured with an excitation at 620 or 615 nm and an emission from 630-750 nm. No fluorescence was detected from December 1993 to June 1994. This was true for turbid, filtered and concentrated samples (80 x, 1000x). A clear phycocyanin fluorescence with a maximum at 642 nm was seen for the first time in a concentrated sample of July 1994 at both excitation wavelengths (Fig. 9.6). The phycocyanin fluorescence could be determined until September 1994. Repeated experiments in 1995 with 80-fold concentrated samples were in agreement with the results of 1994.

Yellow substances are a conglomerate of heterogeneously composed decomposition products of organic matter. The fluorescence spectra were registered with an excitation at 350 nm and an emission from 360-600 nm. An intensive yellow substance fluorescence was found throughout the year. The fluorescence occurred almost through the whole emission range with maxima at 400 and 440 nm (Fig. 9.7a). The spectra demonstrate that the filtration does not diminish the emission remarkably when compared with the turbid water sample. The 20-fold concentrated sample showed an increased and modified spectrum.

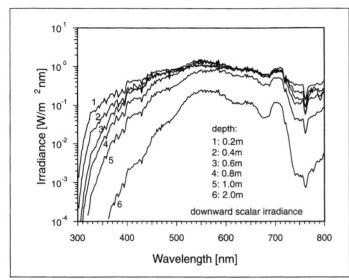

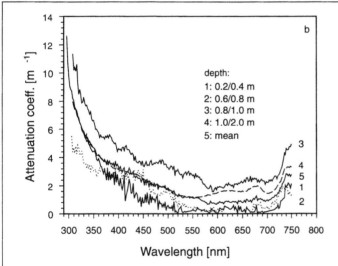

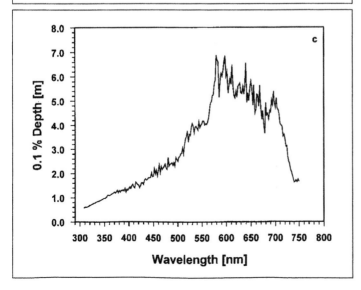

*Fig. 9.3. (a) Spectral distribution of radiation in the water column at the location Erlangen harbor at different depths during local noon under clear sky, measured by a double monochromator spectroradiometer. (b) Attenuation coefficients corresponding to different depth intervals. (c) Water depth at which the surface irradiance decreased to 0.1% of the incident in dependence of the wavelength.*

*Fig. 9.4. (a) Chlorophyll emission spectrum of an 80-fold concentrated water sample taken at the location Möhrendorf in April 1994. (b) Chlorophyll emission spectrum of a 65-fold concentrated water sample taken at the location Möhrendorf in July 1994.*

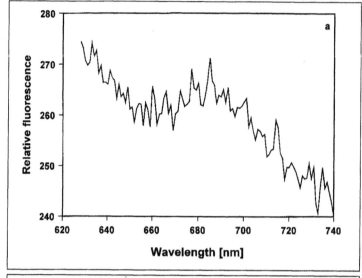

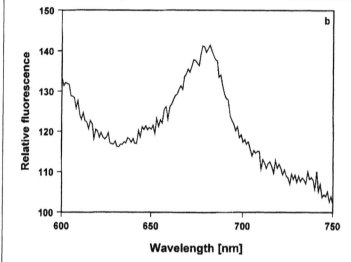

*Fig. 9.5. Absorption spectra of a 25-fold (broken line) and a 33-fold (continuous line) concentrated water sample taken at the location Möhrendorf in June 1994.*

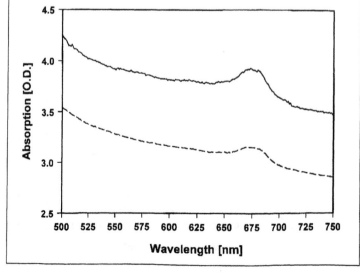

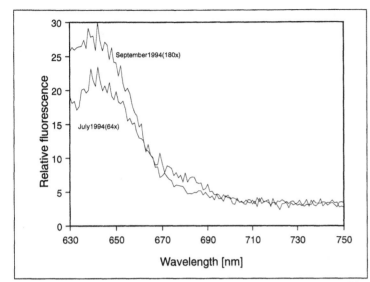

Fig. 9.6. Phycocyanin emission spectra with an excitation wavelength of 615 nm of concentrated water samples taken at the location Möhrendorf. The lower curve corresponds to the 65-fold concentrated sample from July 1994 and the upper curve to the 180-fold concentrated sample from September 1994.

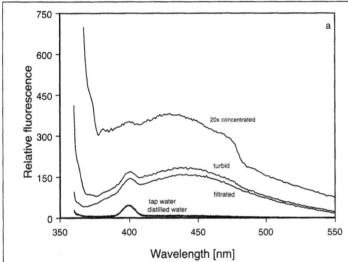

Fig. 9.7. (a) Yellow substance emission spectra of water samples taken at the location Möhrendorf. Curves from top to bottom: 20-fold concentrated sample, turbid sample, filtered sample, control spectra of tap water and distilled water. (b) Relative fluorescence at 440 nm, determined from yellow substance emission spectra of water samples taken at the location Möhrendorf.

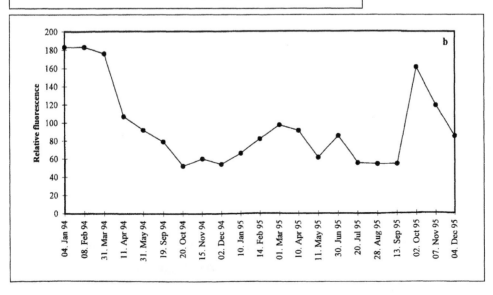

Control spectra of tap water and distilled water were identical to each other. They did not show the broad fluorescence spectrum, but a maximum at 400 nm, which corresponds to the small maximum on top of the broad spectrum of samples from the Main-Donau-Canal. The relative fluorescence at 440 nm was determined and plotted against the month in order to reveal an annual cycle (Fig. 9.7b).

The excitation wavelengths for synchronous spectra extent from 250-750 nm, the resulting emission spectrum from 265-765 nm. Figure 9.8a shows the synchronous spectrum of a turbid and a concentrated sample. Two strong maxima at

362 and 371 nm occurred. Excitation maxima at 336 and 343 nm were found for the emission at 362 nm (Fig. 9.8b). The excitation spectrum for the emission peak at 371 nm indicates a distinct maximum at 357 nm (Fig. 9.8c). Figure 9.8d shows the emission spectra which were obtained by excitations at 336, 343 and 357 nm. It can be seen that only the peak at 362 nm occurred. The same is true when the two weak maxima of the excitation spectra, 319 and 328 nm, were used for excitation (Figs. 9.8b,c). Control spectra with tap water from Erlangen and distilled water were identical to each other. They did not show the emission maxima of the synchronous

*Fig. 9.8. (a) Synchronous spectra of water samples taken at the location Möhrendorf. Curves from top to bottom: 20-fold concentrated sample, turbid sample, control spectra of tap water and distilled water. (b) Excitation spectrum for the emission peak 362 nm. (c) Excitation spectrum for the emission peak 357 nm. (d) Emission spectra by excitation wavelengths of 336 nm, 343 nm and 357 nm.*

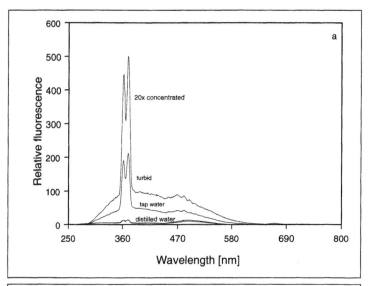

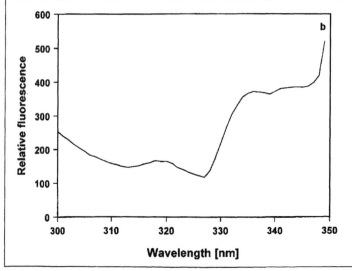

spectrum (Fig. 9.8a). Synchronous spectra were registered for each month.

## BIOLOGICAL PARAMETERS OF THE MAIN-DONAU-CANAL

The plankton analysis documents the total number of cells, the cell number of motile organisms and the most important alga families and genera. The cell numbers were counted from fluorescence-stained organisms in order to count only living organisms and to exclude detritus particles. Fluorescence arose from the acridine orange stained RNA as well as the autofluorescence of chlorophyll present in the plastids which are excited by the same wavelength range. The population density increased in both years with the beginning of the vegetation period (Fig. 9.9a). The density of motile cells in both years is shown in Figure 9.9b. The distribution of plankton organisms in the water column was analyzed in March 1995 at the location Erlangen harbor (Fig. 9.9c). There were only slight differences between the population densities at the different depths. The density of the total cell number was largest near the surface. The Secchi disc transparency measured in parallel was 95 cm.

Algal families with prominent genera at the location Möhrendorf are documented through two vegetation cycles (Fig. 9.10). A bloom of 62% *Asterionella formosa* occurred in the diatom population appearing in spring 1994, but not in spring 1995. The Cyanophyceae bloom in autumn 1994 was almost entirely composed of *Microcystis flos-aquae*. In autumn 1995 the composition of Cyanophyceae was, similar to the diatoms, more heterogeneous. Only 33% of the Cyanophyceae belonged to *Microcystis* in August 1995. Predominant species were *M. flos-aquae* and *M. viridis*. *Aphazinomenon flos-aquae* developed a bloom of 85% in the Cyanophyceae population, causing a second peak in October 1995. Besides this, a bloom of *Nitzschia acicularis* increased the diatom population compared to autumn 1994; 92% of the diatoms were represented by *N. acicularis*. Chlorophyceae were

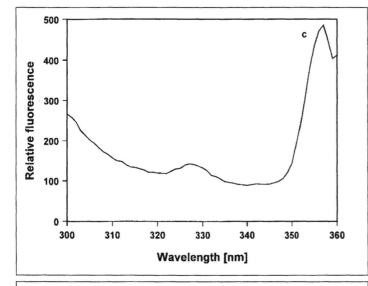

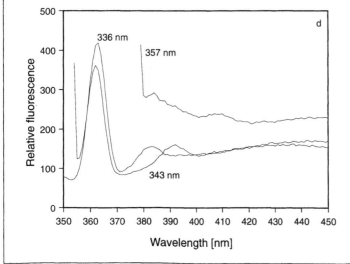

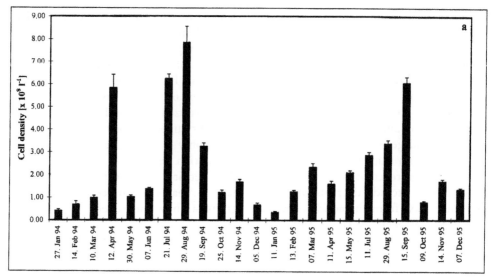

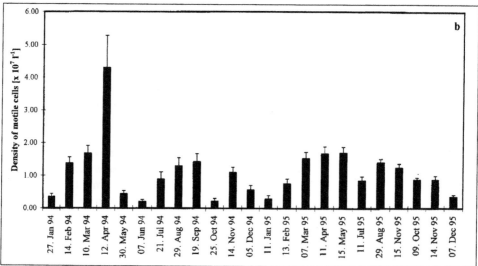

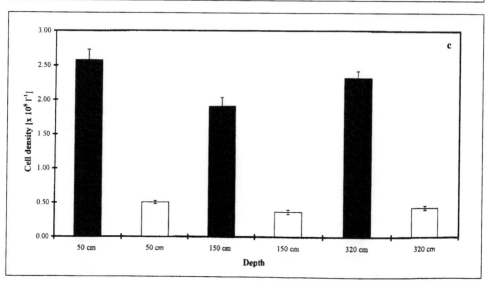

*Fig. 9.9 (opposite). (a) Total number of plankton organisms in water samples taken at the location Möhrendorf in 1994 and 1995. (b) Density of motile cells in water samples taken at the location Möhrendorf in 1994 and 1995. (c) Distribution of total number and density of motile cells in water samples from different depths of the location Erlangen harbor at 27 March 1995.*

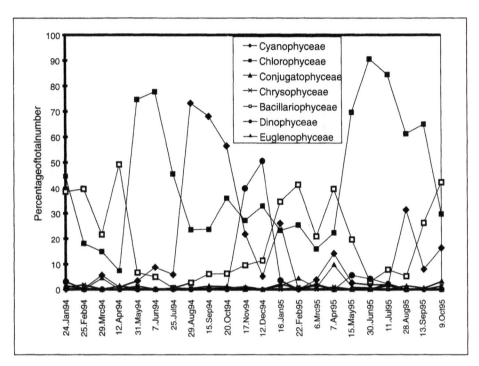

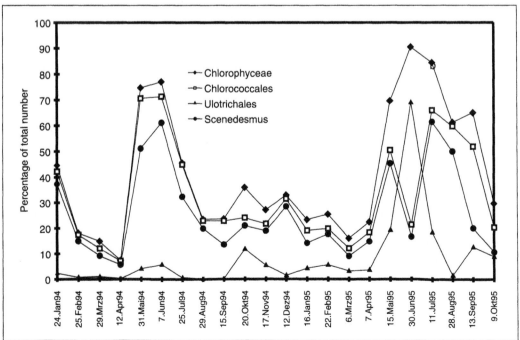

*Fig. 9.10. Phytoplankton composition in the Main-Donau-Canal in 1994 and 1995 at the location Möhrendorf. (a) Algal families and predominant genera. (b) Composition of Chlorophyceae.*

found throughout the years with optima during the summer. The composition is analyzed in Figure 9.10b. *Scenedesmus* occurred in every month with several species. Mass populations appeared during the summer, but were in competition with Ulotrichales, as can be seen in Figure 9.10b. At the end of 1994 a mass population of the dinoflagellate *Peridinium boryanum* occurred in coincidence with results of the year before. *Cladophora* was found as the predominant macroalga during the summer.

## Photosynthesis

The bloom of Chlorococcales at the end of July 1994 in the Main-Donau-Canal enabled photosynthetic measurements with naturally occurring phytoplankton. Oxygen production was measured at different depths of the canal on a sunny day during a heat period. Water temperature was 25°C. The first series of experiments started at 13.30 h (Fig. 9.11a). Oxygen production could be detected down to a water depth of 80 cm. In 100 cm water depth no photosynthetic activity could be determined.

The second series of experiments started at 15.10 h with a concentrated (64 times) sample of organisms diluted 1:1 with fresh canal water (Fig. 9.11b). Due to the advanced time and cloud formation, no oxygen production could be found at a water depth of 50 cm. Similar to the dark control the oxygen concentrations decreased as the result of respiration. Only at the water surface an increase in oxygen concentrations could be found (Fig. 9.11c). The relation between oxygen production and light conditions can clearly be seen. The oxygen production started after the disappearance of a large cloud. Although the sky was clear, no more oxygen production could be detected in the water column after 15.45 h due to decreased light intensities. Photosynthetic measurements were carried out in parallel with spectral measurements.

## DISCUSSION

The transparency of waters is influenced by two factors, dissolved and particulate substances and phytoplankton populations. The Main-Donau-Canal showed a high degree of turbidity in the area of the Seebach inlet, primarily caused by a high amount of suspended particles brought in and continuous mixing by passing ships. The measured Secchi depths fluctuated within the year at each location in the section Bamberg-Nürnberg. There was a clear tendency to increased Secchi depths in 1995, as can be seen from short time measurements. The strikingly low turbidity on 17 April 1995 and 5 May 1995 were measured during holidays when the ship traffic had ceased.

The transparency of waters indicated a high amount of particulate substances. These contribute to the deposits at the bottom if not transported, which eventually may cause silting-up and blocking of the waterway. This is of major importance for stagnant waters.

Essentially important with respect to the absorption within the water body are organic substances, as humus substances, yellow substances and the photosynthetic pigments inside phytoplankton organisms, especially chlorophyll.[19,23,38] The comparison of absorption spectra throughout 1994 and 1995 reveals two predominant peaks in January 1995 and October 1995. These dramatic increases in turbidity corresponded to snow melt or heavy rain falls. The absorption spectra in the whole section Bamberg-Nürnberg confirmed the results of the Secchi disc measurements. The absorption maximum can be seen at the location Seebach inlet. The 0.1% values of incident irradiance are different for different wavelengths in the spectral range. They started with 0.9 m depth in the UV range. The optimum at 600-650 nm was 9 m depth. The maximal depth of the UV-B penetration is characteristic for coastal waters with high concentrations of seston (particulate substances) and gelbstoff

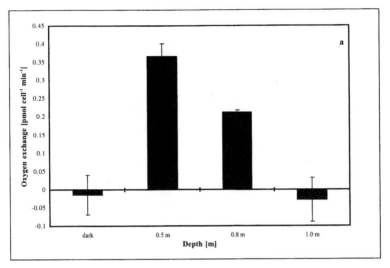

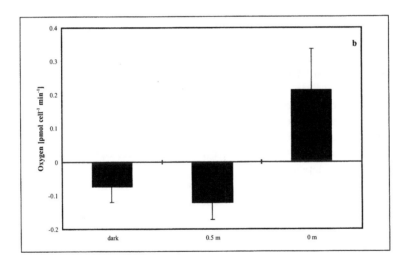

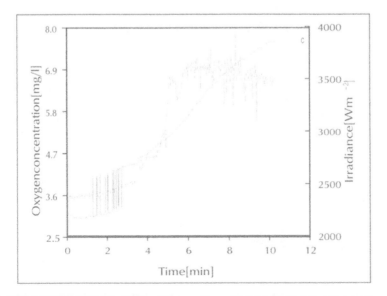

Fig. 9.11. Oxygen production by Chlorococcales at different water depths and in a dark control at the location Erlangen harbor on 27 July 1994. (a) The series started at 13.30 h. (b) The series started at 15.10 h. (c) Oxygen production (continuous line) in relation to irradiance (broken line). Surface experiment from series 11b.

(yellow dissolved organic substances). In contrast, penetration to several tens of meters may occur in clear oceanic waters.[37] The productivity of phytoplanktonic organisms is restricted to the euphotic zone, the top layer of the water column. The lower limit of the euphotic zone equals the depth where the incident PAR (photosynthetic active radiation) irradiance corresponds to the 0.1% level of the surface radiation.[1]

The irradiance above the water can be estimated to be 1160 W/m$^2$ for the whole spectrum at a day with clear sky during zenith sun. The irradiance drastically decreases in the first meters of a water body.[20] Complete spectral energy analysis requires that the number of quanta s$^{-1}$ m$^{-2}$ in the whole range of 350-700 nm are directly measured. After Jerlov[20] optical water types can be derived from such measurements corresponding to the transmittance of downward irradiance through one meter depth. In the case of the Main-Donau-Canal results of the analysis indicated optical water type 10-12, corresponding to extended Jerlov optical water types.[39]

Attenuation is caused by absorption and scattering. The attenuation of solar UV radiation in the water body depends on the wavelength. When the attenuation coefficients of the measured spectra are plotted against the wavelengths, it can be seen that the lowest attenuation in the water column in the visible wavelength range occurs between 0.2-0.4 m, in the UV range from 0.6-0.8 m and the strongest attenuation in the whole spectral range at a depth from 0.8-1.0 m. For comparison, in extreme plankton poor inland waters as Lake Tahoe, U.S., attenuation coefficients of about 0.05 m$^{-1}$ were found; when dense cyanobacteria blooms occur, as in the Nakuru Sea, Kenya, attenuation coefficients can exceed values of 10 m$^{-1}$. In Lake Constance the attenuation coefficient is about 0.2 m$^{-1}$ during a biomass minimum and about 0.8 m$^{-1}$ during the maximum.[27] Ultraviolet radiation is attenuated by high turbidity and yellow substances. The percentage of solar UV at different depths of the water body is important for its effect on phytoplankton. Besides, a harmful effect on phytoplankton organisms, solar UV radiation can cause chemical reactions in many substances in natural waters.[38,40] Growth and metabolism of phytoplanktonic microorganisms depend on solar energy. Close to the surface phytoplankton face high levels of solar irradiation. Most phytoplankton organisms are not adapted to unfiltered sunlight exceeding 1000 W m$^{-2}$. Cells cannot escape detrimental radiation since no photoreceptor active in the UV range was found in the organisms studied so far.[15,26] The situation is even worse in the canal because of frequent artificial mixing of the water mass.

There is growing evidence that many phytoplankton organisms suffer from UV-B stress even at ambient, current levels. Motility, orientation, photosynthesis and nitrogen fixation of phytoplankton organisms are affected by UV-B radiation in most cases studied so far.[15,26,41-44] Furthermore, the general metabolism in microorganisms may be impaired since all aromatic amino acids possess a strong absorption band in the UV-B range. Specifically, proteins are destroyed by UV-B irradiation. High UV-B doses have a harmful effect on membranes, including, for example, increases in ion permeability.[45]

In climate zones with seasonal changes most of the plankton organisms are confronted with bad or lethal environmental conditions.[27] In these periods it is important that enough individuals survive in order to serve as inoculum during improved environmental conditions. Bad weather fronts diminish surface irradiation and temperature. They occur in irregular weekly and monthly cycles. Oscillations of herbivorous zooplankton also exert pressure on the phytoplankton populations. Physical and chemical environmental conditions, predator/prey relations, and sedimentation influence the species composition of the plankton organisms. As mentioned above, UV-B has been shown to have detrimental effects on phytoplankton productivity.

Chances for growth and survival of phytoplankton organisms are impaired if orientation mechanisms as motility or the capability of responses towards external stimuli are inhibited by increased UV-B radiation.[46,47] Exposition to too bright irradiances can photobleach the cells, while exposure to too dim irradiances can reduce the photosynthetic biomass production. Additionally, the consumers are negatively affected by decreased production of primary producers.

Colonization may occur by passive transportation or active migration. Another possibility to change the species composition of phytoplankton can occur when some species are more sensitive than others to unfavorable environmental conditions like UV radiation. Changes in the phytoplankton composition and losses in organic material bear the risk of decreased availability of food for the links that follow in the food chain and the risk that toxic algal blooms like Cyanophyceae can occur; in both cases the food web is negatively affected.[1,48-49] The density of total phytoplankton populations followed the annual cycle with a minimum during winter. The amount of motile cells increased in spring and autumn. Many phytoplankton organisms are able to reach a specific position in the water column since they are motile. The active movements of cells are superposed by wind and waves mixing the water body. Since the euphotic zone exceeds the depth of the canal, plankton organisms are found not only at the water surface but throughout the water column. Furthermore, the Main-Donau-Canal, a stagnant water with continuous ship traffic, causes a specific mixing of phytoplankton organisms in the water column. As a consequence organisms are exposed to solar UV-B irradiation during summer. When critical UV-B values are reached, survival of phytoplankton populations or some species is impaired. The sequence of algae populations in 1995 basically confirmed that found in 1994.[50-52] Differences occurred in the composition of algae blooms. There is a clear tendency to

a more heterogeneous pattern in all algal families in 1995 compared to the previous year before. Coincidental with results from Lugol-fixed samples, Chlorophyceae and Cyanophyceae could be identified spectroscopically.

Photosynthesis plays an important role in primary production. Therefore the transformation of anorganic, dissolved substances in biological particles occurs mainly in the upper irradiated water layer and during seasons with sufficient irradiance.[27] Photosynthesis in the water column depends on the surface irradiation and on the vertical light attenuation. On clear days the surface irradiation can reach $1000 \ W \ m^{-2}$ in the temperate zone. Photosynthesis of phytoplankton needs about $1-10 \ W \ m^{-2}$. Increased UV-B radiation reduces photosynthetic activities; stress factors such as elevated temperature amplify UV-effects.[15,26,53] Limitation of phytoplankton productivity by UV-B increases linearly with increasing dose. Photosynthesis of phytoplankton organisms can be impaired by bleaching of photosynthetic pigments.[13,53] As a consequence the harvesting of solar energy is disturbed. Damage in the reaction center of photosystem II, structural changes in the thylakoids and decreased lipid contents were found. Enzyme activities are affected since UV-B reduces the protein content.[54] Another target of UV-B are nucleic acids.[55] Besides UV-B, other wavelengths in the solar spectrum participate in the harmful effects mentioned previously.

The course of photosynthetic pigment concentrations in the annual cycle could be followed by spectral analysis. The fluorescence data of chlorophyll and phycocyanin emission spectra were in agreement with the vegetation period and bloom of Chlorophyceae and Cyanophyceae. Synchronous spectra revealed that no further photosynthetic pigments could be detected. Fluorescence spectra of yellow substances occurred throughout the year. Particulate and dissolved yellow humic substances originate from breakdown of plant matter.[23] The main source are decomposition products of terrestrial biomass from the soils of the catchment brought in by water. Coincidentally, no seasonal rhythm could be derived from fluorescence spectra.

## ACKNOWLEDGMENTS

This study was enabled by support from the Friedrich-Alexander-University of Erlangen-Nürnberg (Hochschulsonderprogramm II). We are indebted to H. Piazena for supplying the solar spectral distribution data shown in Figure 9.3. Furthermore, we thank K. M. Hartmann for valuable comments on the manuscript and H.-O. Glenk and M. Kraml for contributions concerning phytoplankton determination.

## REFERENCES

1. Häder D-P, Worrest RC, Kumar HD et al. Effects of increased solar ultraviolet radiation on aquatic ecosystems. Ambio 1995; 24:174-180.
2. Damkaer DM, Dey DB. UV damage and photo-reactivation potentials of larval shrimp, *Pandalus platyceros*, and adult euphausiids, *Thysanoessa raschii*. Oecologia 1983; 60:169-175.
3. Kelly JR. How might enhanced levels of solar UV-B radiation affect marine ecosystem? In: Proceedings of EPA/UNEP International Conference on Health and Environment Effects of Ozone Modification and Climate Change. U.S. EPA/UNEP, 1986.
4. Caldwell MM, Madronich S, Björn LO et al. Ozone reduction and increased solar ultraviolet radiation. In: UNEP Environmental Effects Panel Report. 1989:1-10.
5. Tevini M, Teramura AH, Kulandaivelu G et al. Terrestrial plants. In: UNEP Environmental Effects Panel Report. UNEP, 1989:25-37.
6. Tevini M, ed. UV-B Radiation and Ozone Depletion. Effects on Humans, Animals, Plants, Microorganisms, and Materials. Boca Raton, Ann Arbor, London, Tokyo: Lewis Publ, 1993.
7. van der Leun JC. Human health. In: UNEP Environmental Effects Panel Report. UNEP, 1989:11-24.
8. Williamson CE, Zagarese HE. The impact of UV-B radiation on pelagic freshwater ecosystems. In: Kausch H, Lampert W, eds. Advances in Limnology. Arch Hydrobiol Beih 43. Stuttgart: E Schweizerbart'sche Verlagsbuchhandlung, 1994:IX-XI.
9. USEPA (US Environmental Protection Agency). An assessment of the effects of ultraviolet-B radiation on aquatic organisms. In: Assessing the risks of trace gases that can modify the stratosphere. EPA 400/1-87/001C, 1987:1-33.
10. Döhler G, Hagmeier E, Grigoleit E et al. Impact of solar UV radiation on uptake of $^{15}$N-ammonia and $^{15}$N-nitrate by marine diatoms and natural phytoplankton. Biol Phys Pfl 1991; 187:293-303.
11. Häder D-P, Worrest RC, Kumar HD. Aquatic ecosystems. In: UNEP Environmental Effects Panel Report 1989:39-48.
12. Häder D-P, Worrest RC, Kumar HD. Aquatic ecosystems. In: UNEP Environmental Effects Panel Report 1991:33-40.
13. Häder D-P, Worrest RC. Effects of enhanced solar ultraviolet radiation on aquatic ecosystems. Photochem Photobiol 1991; 53:717-725.
14. Worrest RC, Häder D-P. Effects of stratospheric ozone depletion on marine organisms. Environ Conserv 1989; 16:261-263.
15. Häder D-P. Risks of enhanced solar ultraviolet radiation for aquatic ecosystems. In: Round FE, Chapman DJ, eds. Progress in Phycological Research. Vol 9. Bristol: Biopress, 1993:1-45.
16. Cullen JJ, Neale PJ. Ultraviolet radiation, ozone depletion, and marine photosynthesis. Photosynthesis Res 1994; 39:303-320.
17. Holm-Hansen O, Lubin D, Helbling EW. UVR and its effects on organisms in aquatic environments. In: Young AR, Björn O, Moan J et al, eds. Environmental UV Photobiology. New York: Plenum, 1993:379-425.
18. Smith RC, Cullen JJ. Implications of increased solar UVB for aquatic ecosystems. U.S. National Report to the IUGG (1991-1994). American Geophysical Union, 1995.
19. Baker KS, Smith RC. Spectral irradiance penetration in natural waters. In: Calkins J, ed. The role of solar ultraviolet radiation in

marine ecosystems. New York, London:
Plenum Press, 1982:233-246.

20. Jerlov NG. Light. In: Kinne O, ed. Marine
Ecology Vol 1. London, New York: Wiley
1970:95-102.

21. Højerslev NK. Yellow substance in the sea.
In: Calkins J, ed. The role of solar ultraviolet radiation in marine ecosystems. New
York, London: Plenum Press, 1982:
263-281.

22. Gala WR, Giesy JP. Effects of ultraviolet
radiation on the primary production of natural phytoplankton assemblages in Lake
Michigan. Ecotox Environ Safety 1991;
22:345-361.

23. Kirk JTO. Optics of UV-B radiation in
natural waters. Arch Hydrobiol Beih 1994;
43:1-16.

24. Morris DP, Zagarese H, Williamson CE et
al. The attenuation of solar UV radiation in
lakes and the role of dissolved organic carbon. Limnol Oceanogr 1995; 40:1381-1391.

25. Hartmann L. Biologische Abwasserreini-
gung. Berlin, Heidelberg, New York, London: Springer-Verlag, 1989.

26. Häder D-P. Effects of enhanced solar ultra-
violet radiation on aquatic ecosystems. In:
Tevini M, ed. UV-B Radiation and Ozone
Depletion. Effects on Humans, Animals,
Plants, Microorganisms, and Materials. Boca
Raton, Ann Arbor, London, Tokyo: Lewis
Publishers, 1993:155-192.

27. Sommer U. Planktologie. Berlin, Heidelberg, New York, London: Springer-Verlag,
1994.

28. Piazena H, Häder D-P. Penetration of solar
UV irradiation in coastal lagoons of the
Southern Baltic Sea and its effect on phy-
toplankton communities. Photochem
Photobiol 1994; 60:463-469.

29. Häder D-P, Vogel K. Simultaneous tracking of flagellates in real time by image
analysis. J Math Biol 1991; 30:63-72.

30. Koller E, ed. Applied fluorescence technology 1992; 4:30-31.

31. Völkl H, Friedrich F, Häussinger D et al.
Effect of cell volume on acridine orange
fluorescence in hepatocytes. Biochem J
1993; 295:11-14.

32. Glenk H-O. Methoden wissenschaftlicher
Planktonuntersuchung I. Mikrokosmos
1962; 51:178-182.

33. Glenk H-O. Methoden wissenschaftlicher
Planktonuntersuchung II. Mikrokosmos
1962; 51:207-211.

34. Häder D-P, Schäfer J. Photosynthetic oxygen production in macroalgae and phytoplankton under solar irradiation. J Plant
Physiol 1994; 144:293-299.

35. Häder D-P, Schäfer J. In-situ measurement
of photosynthetic oxygen production in the
water column. Environmental Monitoring
and Assessment 1994; 32:259-268.

36. Worrest RC. Review of literature concerning the impact of UV-B radiation upon
marine organisms. In: Calkins J, ed. The
Role of Solar Ultraviolet Radiation in Marine Ecosystems. New York: Plenum Press,
1982:429-457.

37. Smith RC, Prezelin BB, Baker KS et al.
Ozone depletion: Ultraviolet radiation and
phytoplankton biology in Antarctic waters.
Science 1992; 255:952-959.

38. Zepp RG. Photochemical transformations
induced by solar ultraviolet radiation in
marine ecosystems. In: Calkins J, ed. The
Role of Solar Ultraviolet Radiation in Marine Ecosystems. New York, London: Plenum Press, 1982:293-307.

39. Piazena H. Penetration of solar UV and PAR
into different waters of the Baltic Sea and
remote sensing of phytoplankton. This volume.

40. Herndl G. Role of ultraviolet radiation on
bacterioplankton activity. Chapter 8, this
volume. In: Häder D-P (ed.) The Effects of
Ozone Depletion on Aquatic Ecosystems.
Austin: RG Landes Co 1997:129-140.

41. Döhler G. Effect of UV-B radiation
(290-320 nm) on the nitrogen metabolism
of several marine diatoms. J Plant Physiol
1985; 118:391-340.

42. Häder D-P, Häder M. Effects of solar radiation on photoorientation, motility and pigmentation in a freshwater *Cryptomonas*.
Botanica Acta 1989; 102:236-240.

43. Häder D-P, Häder MA. Effects of solar and
artificial radiation on motility and pigmentation in *Cyanophora paradoxa*. Arch
Microbiol 1989; 152:453-457.

44. Häder D-P, Häder MA. Effects of solar
UV-B irradiation on photomovement and
motility in photosynthetic and colorless

flagellates. Environ Exp Bot 1989; 29: 273-282.

45. Imbrie CW, Murphy TM. UV-action spectrum (254-405 nm) for inhibition of a K+ stimulated adenosine triphosphatase from the plasma membrane of *Rosa damascena*. Photochem Photobiol 1982; 36:537-542.

46. Häder D-P, Häder M. Effects of solar radiation on motility, photomovement and pigmentation in two strains of the cyanobacterium, *Phormidium uncinatum*. Acta Protozool 1990; 29:291-303.

47. Häder D-P, Häder M. Effects of UV radiation on motility, photoorientation and pigmentation in a freshwater *Cryptomonas*. J Photochem Photobiol B: Biol 1990; 5:105-114.

48. Nixon SW. Physical energy inputs and the comparative ecology of lake and marine ecosystems. Limmnol Oceanogr 1988; 33:1005-1025.

49. Gucinski H, Lackey RT, Spence EC. Fisheries. Bull Am Fish Soc 1990; 15:33-38.

50. Fott B. Algenkunde. Jena: VEB Gustav Fischer Verlag, 1971.

51. Pascher A. Die Süßwasserflora von Mitteleuropa. Stuttgart: Gustav Fischer Verlag, 1978-1991.

52. Streble H, Krauter D. Das Leben im Wassertropfen. Stuttgart: Kosmos. Franckh'sche Verlagshandlung, 1988.

53. Häder D-P, Worrest RC. Effects of ozone depletion on aquatic ecosystems. In: Ilyas M, ed. Ozone depletion. Implications for the tropics. Penang, Malaysia: University of Science Malaysia; Nairobi, Kenya: United Nations Environment Programme, 1991: 254-270.

54. Döhler G, Biermann I, Zink J. Impact of UV-B radiation on photosynthetic assimilation of $^{14}$C-bicarbonate and inorganic $^{15}$N-compounds by cyanobacteria. Z Naturforsch 1986; 41C:426-432.

55. Karentz D, Bothwell ML, Coffin RB et al. Impact of UVB radiation on pelagic freshwater ecosystems: Report of working group on bacteria and phytoplankton. Arch Hydrobiol 1994; Beiheft 43 (special issue), 31-69.

# CHAPTER 10

# THE EFFECTS OF ULTRAVIOLET-B RADIATION ON AMPHIBIANS IN NATURAL ECOSYSTEMS

Andrew R. Blaustein and Joseph M. Kiesecker

R ecent reports suggest that the populations of amphibians in a wide array of geographic regions and habitats have apparently declined, have experienced range reductions or have displayed unusual embryonic mortality.[1-7] For example, many native, mountain-dwelling frogs (*Rana*) and toads (*Bufo*) of western North America seem to be experiencing one or more of these phenomena.[8-12] Furthermore, many population declines, including some in western North America, have occurred in relatively undisturbed habitats.[1]

Most amphibian population declines are probably due to habitat destruction or habitat alteration,[5,13] the major cause for species loss.[14] Some declines are probably the result of natural population fluctuations.[5,13,15] Other explanations for amphibian population declines and range reductions include disease, pollution, atmospheric changes and introduced competitors and predators.[6,13] Population declines are more apparent in some regions than others;[16] there does not seem to be one explanation for all the declines.

The declines of some species have been especially perplexing because there are no obvious causes and they have occurred in habitats that have not been overtly altered.[1,12,13,17-19] Prominent among declines for which the explanations are not yet clear are disappearances of frogs from higher elevations (>1000 m) in temperate western North America.[2,10-11,20-21] Similar unexplained declines of amphibian populations in other regions of the world have been reported from relatively undisturbed habitats.[17-19,22-23]

---

*The Effects of Ozone Depletion on Aquatic Ecosystems*, edited by Donat-P. Häder.
© 1997 R.G. Landes Company.

Because some of these disappearances have occurred in preserves, protected from direct human intrusion, the causes of species disappearing in these areas may be subtle. The disappearance of amphibians in remote regions suggests a degradation of the environment that could be global in scale.

Since we began studying the behavior and ecology of amphibians in 1979 in Oregon (U.S.), we have documented excessive egg mortality and some population reductions of the Cascades frog (*Rana cascadae*) and western toad (*Bufo boreas*) in relatively undisturbed regions.[2-3,24-25] Moreover, the virtual disappearance of *R. cascadae* in the southern portion of its range in California has recently been well-documented.[10] Similarly, western toads (*B. boreas*) are experiencing declines throughout their range in western North America.[9,26-27] Yet, populations of Pacific treefrogs (*Hyla regilla*) sympatric with both *R. cascadae* and *B. boreas*, seem to be persistent in Oregon and there are no reports of excessive egg mortality in this species.

One hypothesis advanced for the decline of amphibians at several well-separated global locales is hypersensitivity to small increases in UV-B radiation at the earth's surface, that may have already occurred as a result of stratospheric ozone depletion.[28-29] In this chapter we: (1) review the results of our research on the effects of ambient levels of UV-B radiation on amphibian embryos in nature; (2) discuss information on the adverse affects of UV-B on amphibians from relevant laboratory studies; (3) describe some of the defenses amphibians may use to lower their risk to UV-B damage, and (4) speculate on the ecological and evolutionary implications of the detrimental effects of UV-B on amphibians.

## INITIAL LABORATORY STUDIES

Several laboratory studies have provided information showing the potential harmful effects of UV radiation on amphibians.[6,30-33] Worrest and Kimeldorf,[34-35] apparently were the first to address the potential connection between the detrimental

effects of ozone depletion and increasing UV radiation on amphibians. They studied the effects of UV-B radiation on western toads (*B. boreas*) from Oregon. Females from the Oregon Cascade Range were induced to lay eggs which were fertilized in the laboratory. Growth and development of embryos were monitored under different regimes of UV-B radiation.

Tadpoles exposed to enhanced UV-B radiation developed anomalous, concave curvatures of their spine during early development. They also developed abnormally thick, pigmented corneas and areas of hyperplasia in their integument. Moreover, compared with control groups, individuals in UV-B enhanced groups had significantly lower survival.

These results suggest that prolonged exposure to UV-B radiation can induce abnormal growth and development patterns in tadpoles that are in early stages of development. These studies served as the foundation for the field studies that are discussed below.

## FIELD STUDIES

### GENERAL METHODS AND MATERIALS USED IN FIELD STUDIES

We used field experiments to assess the sensitivity of ambient levels of UV-B radiation on developing amphibian embryos in natural habitats.[2] In all of our studies, the methods and materials were similar. We placed newly laid eggs, with their surrounding jelly matrix intact, in enclosures where eggs and embryos were exposed to ambient levels of UV-B radiation or shielded from UV-B radiation.

Enclosures (38 x 38 x 7 cm) and filters (50 x 50 x 7 cm) were made of the same materials and had the same dimensions and light transmission properties in all experiments. Enclosures had clear Plexiglas frames with floors of 1 mm$^2$ fiberglass mesh screen. A UV-B blocking filter made of Mylar was placed over one-third of the enclosures. An acetate filter that transmitted UV-B was placed over another third of the enclosures. The remaining enclosures had no filters. The transmitting properties of the Mylar and acetate used on enclosures were assessed before and after experiments by scanning a UV-B 313 lamp directly with an Optronic 752 spectroradiometer and comparing the transmission with the same lamp covered with acetate and Mylar. The Mylar blocked 100% of UV-B (280-315 nm). The acetate allowed about 80% of UV-B transmission.

In all of our experiments described in this chapter, enclosures were placed in a linear array, in a randomized block design[2] (Fig. 10.1). The placement of filters was randomly assigned to enclosures in all experiments. Enclosures at each site were randomly assigned to three sunlight treatments: unfiltered sunlight, sunlight filtered to remove UV-B, sunlight with a filter that allowed penetration of UV-B (control filter). There were four replicates per treatment in all field experiments. Our experiments were terminated when all the original embryos either hatched or died. All embryos were counted each day and all

were accounted for at the end of each experiment. There was no predation on embryos. Survival was measured as the proportion of hatchlings produced per enclosure. In all experiments, daily temperatures (°C) were taken within enclosures in each treatment.

## FIELD EXPERIMENTS AT RELATIVELY HIGH ELEVATION SITES IN OREGON

Our initial field experiments were designed to compare the effects of ambient levels of UV-B radiation on embryos of *H. regilla*, *B. boreas*, and *R. cascadae* as they developed at natural oviposition sites in the Cascade Range.[2] We chose these species because their eggs are laid in open, shallow water, highly exposed to sunlight.[2,36-37] In fact, the eggs of *R. cascadae* are often laid in such shallow water that they may be exposed partially to air.[36] Moreover, we compared sensitivity to UV-B radiation in two species whose populations are in decline (*R. cascadae* and *B. boreas*) with those

*Fig. 10.1. Enclosures used in field experiments were arranged in a randomized block design at natural oviposition sites amongst egg masses. Enclosures were placed in a linear array along the shore.*

of *H. regilla*, a species whose populations appear to be robust.

## METHODS

Initial field experiments were conducted at several sites in the Oregon Cascade Range (U.S.) from 1220-2000 m elevation in spring 1993.[2] To ensure that any effects were not due to a particular site, we tested *H. regilla*, *R. cascadae*, and *B. boreas* at two sites each, including one where all three species were found together.

We placed 150 newly deposited eggs (<24 h old) in each of 12 enclosures at natural oviposition sites of each species (two sites per species). Enclosures were placed parallel to the water's edge at depths of 5-10 cm, where eggs are naturally laid.[2] For *R. cascadae* and *B. boreas*, 25 eggs from each of six different clutches (total = 150 eggs per enclosure) were placed in each enclosure. Because of their small clutch size, for *H. regilla* we used eggs from more than six clutches and randomly assigned 25

eggs from at least six clutches to each enclosure (total = 150 eggs per enclosure).

## RESULTS

There were no differences in hatching success among the three regimes for *H. regilla*. Thus, *Hyla regilla* embryos were highly resistant to UV-B-containing sunlight (Fig. 10.2; Table 10.1). The hatching success of *B. boreas* and *R. cascadae*, however, was greater under sunlight lacking UV-B than under unfiltered sunlight or under regimes with a control filter (Fig. 10.2).

## FIELD EXPERIMENTS AT LOW ELEVATION SITES IN OREGON

During the winter and spring of 1994, we conducted field experiments similar to those described above to investigate the effects of UV-B radiation on Northwestern salamander (*Ambystoma gracile*) and red-legged frog (*R. aurora*) embryos.[38-39] These studies were conducted at 76 and 183 m

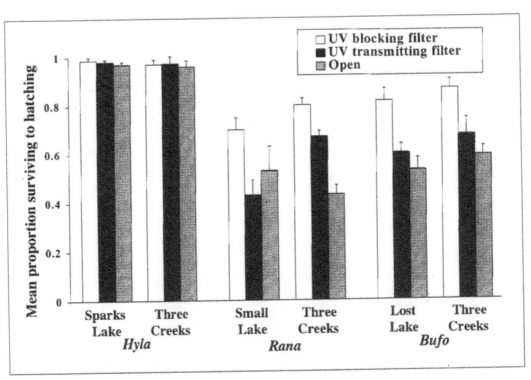

Fig. 10.2. *Effects of UV-B on hatching success (mean ± SE) in* Hyla regilla, Bufo boreas *and* Rana cascadae *at relatively high elevation sites in Oregon.*

in elevation for *R. aurora* and *A. gracile*, respectively. The red-legged frog has disappeared from large portions of its historical range.[36] The population status of the Northwestern salamander is unknown.[38] Both of these species lay eggs in shallow water, exposed to solar radiation.[38-39] However, there are no reports of unusual egg mortality in these species.

## METHODS

We collected whole clutches of *A. gracile* eggs, within 24 h after being laid, from the Oregon Coast Range. Field experiments were conducted at a natural breeding site of *A. gracile* in the Oregon Coast Range. Each enclosure had eggs from eight different clutches. Ten eggs from each of seven clutches and five eggs from an additional clutch were placed in each enclosure (total of 75 eggs per enclosure). Enclosures were placed within small plastic pools (110 cm diameter, 18 cm deep). Within the pools, eggs were immersed in 5-10 cm of natural pond water, a depth at which eggs are

naturally laid.[40] The 12 enclosures were randomly assigned to the three sunlight treatments described above.

For *R. aurora*, we collected freshly laid (<24 h old) clutches of eggs from a site in Benton Co., Oregon. Tests were conducted southwest of Corvallis, OR (U.S.) where *R. aurora* were historically abundant but where they are now rare.[36] We placed 150 eggs (25 eggs from each of six clutches) in each of 12 enclosures. Enclosures were placed in the same types of plastic pools as described for *A. gracile*. Within the pools, enclosures with eggs were immersed in 5-10 cm of well water, a depth at which eggs are often laid.[40] The 12 enclosures were assigned to the three sunlight treatments.

## RESULTS

The hatching success of *A. gracile* was significantly greater under sunlight lacking UV-B radiation than under unfiltered sunlight or under control filter regimes (Fig. 10.3; Table 10.1). There were no differences in hatching success between

---

**Table 10.1. Effects of ambient UV radiation on amphibian species in Oregon and a summary of key ecological characteristics for each species**

| Species | Hatching success under ambient UV-B | UV-B/fungus synergism | Egg laying behavior | Population status |
|---|---|---|---|---|
| Pacific treefrog *Hyla regilla* | no effect | no effect | eggs laid in open shallow water[36-37] | persistent |
| Red-legged frog *Rana aurora* | no effect | experiment not conducted | eggs laid in open shallow water[39] | declining |
| Cascades frog *Rana cascadae* | hatching reduced | yes | eggs laid in open shallow water[36-37] | declining |
| Western toad *Bufo boreas* | hatching reduced | yes | eggs laid in open, often shallow water[36-37] | declining |
| Northwestern salamander *Ambystoma gracile* | hatching reduced | experiment not conducted | eggs laid in open, often shallow water[38] | unknown |

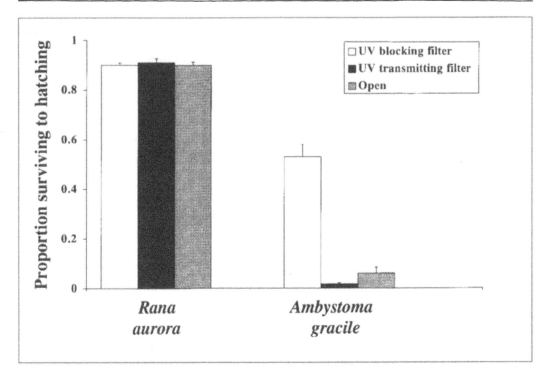

Fig. 10.3. Effects of UV-B on hatching success (mean ± SE) in Rana aurora and Ambystoma gracile at relatively low elevation sites in Oregon.

unfiltered regimes and regimes with control filters that allowed transmission of UV-B[38] (Fig. 10.3).

Unlike *A. gracile*, *R. aurora* embryos were not affected by ambient levels of UV-B radiation. In *R. aurora*, there were no significant differences in hatching success among the three sunlight regimes (Fig. 10.3).

## SYNERGISM OF UV-B RADIATION WITH OTHER AGENTS

The experiments described above tested the effects of UV-B radiation alone on developing embryos. Obviously, in nature, amphibians are exposed to numerous biotic and abiotic agents that may act alone or in conjunction with one another. For example, in Oregon, at least one other factor, the fungus, *Saprolegnia ferax*, contributes to amphibian egg mortality.[3] In a series of field experiments similar to those described above, we tested the hypothesis that there is a synergism between UV-B

radiation and *Saprolegnia*. This hypothesis would be supported if the effects of both UV-B radiation and *Saprolegnia* together were greater than those of either factor alone.

## SYNERGISM OF UV-B RADIATION AND *SAPROLEGNIA*

### Methods

Experiments investigating the effects of UV-B radiation and *Saprolegnia* were conducted at three natural oviposition sites in the central Oregon Cascade Range.[25] Tests of all species were conducted at two sites, including one site where all three species are found together. The sites were from 1190-2000 m in elevation. Our standard enclosures were placed in small plastic pools (as described above) so that we could control *Saprolegnia* densities. Within the pools eggs were immersed in 5-10 cm of natural lake water. Pools with enclosures were placed in a linear array parallel to the water's edge in a 2 x 3 randomized block design with three sunlight treatments crossed with two fungal treatments.[25] There were four replicates for each treatment. Thus, there were 24 enclosures at each site.

For *R. cascadae* and *B. boreas*, 25 eggs from each of six different clutches (total = 150 eggs per enclosure) were placed in each enclosure. Because of their small clutch size, for *H. regilla* we used eggs from more than six clutches and randomly assigned 25 eggs from at least six clutches to each enclosure (total = 150 eggs per enclosure). Filters were placed over enclosures in a manner identical to that described above.

*Saprolegnia* was cultured in the laboratory on 20 ml corn meal agar in standard Petri dishes. Using the standardized culture protocol, boiled hemp seeds were added to cultures and cultures were allowed to incubate at 20°C for approximately 168 h.[25] In the pools where *Saprolegnia* was added, we introduced three hemp seeds laden with *Saprolegnia*.

In the first synergism experiment, we placed 150 newly deposited eggs (<24 h old) in each of 24 enclosures at natural oviposition sites of each species. The enclosures at each site were randomly assigned to the three sunlight treatments. We randomly added *S. ferax* to half of the enclosures in each sunlight treatment. To the remainder of the enclosures, we added the antifungal agent, Malachite green (nontoxic to amphibian eggs[25]), to remove any *Saprolegnia* that may be present naturally.[25] Enclosures were placed in plastic pools as described previously. Within the pools, eggs were immersed in natural lake water.

We conducted a second experiment simultaneously with the first *Saprolegnia* experiment using identical techniques except that enclosures were placed directly into the lake. Thus, in this experiment, embryos were exposed to all three sunlight regimes and natural levels of *Saprolegnia*.

As in all our experiments, these tests ended when all embryos either hatched or died and survival was measured as the proportion of hatchlings produced per enclosure.

## Results

In the first *Saprolegnia* experiment we found a significant UV-B effect by itself (Fig. 10.4). However, the effect is secondary to the interaction effect of UV-B radiation and fungus. All three species had reduced hatching success in the presence of *Saprolegnia*. However, with *Saprolegnia* present, UV-B enhanced this effect in *Bufo* and *Rana* (Fig. 10.4). *Hyla regilla* hatching success was not affected by UV-B radiation and its hatching success was only reduced in the presence of *Saprolegnia*.

In the second *Saprolegnia* experiment, the hatching success of *B. boreas* and *R. cascadae* was also greater in regimes shielded from UV-B radiation.[25] *Hyla regilla* hatching success did not differ among the regimes.

## SYNERGISM BETWEEN pH AND UV-B RADIATION

Many workers have shown that low pH can decrease the survival of amphibian embryos and can have other adverse affects on amphibians.[41-43] For example, Harte and Hoffman[42] suggested that episodic acidification may have detrimental affects on tiger salamander (*Ambystoma tigrinum*) populations in Colorado. Thus, potentially acidification may contribute to amphibian population declines in areas where acid pollution occurs. Moreover, there may be synergistic effects between low pH and other factors such as UV-B radiation. Indeed, a recent study by Long et al[44] suggests that there is potential for such a synergism to affect amphibian embryos, and perhaps, amphibian populations.

Long et al[44] studied eggs of the leopard frog (*Rana pipiens*) exposed to combinations of three levels of UV-B and pH. The experiments were conducted outdoors (250 m elevation) so that there were natural levels of UV-A and visible light for photorepair. Natural UV-B was supplemented with artificial sources of UV-B. One regime had no UV-B, another regime had UV-B levels simulated for high elevations (intermediate UV-B level) and a third regime had UV-B levels simulated

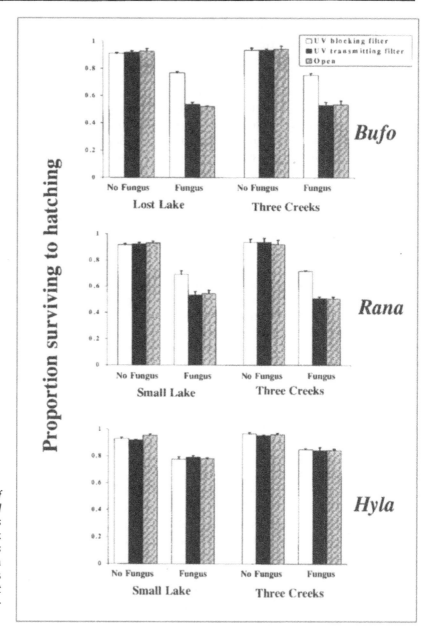

Fig. 10.4. Effects of UV-B radiation and manipulated amounts of Saprolegnia ferax on hatching success (mean ± SE) in Hyla regilla, Bufo boreas and Rana cascadae at relatively high elevation sites in Oregon.

to approach levels predicted with a 30% loss in ozone (high UV-B level). These UV levels were crossed with pH levels of 4.5, 5, and 6.

With low pH and simulated intermediate and high UV-B levels, hatching success was significantly lower when compared with the other regimes. There was no increase in mortality with either UV-B or pH alone.

## DISCUSSION

The results of the laboratory and field experiments described in this chapter suggest that prolonged exposure to UV-B radiation adversely affects the embryos of certain amphibian species but not others. Thus, under ambient conditions, in the field, embryos of *R. cascadae*, *B. boreas*, and *A. gracile* had lowered hatching success in treatments where they were subjected to

UV-B radiation compared with treatments that were shielded from UV-B.[2,38] Embryos of *H. regilla* and *R. aurora* hatched at about the same frequencies under UV-B and in shielded enclosures.[2,39] For *R. cascadae* and *B. boreas* the UV effect is enhanced in the presence of *Saprolegnia*.[25] Under seminatural conditions, low pH and intermediate and high UV-B levels decreased hatching success in *R. pipiens*.[44] The three species studied in our initial field experiments[2] were examined at several sites at various elevations with similar results, which showed that *B. boreas* and *R. cascadae* embryos were affected by UV-B but *H. regilla* embryos were not. Although there is potential for greater UV effects at higher elevations,[45] hatching success in one species, *A. gracile*, was reduced at low elevations.[38] However, hatching success of *R. aurora* embryos was not hampered by UV-B at low elevations.[39] Our results suggest that UV-B radiation is an unlikely contributor to population declines of *R. aurora* in Oregon unless this species is adversely affected by UV-B in larval or postmetamorphic stages.

## AMPHIBIANS AS BIOINDICATORS OF ENVIRONMENTAL STRESS

Under many conditions, amphibians may be good indicators of environmental stress.[13] They have permeable, exposed skin (not covered by tough scales, hair or feathers) and eggs (not covered by hard or leathery shells) that may readily absorb substances from the environment. The complex life cycles of many species potentially exposes them to both aquatic and terrestrial environmental agents. Exposed skin and eggs may make amphibians especially sensitive to the adverse affects of UV-B radiation. Moreover, many amphibians are sensitive to pollutants and acid conditions that may act synergistically with UV-B radiation.[25,44]

## FACTORS AFFECTING EMBRYO RESISTANCE TO UV-B RADIATION

The species that we have studied all lay their eggs in open, shallow water, exposed to sunlight.[2,36] However, the embryos of some species seem to be more resistant to UV-B radiation than others. Several factors may allow the embryos of certain species to be relatively resistant to UV-B radiation.

## EGG PIGMENTATION AND JELLY COAT

The presence of melanin in eggs exposed to sunlight may protect them to some degree from the harmful effects of UV-B radiation.[30] Darkly pigmented eggs may be more resistant to UV-B radiation than more lightly colored eggs.[30]

The jelly matrix that surrounds the eggs of amphibians may protect the eggs from desiccation and predators.[6,30] It may also afford some protection from UV-B radiation. Thicker jelly coats may be more protective than thinner coats. Of course, the jelly coats of different species may have different light penetrance properties, therby, making some species more resistant than others to UV-B, regardless of the thickness of the jelly matrix. Thus, *B. boreas*, *R. cascadae*, and *A. gracile* have relatively thick jelly coats but show less resistance to UV-B radiation than *H. regilla*, whose jelly coat is relatively thin. Therefore, in some species, the thickness of the coat alone may not indicate how resistant a species is to UV-B.

## DNA REPAIR

When UV radiation penetrates the cells of an organism, photoproducts may form that can lead to mutations or cell death. The embryos of some amphibian species may be more resistant to UV-B radiation because they have greater amounts of photoreactivating enzyme, photolyase.[2,39] Photolyase can remove some of the harmful photoproducts. Species with high levels of photolyase probably can remove DNA damaging photoproducts more efficiently than those with lower levels.[2] Indeed, the embryos of *H. regilla* and *R. aurora*, species that are relatively resistant to UV-B, have much greater photolyase activities than do *R. cascadae*, *B. boreas*, and *A. gracile*, species with relatively little resistance to UV-B.[2,38-39] Even species that have relatively long developmental times and are exposed to prolonged periods of solar radiation, such as *R. aurora*, may be able to cope with UV B due to high levels of photolyase activity.[2,39]

DNA repair via photolyase targets specific DNA-damaging photoproducts. Other photoproducts may be removed by excision repair, a complimentary mechanism that may remove photoproducts that are not removed by photolyase.[46-47] Both mechanisms may be used simultaneously. However, excision repair is poorly understood in amphibians.[47]

## BEHAVIOR AND ECOLOGY

Behavioral and ecological attributes may also play roles in limiting UV damage to amphibians. For example, many salamander species lay their eggs underground, in logs, or in crevices where they are not subjected to high levels of sunlight.[2] Many frogs and toads lay their eggs in muddy water where UV penetration may not be significant.[30,40] Some amphibian species may lay their eggs in relatively deep water where penetration of UV radiation is not high.[48] In the clear, high mountain lakes and ponds of Oregon, where we have conducted our research, the potential for UV-B penetration is greater than in relatively turbid sea level ponds and lakes where many eastern North American species and some western species lay their eggs.[30,36,40] Thus, the adverse effects of UV-B on developing embryos may be more pronounced in relatively clear bodies of water.

## ECOLOGICAL CONSEQUENCES OF AMPHIBIAN LOSSES

The loss of amphibians could profoundly affect certain ecosystems because they are integral components of many ecosystems. In some ecosystems, they may comprise the highest fraction of vertebrate biomass.[49] Through their trophic dynamics within ecological communities, a loss of amphibians could potentially have an impact on a variety of other organisms.[6] Adult amphibians are important carnivores

and prey species in many ecosystems.[30,50] Larval amphibians can be important herbivores as well as prey in aquatic habitats.[51-53]

Larval amphibians influence both the physical and biological parameters of lakes and ponds as they move about and forage.[52,54] Tadpoles are important regulators of primary production.[6,51-52] As tadpole numbers fluctuate, shifts in nutrient cycling, algal standing crops, and suspended particle concentrations may change.[52] Therefore, declines in numbers of larval amphibians may lead to significant physical and biological changes in some ecosystems.

Due to interspecific differential sensitivity of amphibians to UV-B radiation, there is a potential for the loss of some species within aquatic communities and not others. As the larvae of certain species disappear, significant changes in an ecosystem may occur. The community changes will depend upon the species that are affected by UV-B because the larvae of different species have different predators and feed on different food items.

In Oregon, mortality rates of *R. cascadae* and *B. boreas* eggs seem to have been no more than 10% from the 1950s to mid 1980s.[55] Since the mid 1980s mortality of eggs in these species has been more than 50% at most of our study sites and may often be over 80%.[55] Presumably, egg mortality in *B. boreas* and *R. cascadae* is at least in part due to UV-B radiation. Mortality of *H. regilla* eggs, however, has remained at less than 10% since the 1950s.[55] Therefore, in Oregon ecosystems, where these species co-occur, there is great potential for change. Significant mortality of eggs in the UV-B sensitive species, *R. cascadae* and *B. boreas*, ultimately leads to declining numbers of larvae of these species. This may initially lead to an increase in larval *H. regilla*, whose eggs are relatively resistant to UV-B radiation as potential competitors (*B. boreas* and *R. cascadae*) decline in numbers. We have no evidence that *B. boreas* and *R. cascadae* larvae are superior competitively to *H. regilla* larvae.

However, several studies indicate that ranids and bufonids are superior competitors to hylids,[56-57] suggesting that this scenario is possible.

Other community changes are possible if *B. boreas* and *R. cascadae* decline in numbers. *Bufo boreas* larvae are unpalatable to many vertebrate predators.[58] If *B. boreas* larvae decline in numbers, the larvae of a highly palatable species, *H. regilla*[58] may persist or even increase (as discussed above). This could lead to an influx of predators that are attracted to increasing numbers of palatable *H. regilla*. It is possible that many of the ponds and lakes in Oregon could change from a two or three species amphibian system to a system containing only *H. regilla*. However, the population and community dynamics of such changes are complex and are difficult to predict. Indeed, increasing numbers of *H. regilla* may be regulated by increasing predation pressure.

Physical parameters of the lakes and ponds could change as species composition of larvae change. For example, the large schools of *B. boreas* churn water and move substrate as they forage for food.[54] The disappearance of these large schools could affect foraging and other aspects of the biology of aquatic invertebrates, fish and plants. Thus disappearance of one or more species could potentially affect a wide variety of community dynamics. Indeed, recent studies in other systems show how sensitivity to UV-B in one species may lead to profound ecosystem changes.[59]

## CONCLUSIONS

We suggest that amphibian eggs and embryos of certain species are presently being harmed by UV-B radiation in nature.[2,38-39] Species that lay their eggs in open shallow water exposed to solar radiation are at greatest risk of UV damage (Table 7.1). Continued mortality of eggs and embryos could ultimately affect amphibians at the population level. However, we do not have enough long-term data to unequivocally state that egg mortality has led directly to population declines. UV-B radiation may act alone or in synergism with other agents such as pesticides, heavy metals, acid conditions or pathogens. However, synergistic interactions of UV-B with other agents are only poorly understood in amphibians in natural systems. For example, in nature, it is possible that the effects of pathogens, such as *Saprolegnia*[3,25] or the bacterium *Aeromonus hydrophila*, another common amphibian pathogen,[9] may be enhanced when defense systems are weakened by stressors.[3,9] One source of stress, UV-B radiation, has well-documented effects that weaken disease defense systems.[60-61] Synergistic interactions between UV-B and acid conditions[44] may also contribute to amphibian declines.

Although UV-B radiation may contribute to some amphibian population declines, obviously, not all amphibians will be affected by UV-B radiation. Behavioral, anatomical, biochemical and ecological attributes of amphibians may limit the exposure of some amphibian species to solar radiation. For example, UV-B radiation is an unlikely cause for declines in species that live or lay their eggs in dense forests or in deep water where they are shielded from solar radiation. Species that have relatively rapid developmental rates may be exposed to relatively short bouts of exposure to solar radiation and may not be adversely affected by UV-B radiation. Species with high capacities to repair UV-induced DNA damage may be less prone to egg mortality or population declines.

If projected increases in UV-B radiation occur[28-29,62] this could potentially lead to increased mortality in amphibian embryos as they develop in nature. Sustained small increases or temporary fluctuations in UV-B may affect especially sensitive species. There may be increased selection pressure on amphibians to evolve efficient repair mechanisms or behaviors that minimize their exposure to UV-B. However, anthropogenic changes in the environment are occurring at such rapid rates that there may not be time for organisms to display adaptations to these changes.

Many factors, including habitat destruction, pollution, disease, introduction of exotic organisms and natural population fluctuations may all play significant roles in amphibian population declines. UV-B radiation is one agent that may act alone or in conjunction with any of these agents to adversely affect amphibian populations.

## ACKNOWLEDGMENTS

Financial support was provided by the National Science Foundation (U.S.) DEB-9423333 and U.S. Forest Service. We thank Jeff Jeffries, Lisa Fremont, Tom Doyle and Lars Thorwald for their help.

## REFERENCES

1. Blaustein AR, Wake DB, Sousa WP. Amphibian declines: Judging stability, persistence, and susceptibility of populations to local and global extinctions. Conserv Biol 1994; 8:60-71.

2. Blaustein AR, Hoffman PD, Hokit DG et al. UV repair and resistance to solar UV-B in amphibian eggs: a link to population declines? Proc Nat Acad Sci USA 1994; 91:1791-1795.

3. Blaustein AR, Hokit DG, O'Hara RK et al. Pathogenic fungus contributes to amphibian losses in the Pacific Northwest. Biol Conserv 1994; 67:251-254.

4. Barinaga M. Where have all the froggies gone? Science 1990; 247:1033-1034.

5. Pechmann JHK, Wilbur HM. Putting amphibian decline populations in perspective: natural fluctuations and human impacts. Herpetologica 1994; 50:65-84.

6. Stebbins RC, Cohen NW. A Natural History of Amphibians. Princeton: Princeton University Press, 1995:1-316

7. Wake DB. Declining amphibian populations. Science 1991; 253:860.

8. Bradford DF. Mass mortality and extinction in a high-elevation population of *Rana muscosa*. J Herpetol 1991; 25:174-177.

9. Carey C. Hypothesis concerning the causes of the disappearance of boreal toads from the mountains of Colorado. Conserv Biol 1993; 7:355-362.

10. Fellers GM, Drost CA. Disappearance of the Cascades frog *Rana cascadae*, at the southern end of its range. Biol Conserv 1993; 65:177-181.

11. Kagarise Sherman C, Morton ML. Population declines of Yosemite toads in the eastern Sierra Nevada of California. J Herpetol 1993; 27:186-198.

12. McAllister KR, Leonard WP, Storm RM. Spotted frog (*Rana pretiosa*) surveys in the Puget Trough of Washington, 1989-1991. Northwest Nat 1993; 74:10-15.

13. Blaustein AR. Chicken little or Nero's fiddle? A perspective on declining amphibian populations. Herpetologica 1994; 50:85-97.

14. McNeely JA, Miller KR, Reid WV et al. Conserving the World's Biodiversity. Washington, D.C. International Union for the Conservation of Nature, World Resource Institute, CI, World Wildlife Fund-U.S., The World Bank 1990:1-193.

15. Pechmann JHK, Scott DE, Semlitsch J et al. Declining amphibian populations: the problem of separating human impacts from natural fluctuations. Science 1991; 253:892-895.

16. Blaustein AR, Wake DB. Declining amphibian populations: a global phenomenon? Trends Ecol Evol 1990; 5:203-204.

17. Crump ML, Hensley FR, Clark KL. Apparent decline of the golden toad: underground or extinct? Copeia 1992; 1992:413-420

18. Tyler MJ. Declining amphibian populations—a global phenomenon? An Australian perspective. Alytes 1991; 9:43-50.

19. Richards SJ, McDonald KR, Alford RA. Declines in populations of Australia's endemic tropical rainforest frogs. Pac Conserv Biol 1993; 1:66-77.

20. Corn PS, Fogleman JC. Extinction of montane populations of the northern leopard frog (*Rana pipiens*) in Colorado. J Herpetol 1984; 18:147-152.

21. Bradford DF. Allotopic distribution of native frogs and introduced fishes in high Sierra Nevada lakes of California: implication of the negative effect of fish introductions. Copeia 1989; 1989:775-778.

22. Semb-Johansson A. Padden (*Bufo bufo*)-Et stebarn inorsk zoologi. Fauna 1989; 42:174-179.

23. La Marca E, Reinthaler HP. Population changes in *Atelopus* species of the Cordillera de Merida, Venezuela. Herpetol Rev 1991; 22:125-128.

24. Blaustein AR, Olson DH. Declining amphibians. Science 1991; 253:1467.

25. Kiesecker JM, Blaustein AR. Synergism between UV-B radiation and a pathogen magnifies embryo mortality in nature. Proc Nat Acad Sci USA 1995; 92:11049-11052.

26. Federal Register. Endangered and threatened wildlife and plants: animal candidate review listing as endangered or threatened species proposed rule. Department of the Interior, Fish and Wildlife Service, Part VIII Washington, D.C., USA 1991; 56: 58804-58836.

27. Olson DH. Ecological susceptibility of amphibians to population declines. In: Kerner HM, ed. Proceedings of the Symposium on Biodiversity of Northwestern California. Univ California, Berkeley: Wildland Res Center, 1992:55-62.

28. Worrest RC, Grant LD. Effects of Ultraviolet-B radiation on terrestrial plants and marine organisms. In: Jones RR, Wigley T, ed. Ozone Depletion: Health and Environmental Consequences. New York: John Wiley and Sons, Ltd., 1989.

29. Zurer PS. Ozone depletion's recurring surprises challenge atmospheric scientists. Chem Engin News 1993; 71:8-18.

30. Duellman WE, Trueb L. Biology of Amphibians. New York: McGraw-Hill, 1986: 1-670.

31. Higgins C, Sheard C. Effects of ultraviolet radiation on the early larval development of Rana pipiens. J Exp Zool 1926; 46:333-343.

32. Elinson RP, Pasceri P. Two UV-sensitive targets in dorsoanterior specification of frog embryos. Development 1989; 106:511-518.

33. Scharf SR, Gerhart JC. Determination of the dorsal-ventral axis in eggs of Xenopus laevis: Complete rescue of UV-impaired eggs by oblique orientation before first cleavage. Dev Biol 1980; 79:181-198.

34. Worrest RC, Kimeldorf DJ. Photoreactivation of potentially lethal UV-induced damage to boreal toad (Bufo boreas boreas) tadpoles. Life Sci 1975; 17:1545-1550.

35. Worrest RC, Kimeldorf DJ. Distortions in amphibian development induced by ultraviolet-B enhancement (290-315 nm) of a simulated solar spectrum. Photochem Photobiol 1976; 24:377-382.

36. Nussbaum RA, Brodie ED, Jr. Storm RM. Amphibians and Reptiles of the Pacific Northwest. Moscow: University Press of Idaho, 1983:1-332.

37. O'Hara RK. Habitat Selection Behavior in Three Species of Anuran Larvae: Environmental Cues, Ontogeny, and Adaptive Significance. Ph.D. Thesis, Oregon State University, Corvallis. 1981:1-146.

38. Blaustein AR, Edmond BJ, Kiesecker JM et al. Ambient ultraviolet radiation causes mortality in salamander eggs. Ecol App 1995; 5:740-743.

39. Blaustein AR, Hoffman PD, Kiesecker JM et al. DNA repair activity and resistance to solar UV-B radiation in eggs of the red-legged frog. Conserv Biol 70:1398-1402.

40. Stebbins RC. Amphibians and Reptiles of Western North America. New York: McGraw-Hill, 1954:1-537.

41. Dunson WA, Wyman RL. Symposium: amphibian declines and habitat acidification. J Herpetol 1992; 26:349-442.

42. Harte J, Hoffman E. Possible effects of acidic deposition on Rocky Mountain population of the tiger salamander Ambystoma tigrinum. Conserv Biol 1989; 3:149-158.

43. Kiesecker JM. pH induced growth reduction and its effects on predator-prey interactions between Ambystoma tigrinum and Pseudacris triseriata. Ecol App 6:1325-1331.

44. Long LE, Saylor LS, Soule' ME. pH/UV-B synergism in amphibians. Conserv Biol 1995; 9:1301-1303.

45. Caldwell MM, Robberecht R, Billings WD. A steep latitudinal gradient of solar ultraviolet-B radiation in the Arctic-alpine life zone. Ecology 1980; 61:600-611.

46. Friedberg EC, Walker GC, Siede W. DNA Repair and Mutagenesis. Washington, D.C.: American Society for Microbiology, 1995: 1-698.

47. Hays JB, Ackerman EJ, Pang Q. Rapid and apparent error-prone excision repair of nonreplicating UV-irradiated plasmids in Xenopus laevis oocytes. Mol Cell Biol 1990; 10:3505-3511.

48. Smith RC, Baker KS. Penetration of UV-B and biologically effective dose-rates in natural waters. Photochem Photobiol 1979; 29:311-323.

49. Burton TM, Likens GE. Salamander populations and biomass in the Hubbard Brook experimental forest, New Hampshire. Copeia 1975; 1975:541-546.

50. Stewart MM. Climate driven population fluctuations in rain forest frogs. J Herpetol 1995; 29:437-446.

51. Dickman M. The effect of grazing by tadpoles on the structure of a periphyton community. Ecology 1968; 49:1188-1190.

52. Seale DB. Influence of amphibian larvae on primary production, nutrient flux, and competition in a pond ecosystem. Ecology 1980; 61:1531-1550.

53. Morin PJ, Lawler SP, Johnson EA. Ecology and breeding phenology of larval *Hyla andersonii*: the disadvantage of breeding late. Ecology 1990; 71:1590-1598.

54. Wilbur HM. Density-dependent aspects of growth and metamorphosis in *Bufo americanus*. Ecology 1977; 58:196-200.

55. Kiesecker JM, Blaustein AR. Influences of egg laying behavior on pathogenic infection of amphibian eggs. Conserv Biol (in press).

56. Wilbur HM, Alford RA. Priority effects in experimental pond communities: responses of *Hyla* to *Bufo* and *Rana*. Ecology 1985; 66:1106-1114.

57. Morin PJ, Johnson EA. Experimental studies od asymmetric competition among anurans. Oikos 1988; 53:398-407.

58. Peterson JA, Blaustein AR. Relative palatabilities of anuran larvae to natural aquatic insect predators. Copeia 1992; 1992:577-584.

59. Bothwell ML, Sherbot DJM, Pollock CM. Ecosystem response to solar ultraviolet-B radiation: influence of trophic interactions. Science 1994; 265:97-100.

60. Tevini M. UV-B Radiation and Ozone Depletion: Effects on Humans, Animals, Plants, Microorganisms, and Materials. Boca Raton: Lewis Publishers, 1993:1-248.

61. Orth AB, Teramura AH, Sisler HD. Effects of UV-B radiation on fungal disease development in *Cucumis sativus*. Am J Bot 1990; 77:1188-1192.

62. Kerr JB, McElroy CT. Evidence for large upward trends of ultraviolet-B radiation linked to ozone depletion. Science 1993; 262:1032-1034.

# CHAPTER 11

# IMPACTS OF UV-B IRRADIATION ON RICE-FIELD CYANOBACTERIA

Rajeshwar P. Sinha and Donat-P. Häder

## INTRODUCTION

Cyanobacteria are a primitive group of Gram-negative prokaryotes but are the only group of bacteria to possess higher plant-type oxygenic photosynthesis. A majority of these organisms is capable of fixing atmospheric nitrogen either in the free-living state or in symbiotic organization with a diverse group of plants.[1] Cyanobacterial populations occupy an important place in both aquatic as well as terrestrial ecosystems due to their inherent capacity to fix atmospheric nitrogen, with the help of the enzyme complex nitrogenase, into ammonium ($NH_4^+$), a form through which nitrogen enters into the food chain.[2] Cyanobacteria fix over 35 million tons of nitrogen annually which is thus available for use by higher plants.[3] In several instances the availability of nitrogen has been found to be the limiting factor for productivity in natural habitats.[4]

$N_2$-fixing cyanobacteria form a prominent component of microbial populations in wet land soils, especially in rice paddy fields, where they significantly contribute to fertility as a natural biofertilizer.[5,6] The agronomic potential of cyanobacteria has long been recognized as a promising biofertilizer for wet land soils.[7] In nature the process of nitrogen fixation is of utmost importance because it counterbalances losses of combined nitrogen from the environment by denitrification.[8] Rice is one of the major staples in the diet of billions of people and requires a large amount of nitrogenous fertilizers for its growth and productivity. Industrial production of nitrogenous fertilizers is highly expensive, and in rice growing countries the availability of nitrogenous fertilizers is a major factor limiting crop yield. Thus cyanobacteria are of even greater importance as far as rice cultivation and other sustainable agricultural practices are concerned.[9]

*The Effects of Ozone Depletion on Aquatic Ecosystems*, edited by Donat-P. Häder.

Cyanobacteria are thought to have survived from a wide spectrum of global environmental stresses such as heat and cold, salinity, drought, photo-oxidation, osmotic and UV stress. The cosmopolitan colonization of cyanobacteria demand high variability in adapting to diverse environmental factors.[10,11]

Recent findings suggest that ultraviolet-B (UV-B; 280-315 nm) radiation at the earth's surface is increasing due to continued depletion of stratospheric ozone, caused by anthropogenically released atmospheric pollutants such as chlorofluorocarbons (CFCs), chlorocarbons (CCs) and organobromides (OBs).[12-15] Despite international efforts to minimize the problem, it is thought that ozone depletion will intensify and spread to a broader range of latitudes throughout the next century. UV-B radiation has been reported to cause deleterious effects in a number of biological systems including cyanobacteria.[16,17] In cyanobacteria, processes such as growth, survival, pigmentation, motility, as well as enzymes of nitrogen metabolism and $CO_2$ fixation have been reported to be susceptible to UV-B.[17-23] In this chapter physiological and biochemical aspects of rice-field cyanobacteria under UV-B irradiation are discussed.

## GROWTH AND SURVIVAL

Growth and survival of various rice-field cyanobacteria have been reported to be severely affected following UV-B irradiation for different durations.[18,24,25] The percentage survival of several rice-field cyanobacteria following UV-B irradiation are listed in Table 11.1. Depending upon the species, growth and survival stopped within 120-150 min of UV-B irradiation. Cyanobacterial strains such as *Scytonema* sp. and *Nostoc commune*, filaments of which are embedded in mucilagenous sheath, have been reported to be more tolerant against UV-B irradiation in comparison to cyanobacteria such as *Anabaena* sp. and *Nostoc* sp., which do not contain such covering.[18] Growth and survival of any organism depends exclusively on the genetic machinery

of that organism. Complete killing of the organisms following UV-B irradiation could be due to damage in the basic genetic material, DNA. A number of workers are of the opinion that the cellular constituents absorbing radiation between 280-315 nm are destroyed by UV-B irradiation, which may further affect the cellular membrane permeability and protein damage, and eventually result in the death of the organism.[16,18,22,25,26]

## PIGMENTATION

A gradual decrease in the pigment content with increasing UV-B exposure time has been reported in a number of cyanobacteria. Phycobiliproteins have been shown to be bleached more drastically and rapidly than any other pigments such as chlorophyll or carotenoids.[18,24,25] The impact of UV-B on pigmentation of various rice-field cyanobacteria has been summarized in Table 11.2 and is shown in Figure 11.1.

Absorption spectra of the organisms have been shown to possess four major peaks at 437, 485, 620 and 672 nm in almost all strains, with a steady decline in absorption after increasing UV-B exposure time. However, in comparison to other peaks there was a drastic decline at 620 nm, suggesting once again that the accessory pigment phycocyanin is bleached more rapidly and dramatically than Chl *a* (437 and 672 nm) or the carotenoids (485 nm).[18,20]

The intensity of phycobiliproteins separated by sucrose density gradient ultracentrifugation have been shown to be decreased within 3-4 h of UV-B irradiation, depending upon the species type.[27] The phycobiliproteins of the cyanobacterium *Anabaena* sp. separated by sucrose density gradient ultracentrifugation with increasing UV-B exposure time are illustrated in Figure 11.2. Absorption spectra of the phycocyanin (phycobiliprotein) revealed a typical peak at 620 nm and showed gradual decline in absorbance with increasing UV-B exposure time. Fluorescence excitation at 620 nm resulted in an emission at

640-650 nm, depending upon the organisms type, which first showed a slight increase in fluorescence and second, a shift towards shorter wavelengths, and thereafter a gradual and steady decline in fluorescence by a further shift towards shorter wavelengths with increasing durations of UV-B exposure.[19,20,22] The transient rise in the fluorescence after short UV-B exposure may be due to an uncoupling of the energy transfer between phycocyanin and Chl *a*, which results in a higher fluorescence rate. This process is paralleled by the photodestruction of the chromophores in the phycobilisomal complex, which in turn lowers the fluorescence rate and decreases the absorption. After an increase in UV-B exposure time both processes compensate

### Table 11.1. Impact of UV-B on percent survival of various rice-field cyanobacteria[1]

| Organisms | Percent survival UV-B exposure time (min) | | | | | |
|---|---|---|---|---|---|---|
| | 0 | 30 | 60 | 90 | 120 | 150 |
| Anabaena sp. | 100 | 68.0 | 50.5 | 10.5 | 2.80 | 0.0 |
| Nostoc carmium | 100 | 65.5 | 45.8 | 10.0 | 2.50 | 0.0 |
| Nostoc spongiaeforme | 100 | 80.0 | 30.5 | 10.5 | 0.00 | 0.0 |
| Nostoc muscorum | 100 | 70.0 | 25.5 | 08.0 | 0.00 | 0.0 |
| Nostoc commune | 100 | 80.0 | 60.5 | 50.5 | 10.5 | 0.0 |
| Nostoc sp. | 100 | 50.0 | 15.0 | 03.0 | 0.00 | 0.0 |
| Scytonema sp. | 100 | 80.5 | 58.5 | 35.8 | 12.5 | 0.0 |

[1]The values are representative of three separate but identical experiments. S.D. was consistently less than 10% of

### Table 11.2. Impact of UV-B on pigmentation of various rice-field cyanobacteria[1]

| UV-B exposure time (min) | 0 | 30 | 60 | 90 | 120 | 0 | 30 | 60 | 90 | 120 | 0 | 30 | 60 | 90 | 120 |
|---|---|---|---|---|---|---|---|---|---|---|---|---|---|---|---|
| Organisms | Chl a (µg/mL) | | | | | Carotenoid (µg/mL) | | | | | Phycobiliproteins (µg/mL) | | | | |
| Anabaena sp. | 8.0 | 7.2 | 5.0 | 2.5 | 1.2 | 2.5 | 2.0 | 1.8 | 1.2 | 0.5 | 75.0 | 60.0 | 40.5 | 18.0 | 6.0 |
| Nostoc carmium | 7.5 | 6.5 | 4.8 | 2.0 | 0.8 | 2.9 | 2.0 | 1.5 | 0.8 | 0.3 | 60.5 | 50.0 | 30.2 | 10.2 | 5.0 |
| Nostoc spongiaeforme | 6.3 | 5.0 | 4.6 | 1.3 | 1.0 | 2.2 | 1.9 | 1.8 | 0.4 | 0.2 | 56.0 | 39.9 | 31.0 | 11.2 | 5.2 |
| Nostoc muscorum | 5.2 | 4.5 | 1.0 | 0.8 | 0.1 | 1.5 | 1.0 | 0.6 | 0.2 | 0.1 | 40.0 | 30.5 | 20.8 | 8.5 | 4.5 |
| Nostoc commune | 6.5 | 6.0 | 5.2 | 2.5 | 1.5 | 2.3 | 2.0 | 1.7 | 1.2 | 0.8 | 70.0 | 60.0 | 50.5 | 30.0 | 10.5 |
| Nostoc sp. | 6.8 | 5.5 | 4.3 | 1.2 | 0.5 | 2.0 | 1.2 | 0.8 | 0.2 | 0.1 | 60.0 | 40.5 | 20.2 | 9.8 | 3.5 |
| Scytonema sp. | 7.8 | 6.5 | 6.0 | 5.2 | 3.5 | 2.5 | 2.3 | 2.0 | 1.5 | 0.8 | 72.0 | 65.0 | 55.5 | 40.2 | 18.0 |

[1]The values are representative of three separate but identical experiments. S.D. was consistently less than 10% of means.

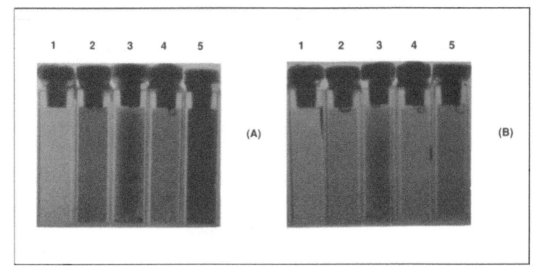

Fig. 11.1. (A) Exponentially growing nitrogen-fixing cyanobacterial strains immobilized in agar-agar (0.5%) and sealed in a quartz cuvette. Lane 1: only agar-agar, lane 2: Nostoc carmium, lane 3: Nostoc commune, lane 4: Scytonema sp. and lane 5: Nostoc sp. (B) Same as in A, after 3 h of UV-B irradiation.

Fig. 11.2. Sucrose density gradient (5-40% w/v sucrose in 0.75 M phosphate buffer, pH 7.0) pattern of phycobiliproteins from Anabaena sp. after different times of UV-B irradiation. Lane A: Unirradiated control (0 h), lane b: 1 h, lane c: 2 h and lane d: 3 h of UV-B irradiation.

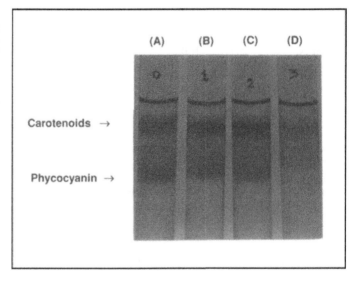

each other in terms of fluorescence rates, while the absorption is already much decreased.[19,20,22]

Thus absorption and fluorescence data of phycobiliproteins indicate that pigmented proteins are one of the main targets of UV-B in cyanobacteria. The drastic decline in both absorption and fluorescence as well as a shift of the fluorescence peak towards shorter wavelengths are indicative of a partial disruption of the intact phycobilisomal complex and an im-

paired energy transfer from the biliproteins to the photosystems due to UV-B irradiation.[19,20,22]

## PHYTOSYNTHESIS

Like all photoautotrophs, cyanobacteria are dependent upon solar radiation as their primary source of energy in their natural environment. The cyanobacteria, along with several diverse groups of organisms, are capable of utilizing $CO_2$ as their sole source of carbon. Ribulose 1,5-bisphosphate carboxylase (RuBisCO) is the primary $CO_2$-fixing enzyme and the most abundant protein found on earth.[28] The control of RuBisCO biosynthesis is strongly influenced by the prevailing light environment.

UV-B-induced inhibition of photosynthetic activities have been shown in a number of rice-field cyanobacteria.[21,22] RuBisCO activity has also been reported to be severely effected by UV-B treatment.[22] UV-B-induced inhibition of $^{14}CO_2$ uptake in cyanobacteria may be due to the reduced supply of ATP and NADPH resulting from the destruction of the photosynthetic apparatus following UV-B irradiation.[21,23] It is known that UV-B irradiation causes an inactivation of PSII reaction centers.[29] A disruption of the cell membrane and/or disturbance in the thylakoid integrity as a result of UV-B radiation may partly or wholly destroy the components required for photosynthesis and may thus affect the rate of $CO_2$ fixation.[22]

Inhibition of RuBisCO activity by UV-B radiation could be due to protein destruction or enzyme inactivation. This notion is supported by the fact that crude enzyme extracts of several cyanobacteria, when irradiated with UV-B, contained less proteins than those of the unirradiated control.[18,21,22] Protein may undergo a number of modifications during UV-B irradiation, including photodegradation, increased aqueous solubility of membrane proteins, and fragmentation of peptide chains, leading to inactivation of proteins (enzymes) and disruption of their structural entities.[30] It has been demonstrated that both nuclear (chlorophyll *a/b* binding protein and small subunit of RuBisCO) and chloroplast (D1 polypeptide of photosystem II and large subunit of RuBisCO) encoded mRNA transcripts were reduced in response to UV-B exposure.[31,32]

## Enzymes of Nitrogen Metabolism

UV-B-induced inactivation of the nitrogen fixing enzyme nitrogenase has been reported in many cyanobacteria.[21,23,24,26,27] UV-B irradiation of as little as 5 min has been reported to cause considerable inhibition in enzyme activity, with a complete loss within 35-55 min depending upon the species.[21,23,27] The process of nitrogen fixation has also been reported to be depressed during mid-day when visible and UV-B irradiation are highest.[33] Nitrogenase requires continued and abundant supply of suitable reductant and ATP for conversion of $N_2$ to $NH_3$. One possibility of the inhibition may be due to the impairment in the supply of ATP and reductant following UV-B irradiation. However, most cyanobacteria contain specific levels of endogenous reductant and ATP, which in turn support nitrogenase activity for a limited period.[1,21,23] This is also supported by the fact that when cells were treated with DCMU (3-(3,4-dichlorophenyl)-1,1-dimethylurea) or incubated in the dark, the complete loss of nitrogenase activity occurs only after 6-8 h. When such cultures were transferred to light in DCMU free medium, the activity started appearing and by 3 h it was almost comparable to that in control cultures. In contrast, the UV-B-treated cultures failed to restore activity, even with glucose treatment, up to at least 3 h; this may be largely due to the fact that nitrogenase is not present in the cell. Thus it seems that the loss of nitrogenase activity following UV-B irradiation is not related to the instant impairment of reductant/ATP supply, but rather appears to be a novel phenomenon.[21,23,27]

More or less, UV-B irradiation of cyanobacterial strains causes deleterious effects on many metabolic processes, however, it was surprising to find stimulation of nitrate reductase (NR; the other important enzyme of nitrogen metabolism) activity in all nitrogen fixing cyanobacterial strains studied so far.[18,23] The observed stimulation of NR activity could not be explained as yet. Whether it is due to an artifact caused by changes in cellular membrane permeability following UV-B irradiation and thus leading to higher nitrate uptake or whether UV-B has a direct effect on $NO_2^-$ accumulation in the cells needs thorough investigation.[18,23] Inhibition in the activity of ammonia assimilating enzyme glutamine synthetase (GS) has also been reported in a number of rice-field cyanobacteria following UV-B irradiation. However, there was no complete inhibition even after long hours of UV-B irradiation.[18,23,27] To date, it is not well understood to what extent the effect of UV-B on the enzymes of nitrogen metabolism results from a direct impact on enzyme activity or whether enzyme synthesis is affected via RNA damage or from some other as yet unknown reasons.

Differentiation of vegetative cells into heterocysts has been reported to be also severely affected by UV-B irradiation in a number of cyanobacteria.[21,34] Processes such as heterocysts differentiation and nitrogen fixation are metabolically very expensive, and during extreme stress conditions may require cellular resources for their maintenance. Since heterocysts are important primarily in supplying fixed nitrogen to the vegetative cells, apparently vegetative cells growing in areas without external nitrogen sources may be seriously affected. Most probably the C:N ratio is severely affected following UV-B irradiation, which in turn affects the spacing pattern of heterocysts in a filament.[21]

## PROTEIN SYNTHESIS AND TOTAL PROTEIN PROFILE

By using SDS-PAGE techniques possible changes in the total protein profile of a number of rice paddy field $N_2$-fixing cyanobacteria have been demonstrated following increasing UV-B exposure time.[18,21,23] A gradual decrease in the intensity of all protein bands has been reported with increasing UV-B exposure times, and depending upon the species, these bands were completely eliminated within 2-3 h of UV-B exposure. Protein bands at around 20 kDa (α and β monomers of phycocyanin involved in PSII) have been reported to be severely affected even after a short exposure to UV-B in many cyanobacteria. A prominent protein band of approximately 55 kDa was less affected even after long durations of UV-B exposure.[18,21] The SDS-PAGE protein profile of *Scytonema* sp. following increasing UV-B exposure times is shown in Figure 11.3. This also shows a linear decrease in protein content with increasing UV-B exposure times and a severe effect on the protein bands around 20 kDa. SDS-PAGE protein profile of isolated heterocysts have been shown to possess three prominent bands at around 26, 54 and 55 kDa with a decrease in the first two and complete elimination of the last band following 1 h of UV-B exposure.[21]

SDS-PAGE profiles of isolated phycobiliproteins have been reported to possess three prominent bands ground 20 kDa (αβ subunits of phycocyanin), 45 kDa (rod and rod-core linker polypeptides) and 70 kDa (core-membrane linker polypeptides). Depending on the species all three bands have been reported to be completely eliminated within 3-4 h of UV-B exposure.[19,20,22]

Proteins are thought to be one of the main targets of UV-B, since aromatic amino acids such as tryptophan, tyrosine,

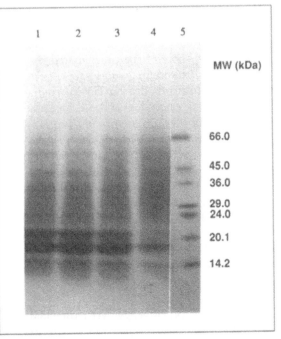

*Fig. 11.3. Vertical SDS-PAGE (gradient 5-15% T) of Scytonema sp. following increasing exposure to UV-B. Lane 1: control, lane 2: 30 min, lane 3: 90 min, lane 4: 150 min, and lane 5: marker proteins.*

phenylalanine and histidine strongly absorb in the UV-B range.[21] Loss of low molecular mass ($\alpha\beta$) monomers and high molecular mass linker polypeptides have been attributed to impart a partial disruption of the intact phycobilisomes and an impaired energy transfer from the phycobiliproteins to the photosynthetic reaction centers due to UV-B irradiation.[19,20,22] Denaturation of various macromolecules, especially proteins, with a consequent loss of structure and function and also changes in cellular membrane permeability leading to the death of the cell have been discussed.[16-18] This may further affect other important processes such as $N_2$-fixation by cyanobacteria via inactivation of the $N_2$-fixing enzyme nitrogenase.[17,21,23]

## PROTECTIVE MECHANISMS

Several cyanobacteria have developed a broad range of adaptive strategies to reduce the deleterious effects of UV-B irradiation. This includes the avoidance of brightly lit habitats, the production of UV screening pigments and a number of scavenging mechanisms that detoxify the highly reactive oxidants produced photochemically.[16] In addition, a number of cyanobacteria have the ability to vary their phycobiliprotein (phycocyanin/phycoerythrin) ratio which allows regulation of the balance of wavelengths of light absorbed by cyanobacteria, a phenomenon known as chromatic adaptation.[35]

A number of naturally occurring compounds have been reported in cyanobacterial cells, which strongly absorb in the UV-A (315-400 nm) or UV-B (280-315 nm) regions. These compounds include cyanobacterial sheath pigments (scytonemin) and mycosporin-like amino acids (MAAs). Scytonemin is thought to act as a photoprotectant against UV irradiation. It mostly occurs in the mucilagenous sheath surrounding the cells and has a broad band absorbance which peaks at about 380 nm.[36,37] MAAs have also been reported to provide protection against UV irradia-

tion, and absorb maximally in the range between 310-360 nm.[38] UV-A/B protecting pigments have been reported earlier from the terrestrial cyanobacterium *Nostoc commune*.[39] A brown *Nostoc* sp. producing three UV absorbing compounds has been reported to be resistant to UV irradiation.[40] Recently, the shielding role of certain cyanobacterial pigments (a brown-colored pigment from *Scytonema hofmanii* and a pink extract from *Nostoc spongiaeforme*) against UV-B, induced damage have been demonstrated.[41]

## CONCLUSION

A number of $N_2$-fixing cyanobacterial strains have been shown to be sensitive to UV-B irradiation, but to a variable extent. Changes in the cyanobacterial community composition may occur due to the differential sensitivity of individual species which may further expose the habitat to the development of toxic algal blooms. Any substantial increase in the solar UV-B radiation might be detrimental to these ecologically and economically important cyanobacterial communities, which in turn may affect the productivity of higher plants, particularly rice in paddy fields. There might be a setback in the agricultural economy of all countries where cyanobacteria are being considered as an alternate natural source of nitrogenous fertilizers for rice paddies and other crops. It would be worthwhile to study the long-term effects of UV-B on rice-field cyanobacteria under natural conditions.

## ACKNOWLEDGMENTS

This work was financially supported by the DAAD in Germany, the Ministry of H.R.D., and the Council of Scientific and Industrial Research, Govt. of India, New Delhi to R.P. Sinha and by the European Community (EV5V-CT91-0026) and by the BayFORKLIM (DIII 1) to D.-P. Häder. The authors are thankful to S.C. Singh for great help during preparation of the manuscript.

# REFERENCES

1. Stewart WDP. Some aspects of structure and function in $N_2$-fixing cyanobacteria. Annu Rev Microbiol 1980; 34:497-536.

2. Spiller H, Latorre C, Hassan ME et al. Isolation and characterization of nitrogenase-depressed mutant strains of cyanobacterium *Anabaena variabilis*. J Bacteriol 1986; 165:412-419.

3. Häder D-P, Worrest RC. Aquatic ecosystems. UNEP Environmental Effects Panel Report 1989:39-48.

4. Singh RN. Role of Blue-Green Algae in Nitrogen Economy of Indian Agriculture. New Delhi, India: Indian Council of Agricultural Research, 1961:61-82.

5. Venkataraman GS. Blue-green algae: a possible remedy to nitrogen scarcity. Curr Sci 1981; 50:253-256.

6. Sinha RP, Kumar A. Screening of blue-green algae for biofertilizer. In: Patil PL, eds. Proceedings of the National Seminar on Organic Farming. Pune, India, 1992:95-97.

7. De PK. The role of blue-green algae in nitrogen fixation in rice fields. Proc R Soc Lond 1939; 127:121-139.

8. Kuhlbusch TA, Lobert JM, Crutzen PJ et al. Molecular nitrogen emission. Trace nutrification during biomass burning. Nature 1991; 351:135-137.

9. Sinha RP, Häder D-P. Photobiology and ecophysiology of rice field cyanobacteria. Photochem Photobiol 1996; 64:887-896.

10. Stanier RY, Cohen-Bazire G. Phototrophic prokaryotes: the cyanobacteria. Annu Rev Microbiol 1977; 31:225-274.

11. Tandeau de Marsac N, Houmard J. Adaptation of cyanobacteria to environmental stimuli: new steps towards molecular mechanisms. FEMS Microbiol Rev 1993; 104:119-190.

12. Blumthaler M, Ambach W. Indication of increasing solar ultraviolet-B radiation flux in Alpine regions. Science 1990; 248:206-208.

13. Crutzen PJ. Ultraviolet on the increase. Nature 1992; 356:104-105.

14. Kerr JB, McElroy CT. Evidence for large upward trends of ultraviolet-B radiation linked to ozone depletion. Science 1993; 262:1032-1034.

15. Madronich S. The atmosphere and UV-B radiation at ground level. In: Young AR, Björn LO, Moan J, Nultsch W, eds. Environmental UV Photobiology. New York: Plenum Press, 1993:1-39.

16. Vincent WF, Roy S. Solar ultraviolet radiation and aquatic primary production: damage, protection and recovery. Environ Rev 1993; 1:1-12.

17. Häder D-P, Worrest RC, Kumar HD et al. Effects of increased solar ultraviolet radiation on aquatic ecosystems. Ambio 1995; 24:174-180.

18. Sinha RP, Kumar HD, Kumar A et al. Effects of UV-B irradiation on growth, survival, pigmentation and nitrogen metabolism enzymes in cyanobacteria. Acta Protozool 1995; 34:187-192.

19. Sinha RP, Lebert M, Kumar A et al. Spectroscopic and biochemical analyses of UV effects on phycobiliproteins of *Anabaena* sp. and *Nostoc carmium*. Bot Acta 1995; 108:87-92.

20. Sinha RP, Lebert M, Kumar A et al. Disintegration of phycobilisomes in a rice field cyanobacterium *Nostoc* sp. following UV irradiation. Biochem Molecu Biol Intern 1995; 37:697-706.

21. Sinha RP, Singh N, Kumar A et al. Effects of UV irradiation on certain physiological and biochemical processes in cyanobacteria. J Photochem Photobiol B: Biol 1996; 32:107-114.

22. Sinha RP, Singh N, Kumar A et al. Impacts of ultraviolet-B irradiation on nitrogen-fixing cyanobacteria of rice paddy fields. J Plant Physiol 1997; (in press).

23. Kumar A, Sinha RP, Häder D-P. Effect of UV-B on enzymes of nitrogen metabolism in the cyanobacteria *Nostoc calcicola*. J Plant Physiol 1996; 148:86-91.

24. Tyagi R, Kumar HD, Vyas D et al. Effects of ultraviolet-B radiation on growth, pigmentation, $NaH^{14}CO_3$ uptake and nitrogen metabolism in *Nostoc muscorum*. In: Abrol YP, Wattal PN, Gnanam A, Govindjee, Ort DR, Teramura AH, eds. Impact of Global Climatic Changes on Photosynthesis and Plant Productivity. New Delhi: Oxford and IBH Publishing Co, 1991:109-124.

25. Tyagi R, Srinivas G, Vyas D et al. Differ-

ential effect of ultraviolet-B radiation on certain metabolic processes in a chromatically adapting *Nostoc*. Photochem Photobiol 1992; 55:401-407.

26. Newton JW, Tyler DD, Slodki ME. Effect of ultraviolet-B (280-320) radiation on blue-green algae (cyanobacteria), possible biological indicators of stratospheric ozone depletion. Appl Environ Microbiol 1979; 37:1137-1141.

27. Sinha RP. Impacts of UV-B stress on $N_2$-fixing cyanobacteria. Ph.D. thesis, Banaras Hindu University, Varanasi, India, 1995.

28. Ellis RJ. The most abundant protein in the world. Trends Biochem Sci 1979; 4: 241-244.

29. Renger GM, Volker M, Eckert HJ et al. On the mechanism of photosystem II deterioration by UV-B radiation. Photochem Photobiol 1989; 49:97-105.

30. Wilson MI, Ghosh S, Gerhardt KE et al. In vivo photomodification of ribulose-1,5-bisphosphate carboxylase/oxygenase holoenzyme by ultraviolet-B radiation. Plant Physiol 1995; 109:221-229.

31. Jordan BR, Chow WS, Strid A et al. Reduction in *cab* and *psb* A RNA transcripts in response to supplementary ultraviolet-B radiation. FEBS Lett 1991; 1:5-8.

32. Jordan BR, He J, Chow WS et al. Changes in mRNA levels and polypeptide subunits of ribulose 1,5-bisphosphate carboxylase in response to supplementary ultraviolet-B radiation. Plant Cell Environ 1992; 15: 91-98.

33. Peterson RB, Friberg EE, Burris RH. Diurnal variation in $N_2$-fixation and photosynthesis by aquatic blue-green algae. Plant Physiol 1979; 59:74-80.

34. Blakefield MK, Harris DO. Delay of cell differentiation in *Anabaena aequalis* caused by UV-B radiation and the role of photoreactivation and excision repair. Photochem Photobiol 1994; 59:204-208.

35. Tandeau de Marsac N. Occurrence and nature of chromatic adaptation in cyanobacteria. J Bacteriol 1977; 130:82-91.

36. Garcia-Pichel F, Castenholz RW. Characterization and biological implications of scytonemin, a cyanobacterial sheath pigment. J. Phycol 1991; 27:395-409.

37. Garcia-Pichel F, Sherry ND, Castenholz RW. Evidence for a UV sunscreen role of the extracellular pigment scytonemin in the terrestrial cyanobacterium *Chlorogloeopsis* sp. Photochem Photobiol 1992; 56:17-23.

38. Garcia-Pichel F, Wingard CE, Castenholz RW. Evidence regarding the UV sunscreen role of a mycosporin-like compounds in the cyanobacterium *Gloeocapsa* sp. Appl Environ Microbiol 1993; 59:170-176.

39. Scherer S, Chen TW, Boger P. A new UV-A/B protecting pigment in the terrestrial cyanobacterium *Nostoc commune*. Plant Physiol 1988; 88:1055-1057.

40. de Chazal NM, Smith GD. Characterization of a brown *Nostoc* sp. from Java that is resistant to high light intensity and UV. Microbiology 1994; 140:3183-3189.

41. Kumar A, Tyagi MB, Srinivas G et al. UVB shielding role of $FeCl_3$ and certain cyanobacterial pigments. Photochem Photobiol 1996; 64:321-325.

# CHAPTER 12

# STUDIES OF EFFECTS OF UV-B RADIATION ON AQUATIC MODEL ECOSYSTEMS

Sten-Åke Wängberg and Johanne-Sophie Selmer

## INTRODUCTION

The significantly increased interest, during the last years, in the influence of UV-B radiation on ecosystems is due to the finding of a decrease in the stratospheric ozone layer and an increased UV-B radiation reaching the earth's surface.[1] Increased UV-B is considered to be a threat to both terrestrial and aquatic ecosystems. Legitimacy for this fear comes from repeatedly measured damage by UV-B at ambient radiation on biological systems.[2,3] If UV-B increases, these effects will increase and eventually result in a decreased productivity and biological diversity.

Most studies of the UV-B effects on aquatic organisms, however, have been performed either with laboratory cultures or as short-term tests, measuring the acute effect of UV-B on some physiological parameter of the cultures or the natural community. Both of these approaches are limited in the sense that extrapolations either in time or in complexity are necessary in order to apply the data for prediction of ecosystem effects. Some parameters measured for estimating the effects of UV-B on algae are shown in Table 12.1. Similar tables could be created for effects on other groups of organisms or for effects on the aquatic chemistry. The extrapolations from data obtained in limited systems (such as those represented in the upper and left columns of Table 12.1) to long-term community effects include several pitfalls. For example:

- The variation in sensitivity for UV-B between different algal species is substantial,[4,5] and single species tests have been made on a limited number of species only—how they are representative for phytoplankton communities is unknown.

*The Effects of Ozone Depletion on Aquatic Ecosystems*, edited by Donat-P. Häder.
© 1997 R.G. Landes Company.

- Laboratory cultures are usually grown under replete nutrient conditions which differ from most ecosystems. The implication of this regarding the sensitivity of phytoplankton towards UV-B is unclear. Contradicting results, however, with different species, have been reported.[6,7]
- The relevance of measured effects on one physiological parameter for the long-term ecological effects is not only dependent on the parameter measured. In the long-term, the sensitivity can be modified by development of mechanisms protecting against effects from UV-B.[8] In a multispecies community, UV-B sensitive species will eventually be replaced by more tolerant ones, thus limiting the reduction in productivity.

To predict the ecological consequences, the effects on different biological communities must also be compiled together with the effects on the chemistry. The only way to circumvent these problems is to run long-term experiments on an ecological scale, usually called mesocosms or model ecosystems. (In this chapter we will use the general term "model ecosystem" since "mesocosm", together with the terms "microcosm" and "macrocosm", indicate some scale in size which sometimes leads to confusion). "Model ecosystems" are parts of a natural ecosystem that are isolated from the natural environment and are allowed to develop during a significant time with conditions as similar to the natural ones as possible. The model ecosystems are exposed to different treatments (in replicates), and different parameters are recorded to determine how the different treatments affect their development. Research with model ecosystems played a central role in the investigation of eutrophication, acidification and ecotoxicology;[9,10] however, up to now only a limited number of model ecosystems have been developed for studies of UV-B.

Unique for model ecosystems is that information can be gained on how increased UV-B changes the interactions between different organisms and the chemical environment. Model ecosystems are also used for testing whether extrapolations from reduced systems are valid or not. One should, however, remember that model ecosystems are not equal to real ecosystems. When enclosing a part of the ecosystem it is impossible to maintain all natural conditions, since, for example, water movement and larger grazers are usually excluded.

There is a large variation in model ecosystem design (reviewed in several books, e.g. refs. 9 and 10). Running model ecosystems is always expensive in terms of labor and capital, and the investments needed increase with increasing size, exposure time and the number of parameters

---

**Table 12.1. UV-B effects on algae. Examples of parameters studied on different levels of biological organization and on different time scales**

|  | Population | Communities |
|---|---|---|
| Short-term (< one generation) | photosynthesis (Cullen and Lesser[7]) DNA lesions (Karentz et al[5]) nitrogen metabolism (Döhler et al[28]) motility (Ekelund[29]) pigment destruction (Häberlein and Häder[30]) biological weighting functions (Cullen et al[11]) | photosynthesis (Smith et al[31]) nitrogen metabolism (Behrenfeldt et al[32]) |
| Long-term (> one generation) | growth (Jokiel and York[4]) protecting pigments (Wängberg et al[26]) | species composition, productivity growth |

measured. On the other hand, increasing size and running time can increase the ecological realism of the system.

Aquatic model ecosystems can be divided in pelagic and benthic systems alone or in combined systems (e.g. a pelagic water phase over a sediment). As the aim of model ecosystems is to simulate the development of natural ecosystems, special attention has to be paid to those factors that influence the development of natural ecosystems (e.g. nutrients, temperature and light).

Of special interest for studies of UV-B effects is, of course, the radiation regime. Not only the levels of UV-B, but also those of PAR (photosynthetic active radiation) and UV-A radiation are important since they have been shown to modify the effects of UV-B.[11] The optimal way to simulate the increased UV-B due to the reduction of stratospheric ozone layer is to have a radiation source in addition to the natural one, the irradiance of which is modulated by the natural radiation, resulting in a fixed enhanced percentage.[12]

## MODEL ECOSYSTEMS USED FOR ASSESSING EFFECTS OF UV-B RADIATION ON AQUATIC ECOSYSTEMS

As mentioned above, only a limited number of model ecosystems have been installed for studies of the effects of UVR (UV radiation). An overview of those covered in this chapter are given in Table 12.2.

The significant difference in design and sampling protocols make it difficult to compare the model ecosystem experiments presented here. Therefore it is not easy to generalize how these systems can increase our understanding of the ecological effects caused by enhanced UV-B. On the other hand, the variation in model ecosystems is advantageous as different ecological aspects are covered. None of the systems are designed for giving detailed and quantitative values on the level at which UV-B will threaten the productivity or biodiversity of marine ecosystems. Instead, they give

qualitative information on ecosystem changes induced by increased UV-B. As mentioned above, it should be stressed that model ecosystems are not real ecosystems and the design and sampling protocols presented here exclude several important ecological mechanisms that could be targets for UV-B (e.g. effects on motility, nutrient metabolism and water chemistry are only measured occasionally).

## BENTHIC SYSTEMS

Worrest et al[13] studied the colonization and growth of benthic algae on artificial substrates (acrylic plates) in two 720 l simulated ecosystem chambers at Oregon State University Marine Science Center, Newport, Oregon. The chambers had a "stair-step" graduation of the floor and the plates were thus placed at different depths in the chamber (ranging from 10-58 cm) (Fig. 12.1). Samples were taken on the fourth and sixth step which had depths of 34 and 18 cm, respectively.

Irradiation of the chambers came from fluorescent lamps and 60 W incandescent bulbs for PAR and fluorescent sunlamps (Westinghouse FS40) for UVR. The PAR was 10,000 lumens $m^{-2}$ ($\approx$15 W $m^{-2}$) on both chambers while the UV-B was 1.0 W $m^{-2}$ in one chamber and 1.4 W $m^{-2}$ in the other (when all lamps were on), measured at the water surface. All UV lamps were on for 8 h day$^{-1}$, while the illumination from half of them was extended one hour on each side of the 8 h period (giving about 30 and 45 kJ $m^{-2}$ d$^{-1}$, respectively). A flow-through system with water from the Yaquina estuary flowing through each chamber at a rate of 6 l min$^{-1}$ provided the chambers with new organisms, nutrients, etc.

Bothwell et al[14,15] studied the effects of UV-B on artificial substrates (open-cell styrofoam-DB) attached on the bottom of 2-meter long perplex channels at Environment Canada's experimental troughs apparatus (EXTRA), British Colombia. The channel system is further described in Bothwell.[16] The water velocity was $\approx$50 cm s$^{-1}$ and the water depth over

**Table 12.2. Overview of model ecosystems for studies of the effects of UV-B radiation that are covered in this chapter**

| Community studied | Artificial or ambient UV-B | Duration | Monitored components | Reference |
|---|---|---|---|---|
| benthos | artificial | 7 weeks | algal species algal biomass (chl) organic weight | Worrest et al[13] |
| phytoplankton | artificial | 4 weeks | algal species plankton biomass (chl, dw) photosynthesis | Worrest et al[18] |
| benthos | natural | 5 weeks | diatom cell number diatom species composition diatom cell volume chlorophyll, chironomid tubes | Bothwell et al[14,15] |
| phytoplankton | ambient and enhanced | 48 hours | pigments | El-Sayed et al[20] |
| phytoplankton | ambient | 16 days | chlorophyll, algal species | Helbling et al,[21] Villafañe[22] |
| plankton | artificial | 7 days | species composition photosynthesis, nitrogen uptake, chlorophyll, particulate nitrogen, particulate carbon | Wängberg et al,[23] Selmer et al[24] |
| plankton | ambient and enhanced | 14 days | species composition photosynthesis, nitrogen uptake, pigment, particulate nitrogen, particulate carbobacterial activity | Wängberg et al (in prep); Gustavson et al (in prep) |
| benthos | ambient and enhanced | 3 weeks | algal biomass, photosynthesis bacterial activity, pigments | Sundbäck et al[17] |

the foam was 1 cm. The radiation regime was manipulated by different screening films. Mylar-D and acrylic sheets were used to exclude UV-B[15] and all UV radiation, respectively. Neutral screens were used to reduce the total radiation in some treatments. All light treatments were run in triplicates.

In August 1994, intact sediment cores covered with mature microbial mats dominated by diatoms were incubated for four days in flow-through water baths at Tjärnö Marine Biological Laboratory on the west coast of Sweden.[17] The sediments were exposed with either ambient radiation without UV-B (Mylar film) or with ambient radiation with additional UV-B ($\approx$1.1 W m$^{-2}$) for 3 h day$^{-1}$. In July 1995 a 3-week experiment was conducted in the same area, instead using sandy sediments and including ambient radiation as a third treatment (Fig. 12.2). In both experiments, all measurements were done of three cores exposed for UV-B radiation and three controls at each time.

## PLANKTON

Twelve 15 l containers were placed in a fiberglass greenhouse at Oregon State University Marine Center in Newport,

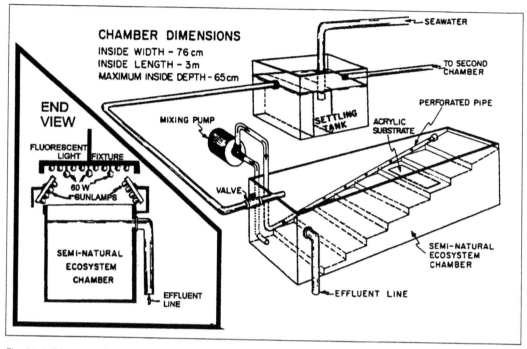

Fig. 12.1. Diagram of seminatural ecosystem chamber (continuous flow-through design) and associated water supply system. Arrangement of radiation sources is illustrated (Worrest et al[13]). Reprinted with permission from Worrest RC et al, Photochem Photobiol 1978; 27:471-478. ©American Society for Photobiology.

Fig. 12.2 Set-up of the model ecosystem for studies of UV-B effects on benthic communities at Tjärnö Marine Biological Station, Sweden. (Photograph by S. Odmark).

Oregon.[18] UV-B was supplied by sunlamps 3 h day$^{-1}$. The UV-B exposure level was manipulated by the use of Mylar and different thickness of cellulose acetate to give four different levels of radiation (0.16; 6.43; 6.86 and 7.61 kJ m$^{-2}$ d$^{-1}$ (290-320 nm)). When weighted with a generalized plant action spectrum[19] these values simulated the following scenarios: No UV-B; ambient; 5.5% and 15.5% ozone depletion. The experiment was run for 27 days, and inflow and outflow ports of the containers allowed approximately 15 l of seawater from the Yaquina Bay to enter each container daily. Thus, the plankton communities within the containers were, at any given time, a composite of those established and growing in the containers and those freshly pumped in.

To study the effects of UV-B and UV-A on Antarctic algae, four containers with three replicate 18 l chambers each, were placed outdoors at Palmer Station (Fig. 12.3).[20] Each chamber was filled with seawater coming from a sea ice hole about 15 m from the shore. The radiation falling into the chambers was individually manipulated for each container so that four treatment levels were established: (1) near ambient; (2) near ambient light, excluding all solar UVR; (3) near ambient visible and UV-A with reduced levels of UV-B; and (4) near ambient light and artificial enhanced UV-B and UV-A. The increased UVR was achieved by FS-20 Westinghouse fluorescent sunlamps located under treatment tank number 4 and was calculated to represent a 3% increase of the biological weighted radiation (photoinhibition action spectra). The system was run for 48 h and the water was kept in motion by air infiltration.

Water samples collected close to Seal Island in Antarctica were placed in 1-l round Chemglass vessels and placed in an on-deck incubator for 16 d in February-March 1991.[21] Chemglass has a good transparency for most of the UV-B range with a 50% cut-off at 295 nm. Six Chemglass vessels were incubated: two controls with no extra filters, two shielded by Mylar film thereby excluding UV-B, and two shielded by plastic foils excluding UV-B and most of the UV-A (50% cut-off at 378 nm). The incubations lasted for 16 d in February-March. In addition to this experiment, water samples were kept in a refrigerator equipped with daylight fluorescence tubes, in polycarbonate bottles. These were to be used in on-deck incubations in tropical waters using the same technique but with only 5 d incubation. The mean daily total UV radiation during the incubations varied between 4 and 14 W m$^{-2}$ in Antarctica, and between 16 and 21 W m$^{-2}$ in the tropical waters (reduced by neutral screens).

A similar approach was used by Villafañe et al[22] with incubations at Palmer

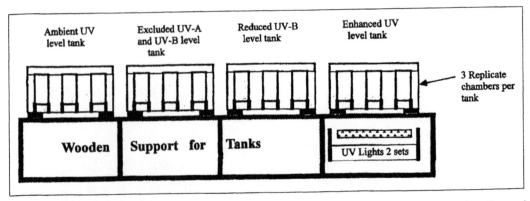

Fig. 12.3. Schematic drawing of the four experimental tanks and chambers used to investigate the effects of UV radiation on Antarctic phytoplankton (El-Sayed et al[20]). Reprinted with permission from El-Sayed SZ, et al. In: Kerry KR, Hempel G, eds. Antarctic Ecosystems–Ecological Change and Conservation. Berlin: Springer Verlag 1990:379-385.

Station and no distinction between UV-A and UV-B since only one filter was used to absorb all UVR. In contrast to Helbling et al,[21] these water samples were kept in quartz vessels and enriched with nutrients. Two experiments were done for 15 and 13 d, respectively, in November-December. The ambient UVR was higher than those by Helbling et al,[21] ranging from 15-50 W m$^{-2}$.

Plankton communities sampled from the upper mixed layer on the Swedish west coast were incubated for one week in 18-l aquaria in an indoor incubation chamber.[23,24] The aquaria were irradiated by daylight fluorescence tubes (20 W m$^{-2}$) for 16 h day$^{-1}$. Two of the aquaria were exposed to UV-B derived from fluorescence tubes with an intensity of about 0.64 W m$^{-2}$, 6 h day$^{-1}$. Each day, before the lighting of the UV-B tubes, half of the water volume in the aquaria was withdrawn and exchanged with filtered sea water (Whatman GF/C glassfiber filters). The water taken out was used for analysis of the plankton community's structure and function. Mixing within the aquaria was achieved by air infiltration.

Large polyethylene enclosures with a diameter of 1.5 meter and a depth of 3.5 m (about 5 m$^3$) were placed in the mesocosm experimental site HEXAGON at Kristineberg Marine Research Station on the Swedish west coast (Wängberg S-Å, Garde K, Gustavson K et al. Effects of UV-B radiation on pelagic ecosystem—a mesocosm study. System set-up, radiation regime and effects on primary producers, manuscript in preparation). The enclosures were sealed at the bottom and filled with sea water from the Gullmar Fjord. UV-B transparent perplex was placed on top of each enclosure. The UV-B exposure of the enclosures was manipulated either by radiation from UV-B fluorescence tubes (Philips TL12) suspended over the opening or by Mylar film

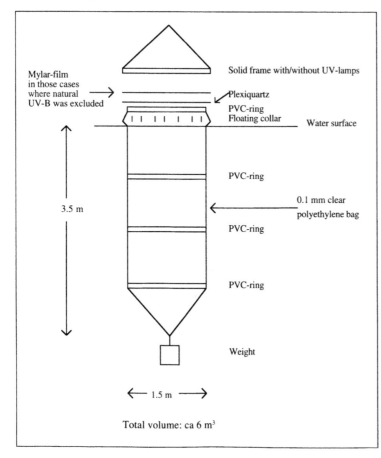

Mylar-film in those cases where natural UV-B was excluded →

Solid frame with/without UV-lamps

Plexiquartz

PVC-ring
Floating collar    Water surface

PVC-ring

3.5 m

0.1 mm clear polyethylene bag

PVC-ring

PVC-ring

Weight

← 1.5 m →

Total volume: ca 6 m$^3$

Fig. 12.4. Schematic drawing of the set-up of the individual pelagic enclosures used for studies of effects of UV-B at HEXAGON, Kristineberg Marine Research Station, Sweden.

Fig. 12.5. Photograph showing the arrangement of radiation source and above-water parts of enclosures used for studies of UV-B effects at HEXAGON, Kristineberg Marine Research Station, Sweden. (Photograph by J.-S. Selmer)

placed on the perplex (Figs. 12.4 and 12.5). Two enclosure experiments were performed, in May and July 1994. In the first experiment one level of enhanced UV-B (1.4 W m$^{-2}$ at the surface centrally under the radiation frame) was used on three replicates, while in the second, two different levels (2.1 and 0.7 W m$^{-2}$) were used in duplicates. The enhanced UV-B was applied for 4 h day$^{-1}$. For both experiments both control enclosures and enclosures with ambient UV-B excluded (by Mylar film) were installed in three replicates. In the first experiment the under-water irradiation levels were measured down to 2 m with a 4$\pi$-sensor coupled to an Optronic 742 spectroradiometer,[25] while in the second, UV-B, UV-A and PAR were measured with cosine corrected sensors. During both experiments the attenuation of UV-B by the enclosed water was high (10% depth around 0.5 meter), but not exceeding that of the water in the Gullmar Fjord outside the HEXAGON. The attenuation for UV-A and PAR was lower and more variable (10% at 1-3 m depth). Water was sampled from the enclosures with an integrating sampler from the surface down to 2 m depth at 9.00 a.m. The highest ambient UV-B was around 0.5 W m$^{-2}$ in the first experiment and 0.8 W m$^{-2}$ in the second.

## REPORTED EFFECTS

As given in the presentation above, most of the model ecosystems run for estimating effects from UV-B have focused on the effects on algae. Only occasionally are effects on other parts of the ecosystem reported.

# EFFECTS ON PRIMARY PRODUCERS

## BIOMASS

In none of the model ecosystems where effects of UV-B on benthic algal communities were studied,[13,15,17] were any immediate effects of the UV-B on the chlorophyll content shown. However, in both Worrest et al[13] and Bothwell et al,[15] effects were found after some weeks when the initial biomass increases were leveling off. In Worrest et al[13] the chlorophyll levels decreased very strongly in the communities exposed to the higher UV-B level (Fig. 12.6), resulting in a significant change in the ratio between organic weight and chlorophyll *a*. In Bothwell et al[15] the UV-B resulted in a continued chlorophyll *a* accumulation even after the radiation had terminated in those communities shielded

from UV-B (Fig. 12.7) (see also below in "Effects on Heterotrophs").

In Bothwell et al[14] the development of chlorophyll and algal cell volumes in the latter phase of the experiment was the same as in Bothwell et al,[15] when excluding all UVR. The growth of the community (measured as increase in chlorophyll content) during the first weeks was, however, significantly higher when UVR was filtered off, resulting in increased chlorophyll content and algal cell volume. The difference during the early phase is explained by higher grazing pressure from chironomides during the experiment presented in Bothwell et al[15] (the density of chironomides were one order of magnitude lower).

In the Antarctic plankton systems at Palmer Station reported by El-Sayed et al,[20] lower chlorophyll concentrations were found after 48 h incubations in all UV exposed chambers when compared with the one shielded from all UVR. There was no difference in reduction of chlorophyll concentration, however, between ambient solar radiation and ambient radiation

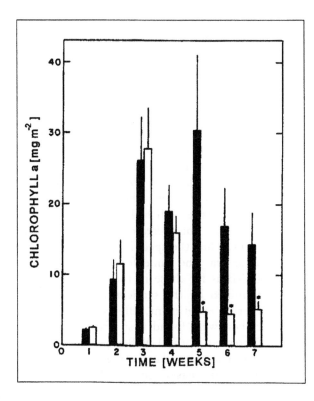

Fig. 12.6. Effects of UV-B intensity on the chlorophyll content of benthic communities in simulated ecosystem chambers (Worrest[13]). Values are from combined samples taken from 26 and 41 cm depths from the two flow-through chambers. Solid bars represent samples taken from the chamber receiving a surface fluence rate of 1 W m⁻² (290-315 nm) and the shaded bars represent samples taken from the chamber receiving a surface fluence rate of 1 W m⁻². Error bars = S.E.; * = p<0.001; all others p>0.05. Reprinted with permission from Worrest RC et al, Photochem Photobiol 1978; 27:471-478. ©American Society for Photobiology.

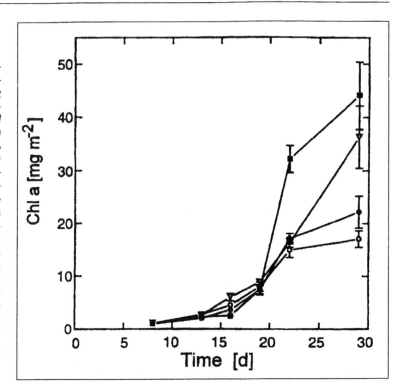

Fig. 12.7. Chlorophyll content of benthic communities established during different radiation regimes in the experimental troughs apparatus (EXTRA) (Bothwell et al[15]). • = 90% PAR; ° = 90% PAR + UV-A; ∇ = 90% PAR + UV-A + UV-B; ■ = 90% PAR + malathion. Addition of the insecticide malathion was done to compare a chemical exclusion of grazers with the effects of UV-B. Reprinted with permission from Bothwell et al. Ecosystem response to a solar ultraviolet-B radiation: Influence of tropic-level interactions. Science 1994; 265:97-100. ©1994 American Association for the Advancement of Science.

excluding UV-B. In the communities with enhanced UVR the chlorophyll levels decreased even further. In contrast, Helbling et al[21] found no differences in chlorophyll content between the various treatments in 1-1 vessels with Antarctic phytoplankton communities incubated on deck for 16 d in Antarctica. The same communities, after storage in a refrigerator and incubation on-deck in the tropics, exhibited less increase in chlorophyll content when exposed to the whole solar spectrum than when UV-B was excluded. The chlorophyll content was even more enhanced if the cultures were shielded from all UVR by plastic foil.

In one of the experiments presented by Villafañe et al[22] the UVR decreased the chlorophyll concentration up to 50% midway through the experiment, however, no differences were found at the end. A reduction in phytoplankton biomass (in mg C m[-3], but calculated from measured algal cell volumes) was found during the whole experiment, thus the chlorophyll content per biomass increased due to UVR exposure. It is not possible to determine

whether the differences in results between this experiment and the Antarctic one[21] were due only to higher radiation in the former. Different absorption by the Chemglass[21] compared to the quartz vessels[22] and the nutrient enrichments could have increased the susceptibility towards UVR in the Antarctic experiment. It can, of course, also be explained by what usually is called the "biological variation."

In the aquarium experiment performed at Kristineberg, a temporary decrease in chlorophyll concentration and continuous lower algal cell volumes were also found when the plankton communities were exposed to UV-B.[23,24] The difference between treatments, established during the first two days, did not increase during the experiment. Thus the specific growth rate of the algae was the same for those exposed to UV-B as for those in the controls and determined by the semicontinuous dilution. In this experiment, effects on particulate carbon were found first after 6 d when it decreased in the UV-B-exposed communities. The particulate nitrogen, on the other hand, increased in the UV-B-exposed cultures during the first days but approached that of the controls on day 6. The particulate C:N ratio decreased during the whole experiment in the UV-B exposed communities, as did the ratio between carbon dioxide fixation and uptake of inorganic nitrogen.

A correlation between the exposure to UV-B and the chlorophyll concentrations was also found in the later part of the 27-d experiment with phytoplankton in Oregon.[18] While the chlorophyll content was almost equal in all containers on day 5, on day 27 it was almost double in the Mylar shielded communities compared with the two containers given the highest UV-B irradiances. All treatments, except the Mylar shielded, showed decreasing biomass with time (measured as ash-free dry weight).

There was no correlation between the pigment content and the exposure treatment for none of the twelve pigments (chlorophylls and carotenoids) measured in the large pelagic mesocosms run in HEXAGON (Wängberg et al, in preparation).

## SPECIES COMPOSITION

When algal communities are enclosed in any system, the changed environmental condition per se will influence the species succession in the community. Thus comparisons should be made on UV-B-induced changes in successional patterns and not on changes in species composition. When a decrease in algal biomass due to UV-B has been recorded it has always been accompanied by changes in the species succession (if measured).

In the benthic communities presented in Worrest et al[13] all communities were dominated by *Melosira* species after two weeks. However, in the communities exposed to high UV-B, *Stauroneis constricta* increased in predominance and after six weeks represented over 50% of the community on both step 4 and step 6 (18 and 34 cm below surface). The communities exposed to the lower UV-B level were more diverse with the highest percentage (20-24%) represented by *Navicula salinicola*

and over 10% consisting of *Synedra fasciculata* and *Fragilaria striatulata* v. *californicum*. The dominance of a few species in the UVR-exposed communities was also reflected in a decreased diversity (Shannon-Weaver information index).

In the channels presented in Bothwell et al[14] the initial community was dominated by *Fragilaria* and *Tabellaria* species but after 3 weeks all communities were dominated by *Fragillaria vaucheriae*. This alga represented high percentages of the total algal cell volume throughout the whole experiment in those communities shielded from UVR. After 35 days it represented 33 and 19% of the cell volume in communities exposed for 90 and 50% of ambient PAR radiation, respectively. Other species with high percentages in those communities were *Nitzschia palea* (21.5% at 90% PAR, <2 at 50% PAR), *Synedra ulna* (21.9% at 90% PAR) and *Melosira varians* (14% at 90% PAR and 42.2% at 50% PAR). In the UVR exposed communities the succession was different resulting in dominance by *Gomphoneis herculeana* (41.5% at 90% PAR+UVR and 28.7 at 50% PAR+UVR) together with *Melosira varians* (24.8 and 32.3% at 90 and 50% PAR+UVR), respectively.

No changes in species composition were found in the mature diatom mats at Tjärnö Marine Biological Laboratory when exposed to UV-B.[17] However, visual observation revealed that the number of *Gyrosigma baltica* cells found on the mat surface decreased during enhanced UV-B exposure during the first three days. This difference remained, however, for only three hours after switching off the UV-B tubes.

In Worrest et al[18] the seed community was dominated by *Chaetoceros* sp. whose dominance even increased in the community exposed to the highest UV-B level. On the other hand, in the community shielded from UV-B, the percentage of the total algal cell volume decreased while *Skeletonema costatum* increased to the same biomass as *Chaetoceros* sp. Among 500 examined diatom species, the number decreased from 37 in the Mylar shielded control to 11 at the highest UV-B.

In both Villafañe et al[22] and Wängberg et al[23] diatoms represented less than 20% of the algal biomass in the seed communities but their contribution increased in the course of the experiments. The succession to a community more dominated by diatoms was, however, enhanced by the UVR in Villafañe et al[22] while it was retarded by additional UV-B in Wängberg et al.[23] Since the designs of the experiments are so diverging (besides the differences in Antarctic and Swedish coastal waters), the different outcome is not surprising. Potentially important factors might be: (1) different radiation regimes; (2) the treatment in the Antarctic experiments was ambient radiation with filters absorbing both UV-A and UV-B, while the treatment in the Swedish experiments were artificial radiation with fluorescence tubes giving mainly UV-B; (3) nutrient conditions differed inasmuch as the Antarctic experiments were fertilized while the Swedish were not. In addition, Antarctic waters usually have a higher nutrient content than Swedish coastal waters.

No changes in species composition were found between the different treatments of Antarctic cultures when incubated in Antarctica by Helbling et al.[21] When incubated in the tropics, however, the ratios of vegetative cells to resting spores of *Chaetoceros* spp. and of vegetative cells to dead cells of *Thalassiosira gracilis* var. *expecta* were lower in the communities exposed to the whole spectrum than in those shielded by Mylar. Shielding the UV-A with plastic increased the proportion of vegetative cells even further. A decreased ratio between nanoplankton ($<20$ μm) and microplankton ($>20$ μm) was also found when shielding the cultures from UVR.

That the size of the algal cell should influence its sensitivity to UV-B has been argued by Karentz et al[5] who also showed experimental evidence for this by measuring DNA lesions in diatoms. These authors concluded that larger algae should have an advantage. In agreement with Karentz et

al,[5] Bothwell et al[14] reported increased mean cell volumes of periphytic communities when exposed to UVR. Wängberg et al,[23] however, showed that the average cell size of the whole phytoplankton community decreased when exposed to UV-B. They noted that the larger mean algal cell volume of the whole community in the controls was dependent on the greater success for diatoms in these communities as compared to those exposed to UV-B. Calculations of the mean cell volume within the Centrales (the algal order with the most number of species), on the other hand, showed an increased mean cell volume during UV-B exposure. This was, however, due to a relative larger success for *Rhizosolenia* than *Chaetoceros* species during UV-B. The smallest of all recorded diatoms, *Cylondrotheca closterium*, also did quite well under UV-B. From model ecosystem experiments it is thus not possible to confirm that increased cell size per se increases the competitive success of the algae during UV-B. This lack of evidence should not be connected with the increases in cell sizes within a single population when exposed to UV-B.[6,26,27]

## PRIMARY PRODUCTION

The only system with benthic communities in which primary productivity has been measured is Sundbäck et al.[17] It was done both as oxygen production in depth profiles from +500 to -2000 μm using microelectrodes and as incorporation of [14]C-labeled carbonate during one hour. All measurements were done with PAR only. Oxygen profiles were run directly after the end of the radiation period on day 2 and 4. On both days it showed a significant decrease of the maximum values in the communities exposed for UV-B as compared to the controls (0.8 and 0.7 compared with 1.2 and 1.1 mM $O_2$). The radiocarbon incorporation experiments were performed four hours after the UV-B treatment on day 2 and showed no differences between the communities exposed to UV-B radiation and the controls. On day 4, on the other hand, the radiocarbon experiment was done at the same time as the oxygen profiles, the incorporation being only 20% in the UV-B exposed communities as compared to the controls. The incorporated carbon was fractionated into different biochemical groups, showing that the UV-B exposed communities allocated significantly larger percentages of the photosynthetic products to proteins and lipids than the controls, while the allocation to polysaccharides and low-molecular-weight compounds was lower.

In the end, both experiments of Villafañe et al[22] showed that carbon dioxide fixation per chlorophyll was higher in the communities exposed to UVR as compared with those shielded from all UVR. In Worrest et al[18] the carbon fixation per chlorophyll was the same during all treatments in the beginning of the experiment. In the last measurement after 26-27 days, it increased in all communities and the increase was highest in the communities shielded from all UV-B, but similar in all communities exposed to UV-B. In Selmer et al[24] the carbon dioxide fixation per chlorophyll was always lower in the communities exposed to UV-B than in the controls. The percentage photosynthate allocated to polysaccharides was significantly higher in the UV-B exposed communities while allocation to low-molecular-weight compounds decreased. The decrease in chlorophyll content in the UV-B exposed communities in Worrest et al[18] and Selmer et al[24] makes the UV-B-dependent reduction in carbon dioxide fixation even greater if expressed per liter.

In Villafañe et al[22] measurements of carbon dioxide fixation were done under different shields absorbing either UV-B or all UVR at the beginning and end of both experiments. Both UV-A and UV-B reduced the carbon dioxide fixation in all communities, except for one that had been exposed for the full spectrum for 13 days. This one community had thus achieved an increased tolerance when the succession pattern was changed. However, the tolerance also increased in the communities shielded from UVR, indicating that a succession towards a more UVR-tolerant community occurred independent of the UVR exposure. Also in Selmer et al[24] incubations were done both with and without UV radiation. The artificial UV-B irradiance used was the same as that in the center of the aquaria during the long-term experiment ($\approx$1 W m$^{-2}$), but as the incubations were done in polystyrene vessels more radiation was coming from the longer wavelengths. The addition of UV-B did not change the carbon dioxide fixation in any treatment.

## Effects on Heterotrophs

Effects of UVR on other organisms than primary producers were only occasionally recorded. Bothwell et al,[15] however, clearly showed that the abundance of chironomides increased when the UVR excluded by selective films. The decreased abundance of these efficient grazers on algae explained the higher algal biomass under UV-B in the later phase of the experiment (Fig. 12.7).

In some cases, however, UV-B exposure has increased the activity of heterotrophic organisms. In the aquarium system at Kristineberg the UV-B increased the number of choanoflagellates,[24] probably due to increased amount of their prey which is mainly bacteria. In the large pelagic enclosures at HEXAGON where no significant effects were found on primary producers, enhanced UV-B significantly increased the bacterial activity (thymidine-incorporation) up to 30-40% during the experiment in May (Gustavson K, Garde K, Selmer J-S et al. The impact of UV-B irradiation on bacterial production in a coastal environment, manuscript in preparation). Increased bacterial activity after UV-B exposure was probably due to increased bacterial substrate availability, inasmuch as UV-B can cause photochemical degradation of humic substances into low-molecular-weight compounds.

## Comparisons Between Short- and Long-Term Effects

One aim of the work with model ecosystems is to test whether predictions on long-term effects made from limited systems are valid or not. Whenever possible, it is therefore of interest to compare the effects of UV-B on limited systems with those found in model ecosystems. Such comparisons, done on three experiments, gave three different results.

Helbling et al[21] have summarized a large number of experiments and shown a correlation between total UVR and the enhancement of carbon dioxide fixation when shielded from different parts of the UVR. The radiation of 8 W m$^{-2}$ during the 16 d incubation in Antarctica coincides with the threshold value where some experiments show no enhancement while others show low enhancement. When the same communities were incubated in the tropics at around 18 W m$^{-2}$, the growth was reduced as predicted from the short-term experiments. This comparison thus validates that the effects on carbon dioxide fixation can predict, at least, the threshold value for long-term effects of UV-B.

The seed communities for both experiments by Villafañe et al[22] showed about 100% increase in carbon dioxide fixation rate when shielded from UVR. The long-term effects on chlorophyll levels and carbon dioxide fixation activity were much smaller, indicating that the effects on carbon dioxide fixation overestimate the long-term effects.

No reduction in carbon dioxide fixation or carbon allocation pattern was found when the seed community for the aquarium experiment at Kristineberg was incubated at the UV-B level that reduced the algal biomass and changed the succession pattern.[23,24] This comparison indicates that effects on carbon dioxide fixation underestimate the long-term effects.

## ACKNOWLEDGMENT

We would like to thank Maureen Jehler for her invaluable correction of the language in this chapter.

## REFERENCES

1. Madronich S. The atmosphere and UV-B radiation at ground level. In: Young AR, Björn LO, Moan J et al, eds. Environmental UV Photobiology. Plenum Press, 1993:1-40.

2. Häder D-P. Risks of enhanced solar ultraviolet radiation for aquatic ecosystems. In: Round FE, Chapman DJ, eds. Progress in Phycological Research. Bristol: Biopress Ltd, 1993:1-45.

3. Holm-Hansen O, Lubin D, Helbling EW. Ultraviolet radiation and its effects on organisms in aquatic environments. In: Young AR, Björn LO, Moan J et al, eds. Environmental UV Photobiology. New York: Plenum Press, 1993:379-425.

4. Jokiel PL, York Jr RH. Importance of ultraviolet radiation in photoinhibition of microalgal growth. Limnol Oceanogr 1984; 29:192-199.

5. Karentz D, Cleaver JE, Mitchell DL. Cell survival characteristics and molecular responses of Antarctic phytoplankton to ultraviolet-B radiation. J Phycol 1991; 27:326-341.

6. Behrenfeld MJ, Hardy JT, Lee II H. Chronic effects of ultraviolet-B radiation on growth and cell volume of *Phaeodactylum tricornutum* (Bacillariophyceae). J Phycol 1992; 28: 757-760.

7. Cullen JJ, Lesser MP. Inhibition of photosynthesis by ultraviolet radiation as a function of dose and dosage rates: results for a marine diatom. Mar Biol 1991; 111: 183-190.

8. Karentz D. Ultraviolet tolerance mechanisms in Antarctic marine organisms. In: Weiler CS, Penhale PA, eds. Ultraviolet radiation in Antarctica: measurements and biological effects. Antarctic research Series 1994; 62:93-110.

9. Grice GD, Reeve MR. Marine mesocosms. Biological and Chemical Research in Experimental Ecosystems. Berlin, New York: Springer Verlag 1982.

10. Ravera O. The "enclosure" method: concepts, technology, and some examples of experiments with trace metals. In: Boudou A, Ribeyre F, eds. Aquatic Ecotoxicology: Fundamental Concepts and Methodologies. Vol. 1. Boca Raton: CRC Press, 1989:250-272.

11. Cullen JJ, Neale PJ, Lesser MP. Biological weighting function for the inhibition of

phytoplankton photosynthesis by ultraviolet radiation. Science 1992; 258:646-650.

12. Björn LO, Teramura AH. Simulation of daylight ultraviolet radiation and effects of ozone depletion. In: Young AR, Björn LO, Moan J et al, eds. Environmental UV Photobiology. New York: Plenum Press, 1993:41-71.

13. Worrest RC, van Dyke H, Thomson BE. Impact of enhanced simulated ultraviolet radiation upon a marine community. Photochem Photobiol 1978; 27:471-478.

14. Bothwell ML, Sherbot D, Roberge AC et al. Influence of natural ultraviolet radiation on lotic periphytic diatom community growth, biomass accrual, and species composition: Short-term versus long-term effects. J Phycol 1993; 29:24-35.

15. Bothwell ML, Sherbot DMJ, Pollack CM. Ecosystem response to a solar ultraviolet-B radiation: influence of tropic-level interactions. Science 1994; 265:97-100.

16. Bothwell ML. Growth rate responses of lotic periphytic diatoms to experimental phosphorus enrichment: the influence of temperature and light. Can J Fish Aquat Sci 1988; 45:261-270.

17. Sundbäck K, Nilsson C, Odmark S et al. Effects of enhanced UV-B radiation on a marine benthic diatom mat. Marine Biology (in press).

18. Worrest RC, Thomson BE, van Dyke H. Impact of UV-B radiation upon estuarine microcosms. Photochem Photobiol 1981; 33:861-867.

19. Caldwell MM. Solar ultraviolet radiation as an ecological factor for alpine plants. Ecol Monogr 1968; 38:243-268.

20. El-Sayed SZ, Stephens FC, Bidigare RR et al. Effects of ultraviolet radiation on Antarctic marine phytoplankton. In: Kerry KR, Hempel G., eds. Antarctic Ecosystems— Ecological Change and Conservation. Berlin: Springer Verlag 1990:379-385.

21. Helbling EW, Villafañe V, Ferrario M et al. Impact of natural ultraviolet radiation on rates of photosynthesis and on specific marine phytoplankton species. Mar Ecol Progr Ser 1992; 80:89-100.

22. Villafañe VE, Helbling EW, Holm-Hansen O et al. Acclimatization of Antarctic natural phytoplankton assemblages when exposed to solar ultraviolet radiation. J Plankt Res 1995; 12:2295-2306

23. Wängberg S-Å, Selmer J-S, Gustavson K. Effects of UV-B radiation on biomass and composition in marine phytoplankton communities. In: Figueroa FL, Jiménez C, Pérez-Llorens JL et al, eds. Underwater light and algal photobiology. Sci Mar 1996; 60 (Suppl.1):81-88.

24. Selmer J-S, Wängberg S-Å, Gustavson K. Effects of UV-B radiation on carbon and nitrogen dynamics in marine planktonic communities. (submitted).

25. Piazena H, Häder D-P. Penetration of solar UV irradiation in coastal lagoons of the southern Baltic Sea and its effects on phytoplankton communities. Photochem Photobiol 1994; 60:463-469.

26. Wängberg S-Å, Persson A, Karlson B. Effects of UV-B radiation on synthesis of mycosporine-like amino acid and growth in *Heterocapsa triquetra* (Dinophyceae). J Photochem Photobiol B 1996; 37:141-146.

27. Buma AGJ, Zemmelink HJ, Sjollema KA et al. Effect of UV-B on cell characteristics of the marine diatom *Cyclotella* sp. Proc Eur Symp on Environm UV Effects (in press).

28. Döhler G, Hagmeier GE, Krause K-D. Impact of solar UV radiation on uptake of $^{15}N$-ammonia and $^{15}N$-nitrate by marine diatoms and natural phytoplankton. Biochem Physiol Pflanzen 1991; 187:293-303.

29. Ekelund, NGA. Influence of UV-B radiation on photosynthetic light-response curves, absorption spectra and motility of four phytoplankton species. Physiologia Plantarum 1994; 91:696-702.

30. Häberlein A, Häder D-P. UV effects on photosynthetic oxygen production and chromoprotein composition in a freshwater flagellate *Cryptomonas*. Acta Protozool 1992; 31:85-92.

31. Smith RC, Prézelin BB, Baker KS et al. Ozone depletion: ultraviolet radiation and phytoplankton biology in Antarctic waters. Science 1992; 255:952-959.

32. Behrenfeld MJ, Lean DRS, Lee II H. Ultraviolet-B radiation effects on inorganic nitrogen uptake by natural assemblages of oceanic plankton. J Phycol 1995; 31:25-36.

# CHAPTER 13

# Solar UV Effects on Benthic Marine Algal Assemblages— Three Case Studies

## Regas Santas

## ABSTRACT

Medium- and long-term effects of solar UV on assemblages of attached filamentous algae and diatoms have been investigated in natural marine environments (Caribbean, Mediterranean seas), and in laboratory mesocosms. In the field, solar UV-B reduced the productivity of assemblages by 40% (Caribbean), while in the laboratory exposure to increased UV-B irradiance caused shifts in species dominance, community composition, and decreased primary productivity. While these UV-B effects are pronounced during early community development, they gradually diminish as succession progresses. Exposure to UV-A did not significantly affect productivity or community composition.

## INTRODUCTION

Incident UV levels at the earth's surface have been increasing due to atmospheric ozone destruction by CFCs. Ultraviolet radiation represents about 7% of the total solar radiation striking the earth's atmosphere and an even smaller proportion of the radiation penetrating to sea surface levels.[1] Wavelengths shorter than 286 nm do not reach the surface of the earth.[2] UV-B radiation (280-315 nm) has adverse effects on biological systems.[3-6] UV-A is less damaging and in some cases beneficial to aquatic primary producers.[7]

*The Effects of Ozone Depletion on Aquatic Ecosystems*, edited by Donat-P. Häder.
© 1997 R.G. Landes Company.

Initial considerations about the impact of increased solar UV-B included agricultural and terrestrial ecosystem primary productivity. However, it was not long before an equal share of scientific concern focused on oceanic primary producers because of their basal position in trophic chains, their regulatory role in gas cycles and in global climate. Since algae are responsible for nearly all of the oceanic primary production, predicting the impact of increased solar UV on oceanic productivity requires the examination of UV effects on algae at the molecular, organism and community level.

## UV-Shielding Substances in Algae

The DNA of algae appears to be poorly shielded as compared to that of higher plants. For example, doses of UV-C radiation (at 254 nm) necessary to kill leaves of higher plants appear to be some four orders of magnitude greater than that required to kill very resistant algae.[8,9] Flavonoids, the highly effective UV screening compounds of higher plants, have not been found in any algal group. These pigments exist in high enough concentrations in the epidermis of plant leaves so that, together with cuticular waxes and other cell-wall components this single-cell layer reduces the incident UV radiation by one to two orders of magnitude.[10] Thus, it appears that higher plants are well adapted to cope with present-day levels of UV through shielding pigments and tissues, as well as other mechanisms.[1] However, although devoid of flavonoids, algae are known to produce other UV absorbing substances. Fogg and Boalch[11] detected excretion of a yellow protein-carotenoid complex by healthy *Ectocarpus* cultures which, in part, accounted for the color found in the culture medium of this organism. Craigie and McLaughlan[12] also report excretion of yellow compounds in the surrounding water by healthy uninjured thalli of *Fucus vesiculosus*. These compounds showed a strong ultraviolet absorption spectrum with a plateau in the 280-260 nm region. The amount of

pigmentation in the medium correlated to the degree of alkalinity, temperature exposure, and to a lesser extent salinity, while there was no difference in the amount of pigment released in light or darkness. The authors suggested that the substance was produced in the physodes and is initially colorless, the yellow color being a result of oxidation shortly after the release of the substance. A number of other physode-possessing brown algae and *Olisthodiscus*, a chrysophycean alga, released substances giving an absorption spectrum similar to that of *F. vesiculosus*. The presence of a flavonol or catechin-type of tannin was suggested in this material, considered as a potential source of yellow, ultraviolet-absorbing substances in seawater. Tsujino and Saito[13] determined the ultraviolet absorption spectra of a perchloric acid extract in 39 species of red, brown, and green algae. All algal extracts tested showed an absorption maximum at 260-270 nm. Red algae indicated another characteristic sharp absorption maximum at 315-330 nm, while green and brown algae did not indicate this second maximum. These ultraviolet absorbing compounds had no relation to phycobilin pigments. Halldal[7] suggested the presence of photosynthetically inactive compounds that screen UV radiation in a green (*Ulva*) and a red (*Trailliella*) alga. Carotenoids are known to absorb UV radiation, thus providing a photoprotection mechanism in copepods,[14] but are effective as a UV screen in algae only when they are not closely associated with the photosynthetic mechanism.[15] Finally, in cyanobacteria, a photoprotective role of the UV-A absorbing extracellular pigment scytonemin was proposed,[16] while mycosporine-like amino acids (MAAs) have often been thought to serve a UV sunscreen role in many phytoplankton and macroalgae.[17]

Despite the presence of shielding pigments, however, UV radiation is harmful to algal cells. Under enhanced UV-B irradiance, photoinhibition of diatom growth[18] and reduction in biomass production, pigmentation, and protein content of microalgae occurs.[19] The photosynthetic activity of *Anacystis nidulans* is inhibited by UV-A.[20]

While UV irradiation is generally considered damaging to a large number of algae, Halldal[7] reports that *Ulva* and *Trailliella* utilize UV-A and a part of the UV-B spectrum in photosynthesis. Photoinhibition occurs at wavelengths shorter than 313 and 302 nm for *Trailliella* and *Ulva*, respectively. UV-B has also been reported to be the component of solar UV radiation responsible for increased algal biomass over longer periods of time[21]—probably the indirect effect of reduced grazing pressure on primary producers.

Studies of UV effects on benthic algal communities are scarce despite the large production potential and ecological significance of benthic algae.[22] The productivity of coral reef algal turfs, a type of benthic community distinguished by its simple morphology and high efficiency in solar energy capture, ranges from 3.4-20.0 (mean = 10) g C m$^{-2}$ d$^{-1}$,[23,24] with yearly estimates ranging from 1,800-4,200 g C m$^{-2}$ yr$^{-1}$. A maximum of 11,680 g C m$^{-2}$ yr$^{-1}$ gross primary productivity for a fringing reef in Hawaii has been reported.[25] This figure rivals the productivity of systems under intense agricultural management.

The effects of solar ultraviolet radiation on biomass production and community development were investigated in algal turf assemblages in the laboratory (case study 1) and in field diatom assemblages of the Caribbean (case study 2) and Mediterranean Seas (case study 3). The waters of the two sites are among the Earth's clearest. UV-B penetration in the East Mediterranean, measured as radiation at 310 nm, is reduced by 14% per meter depth,[26] while for UV-A (at 375 nm) the corresponding value is 5%. The band of the solar spectrum with the lowest attenuation rate is blue (3% per meter depth, measured at 465 nm), while up to 90% of red light is absorbed by the first meter of water. Considering that the absorption rate of UV-B radiation in other areas of the

world can be as high as 90% per meter or more, UV penetration in the Caribbean and the East Mediterranean stands out as very effective.

## CASE STUDY 1—LABORATORY MESOCOSM EXPERIMENTS

Laboratory experiments were carried out at the 3,000 gallon coral reef mesocosm at the Natural History Museum of the Smithsonian Institution in Washington D.C. The mesocosm simulates the natural conditions of a typical Caribbean coral reef, thus providing a suitable environment for housing more than 300 species of marine organisms. The experiments were conducted in the mesocosm's water purification system, the algal turf scrubber (ATS) array. An ATS (Fig. 13.1) consists of a tray, a wave generator, and a plastic substrate (polypropylene screening) for algal spore attachment and growth. Each ATS was divided into two equal trays using an opaque partition positioned parallel to the flow. One half of the tray was used as a UV treatment, while the other served as a control. Assemblages were exposed to the following four treatments: high UV-B (exposure to PAR+UV-A+ UV-B increased by 20% from natural incident levels), low UV-B (PAR+UV-A+ natural levels of UV-B), UV-A (PAR+UV-A) and controls (PAR). After week 6, the UV treatments were reversed over the developing assemblages to investigate: (a) the recovery of assemblages after the elimination of the UV-exposed stress; and (b) the response of already established assemblages to increased UV-B.

### RESULTS

**Biomass**

The biomass productivity data of the ATS units are plotted in Figures 13.2a-c. A pairwise means analysis of laboratory biomass data (Tukey's studentized range test) indicated that the productivity of the assemblage exposed to high UV-B was significantly lower than that of all other

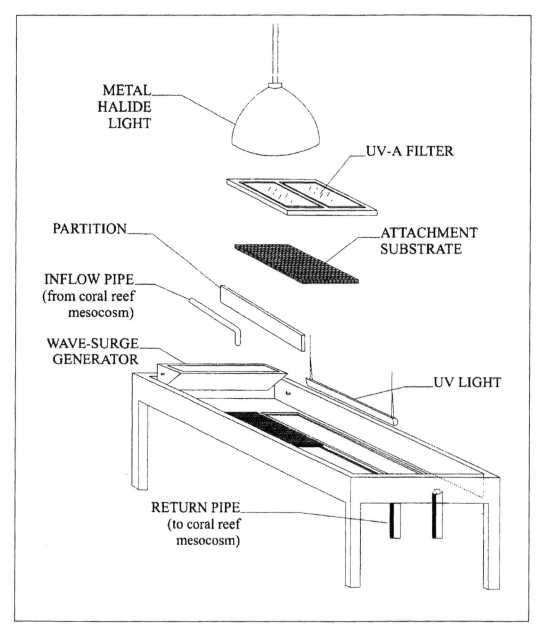

Fig. 13.1. Algal turf scrubber (ATS) used for mesocosm experiments. One half of each ATS was used for exposure to UV radiation, while the other half served as a control (exposure to PAR only).

treatments and controls. More specifically, on the second week of growth, the productivity of the assemblage exposed to high UV-B irradiance was 3.66 g/m²/day, a value lower than one-third of the mean productivity of the controls (12.12 g/m²/day). Mean weekly biomass production of the high UV-B treatment was 55.1% of that of the corresponding control, and significantly different from each one of the controls and the UV-A treatment. Following treatment reversal, the difference in the productivity of the two assemblages gradually decreased in weeks 7 and 8 and was reversed by the end of the experiment (Fig. 13.2a, right). However, the productivity of the two assemblages was not significantly different.

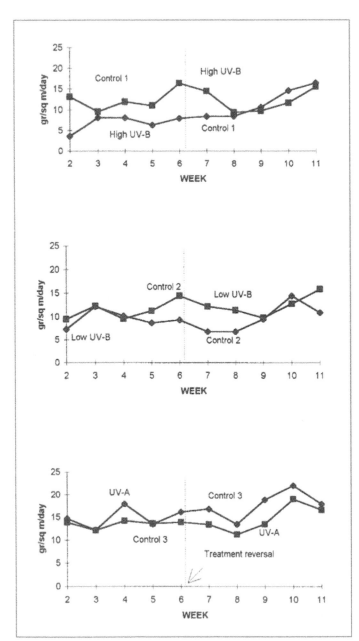

Fig. 13.2. Biomass production coral reef mesocosm assemblages. Primary productivity was lowest under the high UV-B treatment. After the reversal of treatments (week 6, dashed line), there were no significant productivity differences between any two treatments.

A similar trend was observed in the assemblage exposed to low UV-B irradiance. Before treatment reversal (Fig. 13.2b, left), this treatment had a lower biomass production than that of the control. However, the reduction in biomass caused by exposure to low UV-B was less dramatic than that of the high UV-B treatment. Mean weekly productivity of the low UV-B treatment was intermediate between the productivity of the high UV-B treatment and that of the controls, although not significantly different (p>0.05) from either of the two. In the first two weeks following treatment reversal (Fig. 13.2b, right), the productivity of both assemblages declined considerably. The two-week lag in biomass production reversal observed in the high UV-B treated assemblage after treatment reversal is observed here as well.

Finally, the productivity of the UV-A assemblage was not significantly different from that of the controls in either part of the experiment (Fig. 13.2c).

## Community Structure

The initial stages of community development under all treatments were dominated by the green algae *Enteromorpha prolifera* and *Cladophora fuliginosa* (Fig. 13.3a-f). In the high UV-B treatment (Fig. 13.3a, left), the cyanobacterium *Schizothrix calcicola* became increasingly abundant and eventually dominated the assemblage 2 weeks before treatment reversal. The brown alga *Ectocarpus rhodochondroides*, present only in minute quantities in weeks 5 and 6, drastically increased in abundance immediately after cessation of exposure to UV-B (Fig. 13.3a, right). The diatom *Licmophora* sp. was the only other species present in significant quantities after the reversal of treatments.

In the low UV-B treatment, *E. rhodochondroides* appeared in week 3, and became dominant two weeks before treatment reversal (Fig. 13.3c, left). *Licmophora* sp. was present in the second half of the experiment (Fig. 13.3c, right), while *C. fuliginosa* persisted in significant quantities throughout the duration of the experiment.

In the UV-A treatment (Fig. 13.3e) and the controls (Figs. 13.3b, 13.3d, 13.3f), *E. rhodochondroides* remained the dominant species throughout the experiment. Treatment reversal

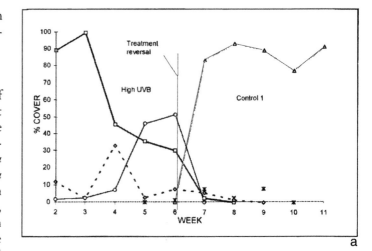

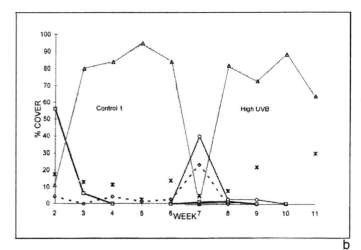

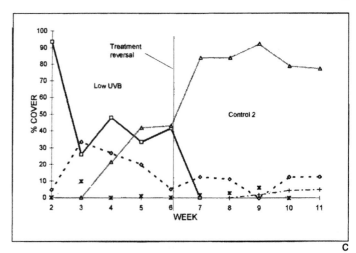

*Fig. 13.3. Abundance of dominant algal species in mesocosm experiment. Each graph represents the development of a single algal assemblage. After week 6 (broken vertical line) the light regime was reversed over the growing assemblages as indicated in the figures.*

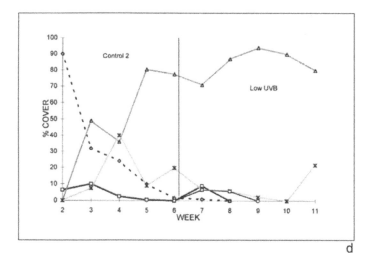

d

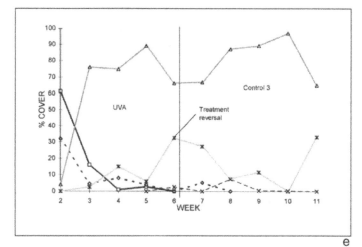

e

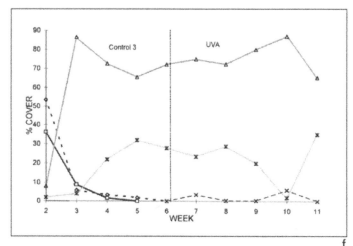

f

—□— *Enteromorpha prolifera*

- -◇- - *Cladophora fuliginosa*

—△— *Ectocarpus rhodochondroides*

- -✖-- *Licmophora sp.*

—○— *Schizothrix calcicola*

- -✖- - *Polysiphonia sp*

- -+- - *Oscillatoria submembranacea*

did not have any dramatic effects on these assemblages, except for some minor differences in the abundance of subdominant species (Figs. 13.3b, 13.3d, 13.3f right).

## CASE STUDY 2— TROPICAL ASSEMBLAGES (CARIBBEAN)

A second experiment was conducted in the lagoon between the coral reef and the island of Grand Turk, Turks and Caicos Islands, lat. 21°2'N, long. 71°3'W. Each experimental unit consisted of a substrate for the attachment and growth of algae (5 mm mesh polypropylene screen) fixed onto a 1/2-inch PVC frame (Fig. 13.4). The units were suspended from PVC rafts at a depth of 1 meter below water surface (Fig. 13.5). Three UV treatments were performed using a combination of Mylar and Plexiglas filters (Fig. 13.4):

1. PAR
2. PAR+ UV-A
3. PAR + UV-A + UV-B

The filters were cleaned regularly every two days to prevent alteration of the transmittance properties due to fouling. Transmittance was periodically checked for replacement of defective filters.

Algal biomass was harvested every seven days by removing the units from the rafts. The units were returned to their original position in the water immediately after harvesting. The collected algal biomass was strained free of salt water, and dried to constant weight at 80°C. To analyze the structure of the diatom assemblage, a part of the harvested biomass was processed for microscope observation using organic matter digestion with $H_2O_2$.

During weeks 2 and 4 (Fig. 13.6) the PAR and PAR+UV-A treatments yielded significantly higher amounts of algae than the PAR+UV-A+UV-B treatment (p<0.05). The productivity of the PAR treatment was more than 40% higher than that of the PAR+UV-A+UV-B treatment until week 4. In the following weeks, the productivity means of the three treatments were not significantly different.

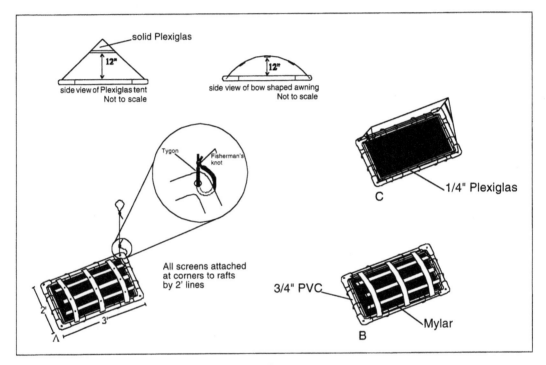

Fig. 13.4. Field enclosures used in case studies 2 and 3.

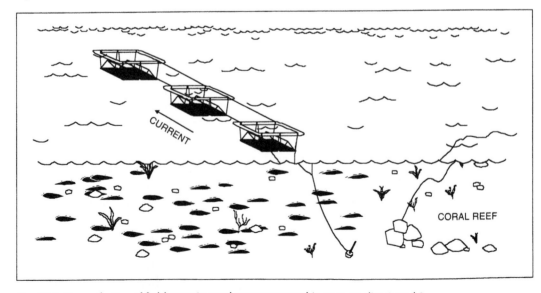

Fig. 13.5. General view of field experimental apparatus used in case studies 2 and 3.

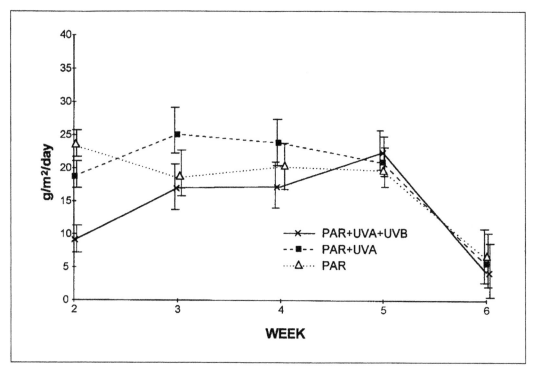

Fig. 13.6. Biomass production of tropical algal assemblages. During the initial stages of succession, the productivity of the assemblages not exposed to UV-B, was 40% higher than under unfiltered solar radiation. After the first four weeks, however, the mean values were not significantly different.

## CASE STUDY 3—
## MEDITERRANEAN ASSEMBLAGES
## (KORINTHOS, GREECE)

The interaction of solar UV with depth and primary succession was studied under the same UV treatments of case study 2, in Saronikos Gulf, Greece (37°58'N, 23°0'E).

Primary succession stages are very clearly reflected in the formation of three distinct assemblage clusters corresponding to the 3 sampling dates (Fig. 13.7). A few days after the placement of the experimental apparatus in the water, a thin layer of debris and bacteria started to develop over the ceramic tiles. One week later, the appearance of a golden-brown color in that layer marked the establishment of pennate diatom species, the early photosynthetic colonizers of newly available substrate in aquatic environments. The occurrence of nonpennate forms was scarce and most likely due to adventitious sedimentation of phytoplankton on the tiles. Pennate

diatoms (Table 13.1) constituted the main taxonomic component of the developing periphytic communities throughout the end of this study. A total of 159 diatom taxa were encountered during the course of the experiment under the 9 treatment combinations (3 depth, 3 radiation regimes).

Exposure to UV-B caused shifts in the species composition of diatom assemblages developing on ceramic tiles during the early stages of primary succession (Figs. 13.8a-b, 13.9a-b). The clear separation of the three UV treatments indicated a discernible UV effect on community structure. A close relation of the replicate assemblage pairs of all treatment combinations was observed. The separation of the different communities became less clear as succession progressed (Figs. 13.8c, 13.9c).

The diatoms *Mastogloia crucicula, N. constricta, N. marginulata* and *N. punctata* were sensitive to both UV-A and UV-B radiation, while *Amphora veneta, N. longissima, Opephora olsenii, Synedra baccilaris* and

**Table 13.1. Common and dominant diatoms, and species indicative of succession stage, UV and depth**

| Common | Dominant |
|---|---|
| Achnanthes brevipes | Cocconeis placentula |
| Amphora coffeaeformis | Mastogloia badjikiana |
| A. inariensis | M. decussata |
| A. ostrearia | M. labuensis |
| A. ovalis | M. lanceolata |
| A. robusta | M. smithi |
| A. ventricosa | Nitzschia constricta |
| Bacillaria paxilifer | Rhopalodia constricta |
| Cocconeis scutellum | Synedra bacillaris |
| Mastogloia corsicana |  |
| M. decipiens | **Succession** |
| M. elliptica | Mastogloia badjikiana |
| M. erythraea | M. labuensis |
| M. paradoxa | Nitzschia lanceolata |
| M. pisciculus | Synedra laevigata |
| M. pussila |  |
| M. punctifera | **UV** |
| Navicula carinifera | Amphora inariensis |
| Nitzschia dissipata | Cocconeis scutellum |
| N. panduriformis | Mastogloia corsicana |
| N. peridistincta | M. decipiens |
| O. olsenii | M. lanceolata |
| Rhaphoneis amphiceros | M. pisciculus |
| R. surirella | Rhopalodia constricta |
| Rhopalodia acuminata |  |
| R. gibberula | **Depth** |
| S. ovata | Amphora robusta |
|  | Cocconeis fluminensis |
|  | Mastogloia erythraea |
|  | M. ovalis |
|  | Nitzschia bilobata |
|  | Rhaphoneis amphiceros |

Reprinted with permission from Limnology & Oceanography (submitted).

S. robusta were excluded by exposure to UV-B, but not UV-A. *Amphora robusta, Cocconeis fluminensis, Mastogloia erythraea, M. ovalis, Nitzschia bilobata,* and *Rhaphoneis amphiceros* were characteristic of different depth assemblages, while *Mastogloia badjikiana, M. labuensis, Nitzschia lanceolata* and *Synedra laevigata* were present during different stages of succession.

## DISCUSSION

The biomass results of the Caribbean field experiment parallel those of the laboratory. The productivity of the UV-B exposed assemblages in weeks 2, 3, and 4 is substantially higher than that of assemblages protected from UV-B. In addition, exclusion of UV-A radiation did not result in significantly different biomass production. However, the productivity of all treatments in the field did not differ significantly from one another after the fourth week, while in the laboratory, productivity differences between the high UV-B treatment and the controls persisted throughout the first part of the experiment (i.e. for 6 weeks). This discrepancy might be due to two facts: (a) the UV/PAR ratio was higher in the laboratory than in the field; and (b) the field communities were highly diversified (over 150 species of filamentous algae and diatoms), while in the laboratory the total number of algal species was lower than 20.

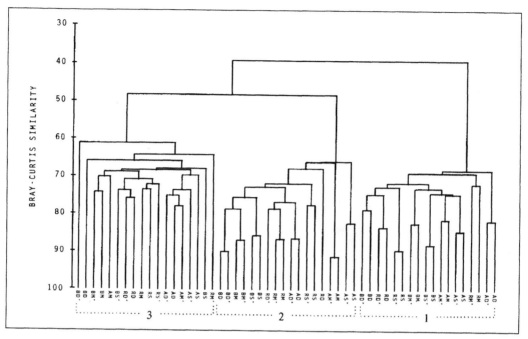

Fig. 13.7. Clustering of Mediterranean assemblages. Reprinted with permission from Limnology & Oceanography (submitted).

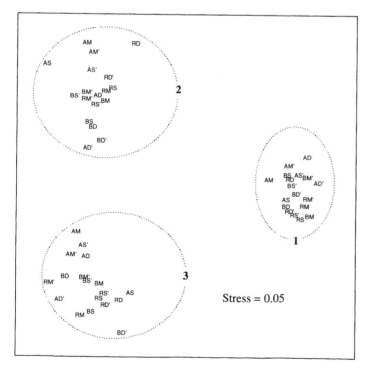

Fig. 13.8. Clustering of diatom assemblages. Primary succession stages are very clearly reflected in the formation of three distinct assemblage clusters corresponding to the 3 sampling dates: 15/9/94 (1); 29/9/94 (2); 13/10/94 (3). The separation of different UV treatments is particularly clear on the second sampling date. A, PAR+UV-A; B, PAR+UV-A+UV-B; R, PAR; S, 0.5 m; M, 1.0 m; D, 1.5 m; AM, AM', replicates. Reprinted with permission from Limnology & Oceanography (submitted).

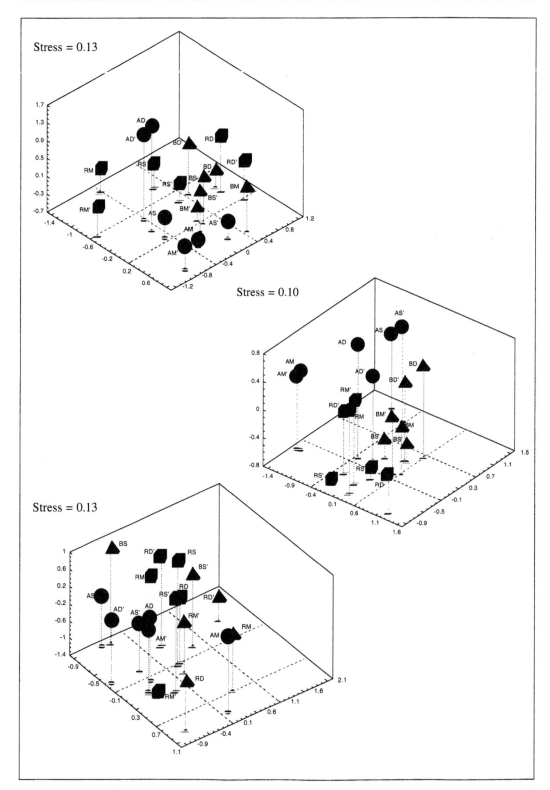

*Fig. 13.9. MDS Ordination: The assemblages grown under different treatments form distinctly separated groups. Cubes, PAR; spheres, PAR+UV-A; pyramids, PAR+UV-A+UV-B. Reprinted with permission from Limnology & Oceanography (submitted).*

In the Mediterranean, the same productivity trends were observed, but the differences were less pronounced than in the Caribbean. This, however, might have been an artifact of the smaller sample area used in the Mediterranean experiment. Exposure to solar UV-B as well as UV-A caused shifts in the species composition of the diatom assemblage during the first few weeks of development. However, as the establishment of the biological community progressed, the differences in community structure between the different light treatments diminished.

In trying to predict the biological effects of increased solar UV irradiance, elaborate systems have been constructed using lamp banks, $O_3$ filters, Plexiglas and/or plastic foil filters, etc. Such systems, including culture chambers, greenhouses and field enclosures, can achieve accurate replication of the desirable *mean* light intensity and spectral composition. However, the inadequate simulation of natural UV *fluctuations* due to cloud cover has been suggested as a possible source of error,[27] leading to overestimates of the predicted damage due to UV. During the 5-month monitoring course of this study, daily doses varied by more than an order of magnitude for the three photosynthetically important bands: 521-12,006 kJ/m² for PAR; 52-1,239 kJ/m² for UV-A; and 0.66-22.5 kJ/m² for UV-B.

The ratio of the different bands of the light spectrum may also be an important factor in determining biological responses: several light-dependent mechanisms such as photoreactivation, photorepair and photoprotection occur simultaneously, but are driven by different wavelengths of the light spectrum, including UV-A and PAR. In field studies, such mechanisms may annihilate or even reverse the impact of UV-B radiation on plant growth and development.[28-30]

## CONCLUSIONS

In conclusion, solar UV-B radiation inhibits community productivity and affects species composition during the early stages of primary succession. However, the fact that such differences do not persist at later successional stages, suggests the presence of adjustment mechanisms against the stress posed by increased solar ultraviolet radiation. The considerable degree of resilience inherent in natural communities should not be ignored when the results of short-term studies carried out in artificial environments are extrapolated to a global scale for predicting climatic changes.

For meaningful data interpretation in enhanced UV-B experiments, in addition to continuous monitoring of irradiance and spectral composition, solar radiation fluctuations should also be simulated and/or accounted for. It is also necessary to examine the interaction of the different radiation bands (UV-A, UV-B and PAR), since photodamage and photorepair mechanisms occur simultaneously.

## Acknowledgment

Supported by grant EV5V-CT94-0425 of EC's DG-XII ('Environment' Programme).

## References

1. Caldwell MM. Plant life and ultraviolet radiation: Some perspective in the history of the Earth's UV climate. Bioscience 1979; 29:520-525.

2. Bener P. Spectral intensity of natural ultraviolet radiation and its dependence on various parameters. In: Urbach F, ed. The Biologic Effects of Ultraviolet Radiation. London: Pergamon Press, 1969; 351-358.

3. Häder D-P. Effects of enhanced solar ultraviolet radiation on aquatic ecosystems. In: Tevini M, ed. UV-B Radiation and Ozone Depletion. Effects on Humans, Animals, Plants, Microorganisms and Materials. Boca Raton, Ann Arbor, London, Tokyo: Lewis Publ, 1993:155-192.

4. Iwanzik W, Tevini M, Dohnt G et al. Action of UV-B radiation on photosynthetic primary reactions in spinach chloroplasts. Physiol Plant 1983; 58:401-407.

5. Murphy TM. Membranes as targets of ultraviolet radiation. Physiol Plant 1983; 58:381-388.

6. Elkind MM, Han A. DNA single-strand lesions due to "sunlight" and UV light: a

comparison of their induction in Chinese hamster and human cells, and their fate in Chinese hamster cells. Photochem Photobiol 1978: 27:717-724.

7. Halldal P. Ultraviolet action spectra of photosynthesis and photosynthetic inhibition in a green and a red alga. Physiol Plant 1964; 17:414-421.

8. Cline MG, Salisbury FB. Effects of ultraviolet radiation on the leaves of higher plants. Radiation Botany 1966; 6:151-163.

9. Nachtwey DS. Linking 254 nm UV effects to UV-B effects via 280 nm equivalents. In: Nachtwey DS, Caldwell MM, Biggs RH, eds. Impacts of Climatic Change on the Biosphere, Part I: Ultraviolet Radiation Effects (CIAP monog. 5). Springfield, VA: US Dept Transp, 1975:3-54.

10. Robberecht R, Caldwell MM. Leaf epidermal transmittance of ultraviolet radiation and its implications for plant sensitivity to ultraviolet-radiation induced injury. Oecologia 1978; 32:277-287.

11. Fogg GE, Boalch GT. Extracellular products in pure cultures of a brown alga. Nature 1958; 181:789-790.

12. Craigie JS, McLaughlan J. Excretion of coloured ultraviolet-absorbing substances by marine algae. Can J Bot 1964; 42:23-33.

13. Tsujino I, Saito T. Studies on the compounds specific for each group of marine alga I. Presence of characteristic ultraviolet absorbing material in Rhodophyceae. Bul Facul Fish, Hokkaido University 1961; 12:49-58.

14. Nelson G, Hairston Jr. Photoprotection by carotenoid pigments in the copepod *Diaptomus nevadensis*. Proc Nat Acad Sci USA 1976; 73:971-974.

15. McLeod GC, McLaughlan J. The sensitivity of several algae to ultraviolet radiation of 2537 A. Physiol Plant 1959; 12:306-309.

16. Garcia-Pichel F, Sherry ND, Castenholz RW. Evidence for an ultraviolet sunscreen role of the extracellular pigment scytonemin in the cyanobacterium *Chlorogloeopsis* species. Photochem Photobiol 1992; 56:17-23.

17. Garcia-Pichel F, Castenholz RW. Occurrence of UV-absorbing, mycosporine-like compounds among cyanobacterial isolates and an estimate of their screening capacity. Applied Environ Microbio 1993; 59: 163-169.

18. Jokiel PL, York RH. Importance of ultraviolet radiation in photoinhibition in microalgal growth. Limnol Oceanogr 1984; 29:192-199.

19. Döhler G. Effect of UV-B radiation on biomass production, pigmentation and protein content of marine diatoms. Z. Naturforsch 1984; 39(c):634-638.

20. Hirosawa T, Miyachi S. Inactivation of Hill reaction by long-wavelength radiation (UV-A) and its photoreactivation by visible light in the cyanobacterium, *Anacystis nidulans*. Arch Microbiol 1983; 135:98-102.

21. Bothwell ML, Sherbot DMJ, Pollock CM. Ecosystem response to solar ultraviolet radiation: Influence of trophic-level interactions. Science 1994; 265:97-100.

22. LaPointe B, Tenore K. Experimental outdoor studies with *Ulva fasciata*. I. Interaction of light and nitrogen on nutrient uptake, growth and biochemical composition. J Exp Mar Biol Ecol 1981; 53:135-152.

23. Lewis JB. Processes of organic production on coral reefs. Biol Rev Cambr Philosoph Soc 1977; 52:305-347.

24. Sournia A. Analyse et bilan de la production primaire dans les recif coralliens. Ann Inst Oceanogr (Paris) 1977; 53:47-74.

25. Gordon MS, Kelly HM. Primary production of an Hawaiian coral reef: a critique of flow respirometry in turbulent waters. Ecology 1962; 43:473-480.

26. Jerlov NG. Ultra-violet radiation in the sea. Nature 1950; 166:111-112.

27. Santas R, Koussoulaki A, Häder D-P. In assessing biological UV-B effects, natural fluctuations of solar radiation should be taken into account. Plant Ecology 1997; 128:1-5.

28. Sinclair TR, N'Diaye O, Biggs RH. Growth and yield of field-grown soybean in response to enhanced exposure to ultraviolet-B radiation. J Environ Qual 1990; 19:478-481.

29. Sullivan JH, Teramura AH, Ziska LH. Variation in UV-B sensitivity in plants from a 3,000-m elevational gradient in Hawaii. Am J Bot 1992; 79:737-743.

30. Miller JE, Booker FL, Fiscus EL et al. Ultraviolet-B radiation and ozone effects on growth, yield, and photosynthesis of soybean. J Environ Qual 1994; 23:83-91.

# CHAPTER 14

# EFFECTS OF UV-B ON CILIATES

Roberto Marangoni, Beatrice Martini and Giuliano Colombetti

## INTRODUCTION

The biological effects of a possible UV-increased irradiation on our planet have been thoroughly investigated over the years, and some of the fundamental processes involved have begun to be understood.

A good amount of the current knowledge in this field derives from studies performed on single cell systems, where the complex interactions between cells and organs typical of higher organisms are not present. It is, therefore, easier to understand the basic molecular processes affected by UV radiation in microorganisms.

A good model system may be represented by ciliates, which are unicellular organisms but at the same time possess a complex morphology and physiology and are, moreover, extremely sensitive to environmental signals; studies performed since the last century have shown that ciliates can perceive a wide variety of stimuli such as: mechanical, optical, thermal, chemical and gravitational stimuli, and react to them by altering their motile behavior. These perception functions are all integrated in a single cell and all participate in controlling the motile behavior; in fact, the final step of a stimulus-induced behavior consists of an alteration of the membrane potential and a modulation of the activity of the motile apparatus. The presence of all these control circuits in the same ciliate cell has lead one author to speak of ciliates as "walking neurons."

It has been known for a long time that the exposure to artificial UV irradiation at high doses can kill the cells; at lower doses, it can destroy (or seriously damage) many of the control circuits and the ultrastructural elements of the cell itself.

This suggests that also environmental UV, the amount of which is increasing as a consequence of the significant depletion of the stratospheric ozone layer,[1-4] can alter the ciliate physiology and lead to decreased chances of survival. Since ciliates belong to early stages of the food chain, it is very important, from an ecological point of view, to understand how they are affected by UV irradiation, and UV-B (280-315 nm) in particular, and if they possess repair mechanisms.

*The Effects of Ozone Depletion on Aquatic Ecosystems*, edited by Donat-P. Häder.
© 1997 R.G. Landes Company.

Many authors have worked on this subject, and have explored a wide range of possible effects of UV exposure (from cell elongation to alterations in the reproduction cycle), often using different investigation methods; in the first part of this chapter we try to give the reader a schematic summary of these works, while in the second part we describe in more detail the most recent studies on the effects of UV-B on motility and photomotility of three diverse Heterotrichida: *Stentor coeruleus, Blepharisma japonicum* and *Fabrea salina*.

## EARLY STUDIES AND GENERAL CONSIDERATIONS

The first studies were mainly aimed at determining the relationship between UV exposure dose and cell survival, and tried to establish the minimum dose sufficient to severely damage the cell.[5-7] Most of these experiments were performed by

determining the UV dose (often using UV-C wavelengths, 200-280 nm) sufficient to determine an immobilization of the cells or a cytolysis: sometimes the minimum dose sufficient to determine an unusual motile pattern, such as rotation, was assumed as a quantitative assessment of cell damage. Together with these dosimetric assays, the role of environmental conditions, such as pH, salt composition of the medium, etc., has been investigated.

These early studies show (see Table 14.1) that low doses of UV irradiation cause an increase in cell speed, while, after a certain dose, the cell speed decreases, the motion pattern becomes anomalous (cells tend to rotate around their cell body) and eventually the cells stop and die.

The resistance against UV irradiation may greatly vary among ciliates: for example, *F. salina* shows a resistance about 10-fold greater than that of *B. undulans*,

**Table 14.1. Resistance to UV-irradiation as a function of environment and physiological state of different protozoa. The dose for 50 % immobilization of P. multimicronucleatum was set at 100 and the other values are relative to this (retyped from Giese AC).[5]**

| Protozoan | Medium or environment | 50% rotation | 50% immobilization | 50% cytolysis |
|---|---|---|---|---|
| Paramecium | Balanced inorganic salt solution | $61 \pm 6.4$ | $100 \pm 2.1$ | |
| | Culture medium | $84 \pm 2.5$ | $114 \pm 0$ | |
| | Distilled water | $72 \pm 5.6$ | $89 \pm 3.8$ | |
| | pH 6.0 | $75 \pm 0$ | $105 \pm 1.8$ | $265 \pm 12$ |
| | pH 7.0 | $65 \pm 0$ | $100 \pm 0$ | $230 \pm 10$ |
| | pH 8.0 | $50 \pm 0$ | $100 \pm 0$ | $205 \pm 10$ |
| | 16°C | $39 \pm 1.7$ | $82 \pm 3.7$ | |
| | 19°C | $51 \pm 4.5$ | $94 \pm 1$ | |
| | 26°C | $79 \pm 2.1$ | $100 \pm 1$ | |
| | 4-day-old culture | $62 \pm 4.7$ | $100 \pm 5.8$ | $220 \pm 12.1$ |
| | 11-day-old culture | $48 \pm 6.7$ | $81 \pm 10.0$ | $172 \pm 3.1$ |
| Blepharisma | pH 6.0 | | $82 \pm 2.5$ | $156 \pm 1.8$ |
| | pH 7.0 | | $58.5 \pm 7.5$ | $101 \pm 5.3$ |
| | pH 8.0 | | $54.5 \pm 7.6$ | $121 \pm 5.2$ |
| Tetrahymena | 1-day-old culture in yeast extract | | $146 \pm 4.4$ | |
| | 8-day-old culture in yeast extract | | $114 \pm 7.0$ | |
| | 3-day-old lettuce culture | | $48.5 \pm 4.2$ | |

which belongs to the same order, Heterotrichida (Table 14.2).

The interaction between UV irradiation and environmental conditions is very complex to understand: it appears, in fact, from Table 14.1, that at certain pH levels or salt concentrations, the cells tolerance to UV increases, even if the possible molecular mechanisms of such an effect remain unknown. They are probably connected with the general "status" of the cell: when a cell is in a good condition, it is less sensitive to UV. This consideration is supported also by observations on the nutritional status and the culture age, also reported in Table 14.1.

Some authors have investigated the relationship between UV-resistance and the presence of pigments in the cell cytoplasm: in some cases (a white mutant of *Blepharisma incertus*, for example) the pigment deprivation produces an increase in UV-vulnerability.[8] These results seem to be in contrast with those obtained by Colombetti et al (unpublished observations) on the apparently colorless strain of *Fabrea salina*, which is more resistant than pigmented protozoa, such as *B. japonicum*, to the same radiation doses. These findings are derived from measures of different parameters, and

are difficult to compare to those of Giese;[5,8] but they indicate, however, that there is no direct relationship between pigment content and UV-resistance.

Structural changes in the cell after UV exposure have been described: cells can elongate, vacuolate or the cell wall can break causing a cytolysis. Particular attention has been given to the so-called retardation effect: it has been noticed, in fact, that cells exposed to UV (and UV-B in particular) strongly reduce the frequency of division.[9] The delay in reproduction has been measured as a function of the UV dose received and of some other environmental factors (e.g. composition of the culture medium, starving period) and also found in this case that all these variables may interact with each other.

In order to try to identify the molecular target of UV irradiation in the ciliates, the action spectrum for the retardation response has been determined in some species; the results are, in certain cases, difficult to interpret: the action spectrum at low doses resembles the absorption spectrum of the cytoplasmic proteins, whereas the action spectrum at large doses is complex, with a contribution surely due to the DNA (Fig. 14.1).[10]

**Table 14.2. Relative resistance of different protozoa to UV-irradiation. The dose for 50% immobilization of P. multimicronucleatum was set at 100 and the other values are relative to this (retyped from Giese AC).[5]**

| Protozoan | Size (μm) | 50% rotation | 50% immobilization |
|---|---|---|---|
| *Tetrahymena glaucomiformia* | 25 x 50 | | 146 ± 4.4 |
| *Colpidium colpoda* | 55 x 88 | | 132 ± 6.8 |
| *Stylonychia curvata* | 48 x 98 | 34 ± 6.4 | 80 ± 4.7 |
| *Paramecium bursaria* | 43 x 105 | 87 ± 6.3 | 103 ± 37.0 |
| *Paramecium aurelia* | 49 x 120 | 92 ± 0.3 | 119 ± 5.7 |
| *Paramecium multimicronucleatum* | | | |
| N. 1 | 51 x 175 | 76 ± 6.9 | 100 ± 4.2 |
| N. 2 | 64 x 213 | 73 ± 22.0 | 116 ± 19 |
| *Paramecium caudatum* | 70 x 173 | 58 ± 4.4 | 101 ± 3.4 |
| *Blepharisma undulans* | 52 x 175 | 43 ± 3.2 | 59 ± 6.1 |
| *Spirostomum ambiguum* | 36 x 369 | 32 ± 3.3 | |
| *Bursaria truncatella* | 200 x 337 | 112 ± 6.3 | 147 ± 14.1 |
| *Fabrea salina* | 169 x 225 | 555 ± 196 | 896 ± 126 |

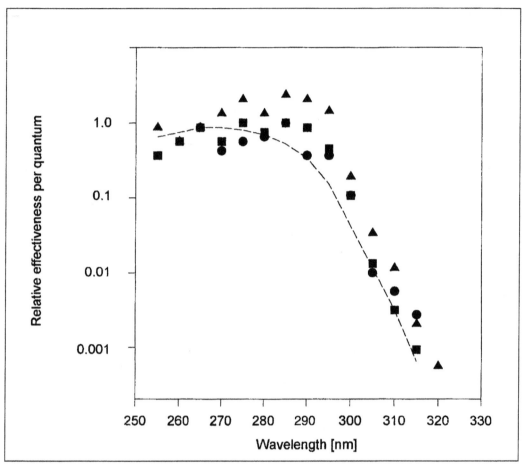

Fig. 14.1. Action spectra of UV-induced photokilling of Tetrahymena as a function of different postirradiation treatments: squares: no treatment, circles: 20 min photoreactivation, triangles: suspension medium containing 0.02% caffeine. The dotted line represents an "average DNA" action spectrum (see ref. 16).

The relationship between UV damage and nucleic acids has been pointed out by different authors; in general it is suggested that the molecular phenomena which take place after UV exposure are complex, and involve both proteins and nucleic acids.[10] Moreover, there are evidences demonstrating that after UV exposure there may be some dark reactions, which, in certain cases (for example in *Didinium nasutum*), may amplify the damage produced by UV itself.[11] In fact, a pulsed UV stimulation, at the same dose as a continuous one, is much more efficient in inducing damages. These dark reactions, being independent of light, are of a thermochemical nature and are therefore strongly dependent on temperature.[9-11]

Probably, although not yet certain, these dark reactions are correlated with the repair mechanisms, and, in particular the photo-repair mechanisms, which operate only in the presence of visible light.[12]

These findings suggest that UV can affect several different physiological functions; this may happen in a somewhat unspecific way, since radiation can affect both proteins and nucleic acids; some researchers, however, have tried to investigate the existence of specific UV-receptors in the ciliates.[13-16] In particular, they tried to characterize the UV sensitivity of some organisms (including little crustacea), in order to establish whether there is a photoreceptor absorbing in the UV region.

These experiments lead to positive results: many microorganisms show a specific UV sensitivity, and can perform an avoiding reaction when exposed to UV rays; moreover, the strength of this reaction seems to be inversely proportional to the relative resistance against UV radiations, i.e. the more resistant the organism, the less evident the reaction (Fig. 14.2).[13] Recent work (Lenci et al, private communication) seems to indicate that also *B. japonicum* has a specific UV-receptor.

## RECENT STUDIES
## ON THE EFFECTS OF UV-B

The effects of UV-B radiation have been studied in detail in three species of ciliated protozoa belonging to the same order Heterotrichida: the colored *Stentor coeruleus* and *Blepharisma japonicum* and the apparently colorless *Fabrea salina*.[17-19] Motile and photomotile behavior have been investigated before and after UV exposure, with the aim of finding possible intracellular targets of UV radiation. Particular attention has been devoted to the role of the intracellular pigments in the sensitivity to UV light.

*B. japonicum* and *S. coeruleus* possess similar pigment granules, located beneath the cell pellicle, which contain two forms of hypericin-like pigments, respectively called blepharismin and stentorin.[20-21] The *F. salina* strain studied in these experiments also contains a hypericin-like pigment, as shown by fluorescence measurements,[22-24] and the pigment granules are very similar to those found in *B. japonicum*, though *Blepharisma* contains a much larger number of them.[24-25] This may explain why *B. japonicum* appears red-pigmented, while *F. salina* appears colorless. In *F. salina* the action spectrum of phototactic reaction[26] together with immunocytochemical studies[27] indicate the presence of another pigment, a rhodopsin-like molecule.

## EFFECT OF UV-B RADIATION
## ON MOTILITY

When *S. coeruleus* cells are exposed to unfiltered solar radiation[17] (the exposure was carried out at 38° northern latitude in Portugal on clear days between 10 and 23 August 1989 starting at 2:00 p.m. local time with a solar radiation fluence rate of about 1140 W/m$^2$) motility is strongly inhibited: after only 4 min exposure (UV-B radiation of about 4.3 W/m$^2$ and UV-A of about 48 W/m$^2$) the percentage of motile cells is reduced to almost zero (Fig. 14.3) and there is a marked decrease in velocity of still motile cells (Fig. 14.4). The inhibition of motility is irreversible: in fact, even after 24 h in darkness or under weak white light no recovery is observed.

When the UV-B component of radiation is removed by inserting a WG 320 cut-off filter (Schott and Gen, Mainz, Germany), the organisms tolerate longer exposures (Figs. 14.3 and 14.4). The insertion of a GG 400 filter (Schott and Gen, Mainz, Germany), which removes all UV radiation, further extends the tolerated exposure time (Figs. 14.3 and 14.4). This indicates that the inhibition of motility is induced by both the UV-B and the UV-A components of solar radiation.

In *B. japonicum*, the exposure to UV radiation impairs cell motility too.[18] Artificial UV irradiation was carried out by means of three TL40W/12 Philips UV-B lamps; cell samples were covered with a cellulose acetate film in order to remove the UV-C component of radiation. Irradiation was performed without additional filter ("UV-B" irradiation) and with a Schott WG1 filter in order to eliminate the UV-B range ("visible" irradiation). In both conditions the total irradiance was kept about the same (~5 W/m$^2$) by means of neutral density filters. In the "UV-B" condition the spectral distribution was about 52% UV-B, 27% UV-A and 21% visible, with UV-B irradiance ~2.5 W/m$^2$ comparable to that in the natural environment, whereas UV-A and visible irradiance are about 30 and 400 times lower than the natural ones. In the "visible" irradiation the spectral distribution was 0% UV-B, 16% UV-A and 84% visible.

"UV-B" irradiation performed for 30 and 60 min caused a gradual rise in the

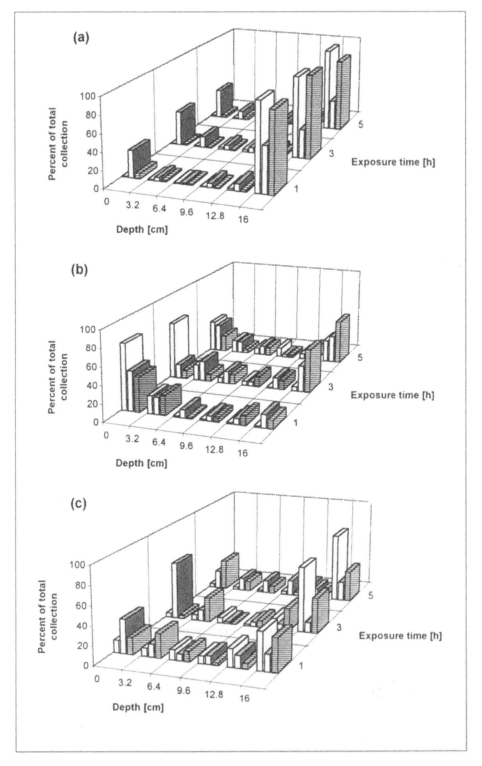

Fig. 14.2. Vertical distribution of the population of three protozoa (a)-Coleps, (b)-Paramecium, (c)-Euplotes, at three different exposure times (1, 3 and 5 h) at 2.5 SU/h. The light gray columns represent the vertical distribution of the protozoa exposed to UV-light; the white ones the dark test cells and the dark gray ones the light control. Redrawn from Barcelo JA et al, Photochem Photobiol 1979; 29:75-83.

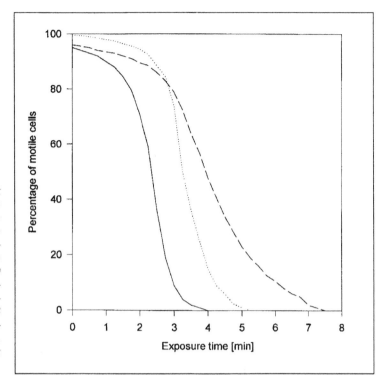

Fig. 14.3. Effect of different illuminations on the fraction of motile cells in a population of S. coeruleus exposed to unfiltered solar radiation (continuous line), solar radiation plus the UV-B cut-off WG320 filter (dotted line), GG400 filtered, UV-deprived, solar radiation (broken line). UV-B fluence rate: 4.3 W/m². Redrawn from Häder DP et al, Photochem Photobiol 1991; 54:423-428.

number of cells that swim following circular trajectories and induced a significant decrease in the cell population speed: from 118 µm/s in dark-kept control samples to 59 µm/s in 30 min "UV-B" irradiated cells and to 7 µm/s in 60 min "UV-B" irradiated samples (Fig. 14.5). However, recent results show that at short UV irradiation times (up to 10 min at 3 W/m²) there is an increase in cell speed (Lenci, private communication). On the contrary, "visible" irradiated cells did not show any movement pattern alteration and a significant reduction of the cell population speed is observable only after 60 min irradiation (Fig. 14.5).

In *F. salina* UV-B irradiation causes a meaningful increase in cell speed,[19] different from the inhibition of motility observed in *S. coeruleus* and *B. japonicum* cells.

Samples were irradiated with two TL40W/12 Philips fluorescent tubes emitting in the UV-B range and one TLD36W/54 Philips fluorescent tube emitting in the visible, with an irradiance about 8 W/m²

in the visible range and about 3 W/m² in the UV range (1 W/m² UV-A, 2 W/m² UV-B). All samples were covered with a cellulose acetate film, as mentioned for *B. japonicum*. Irradiation was performed covering the samples with: a quartz disk transparent to all wavelengths (treated samples), a Schott WG1 filter in order to eliminate the UV-B wavelengths (control 1 samples) and with a Schott GG400 filter transparent to visible radiation only (control 2 samples).

UV-B-induced speed increase is already detectable at short irradiation times (3 min 45 s) and reaches a plateau after one hour of exposure. Control samples, on the contrary, did not show meaningful alteration of their swimming speed (Fig. 14.6). UV-B-induced motion alteration seem to be limited to speed increase. In fact, other motion parameters, such as frequency of directional changes and frequency distribution of the angles of directional changes, did not show any dependence on UV-B irradiation (data not shown).

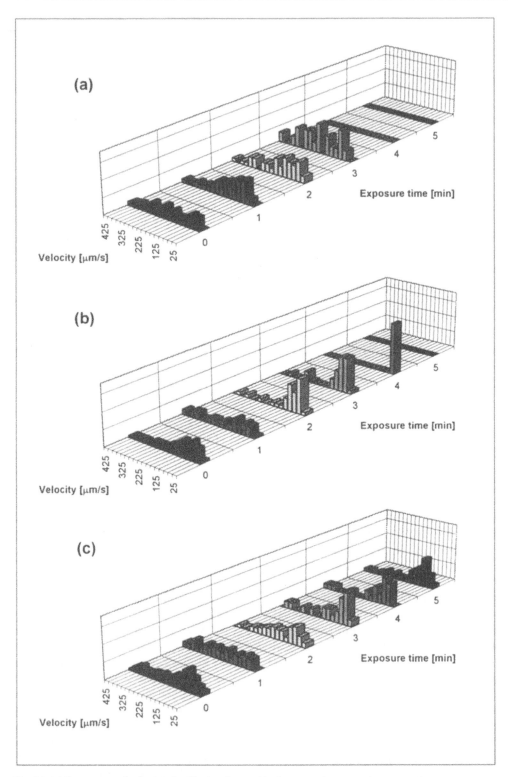

Fig. 14.4. Histograms of velocity distribution (in μm/s) of S. coeruleus as a function of exposure time to (a) unfiltered solar radiation, (b) solar radiation plus a WG 320 UV-B cut-off filter and (c) solar radiation plus a GG400 UV removing filter. UV-B radiation fluence rate : 4.3 W/m². Redrawn from Häder DP et el, Photochem Photobiol 1991; 54:423-428.

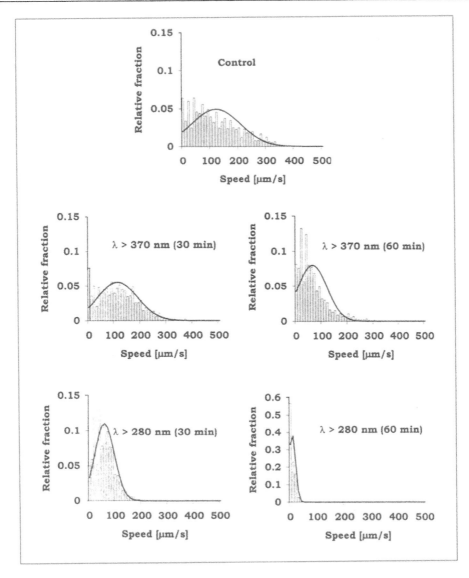

*Fig. 14.5. Histograms of the velocity distribution of samples of red B. japonicum cells under different irradiation conditions ("visible" = λ ≥ 370 nm; "UV-B" = λ ≥ 280 nm) and times (30 and 60 min). Redrawn from Sgarbossa A et al, J Photochem Photobiol B(Biol) 1995; 27:243-249.*

UV-B-induced speed increase is reversible in time, as shown in the 3D histogram reported in Figure 14.7, and the time required for cell speed recovery depends on irradiation time: for short irradiation times (3 min 45 s, 7 min 30 s, 15 min) the speed recovery occurs within about one hour; for longer irradiation times (30 min, 60 min) it occurs within two hours, and for an irradiation lasting 120 min, cell speed resumes its initial value only after three hours.

The cell speed distribution in an irradiated population of *F. salina* has been studied. The analysis of the fraction of cells swimming at a speed ≤0.1 mm/s (slow cells) shows a general reduction of this fraction immediately after the end of irradiation and a subsequent increase. Nevertheless, at the longest irradiation time tested (120 min) the fraction of slow cells as defined above increases immediately after the end of exposure (the relative fraction is 0.05 before irradiation and 0.08

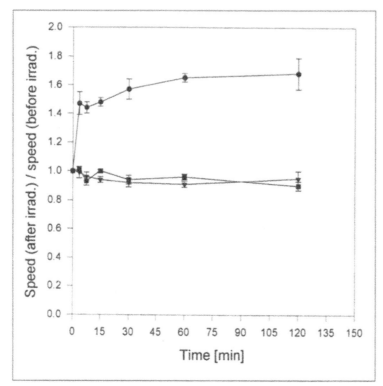

Fig. 14.6. Dose-effect curves of the speed of F. salina cells in: treated samples, irradiated with UV-B, UV-A and visible radiation (circles), control 1 samples, irradiated with UV-A and visible radiation (squares) and control 2 samples, irradiated with visible light only (triangles). The error bars represent the standard error of the mean.

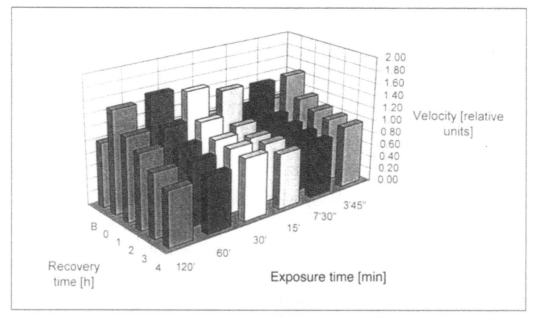

Fig. 14.7. 3D histogram of the speed of UV-irradiated samples of F. salina cells as a function of both the irradiation time and the time elapsed after the end of irradiation. The x-axis represents the time t (min) of UV-irradiation, the y-axis the time T (hours) elapsed after the end of irradiation and the z-axis reports the speed of a sample at (t,T) relative to that of the same sample before irradiation, taken arbitrarily as 1 and shown in the histogram at T = B.

immediately after irradiation); this is clearly visible in Figure. 14.8, where the histogram of cell speed distribution obtained before and after 120 min irradiation are reported. This result could indicate that at long irradiation times a part of the cell population is damaged and swims at a very low speed, whereas the majority of population increases its swimming speed. A possible explanation of this result is the existence of a subpopulation of cells more sensitive to UV-B radiation.

## EFFECTS OF UV-B IRRADIATION ON PHOTOMOTILITY

Photomotile responses of both *B. japonicum* and *F. salina* are impaired by UV-B irradiation.[18-19]

*B. japonicum* reacts to light stimuli exhibiting step-up photophobic responses (upon a sudden increase in light fluence rate the cells stop, turn and start swimming again in a new direction). Action spectroscopy experiments indicate that these responses are mediated by the endogenous pigment blepharismin.[28-30]

The UV-B irradiation causes a specific inhibition of step-up photophobic responses,[18] as shown in Figure 14.9: after 30 min of "UV-B" irradiation, as defined

in the previous section, 50% of still motile cells are unable to respond to photic stimulus and after 60 min irradiation the photoresponsiveness is totally suppressed. This inhibition is specifically determined by the UV-B radiation, in fact, visible irradiation does not affect in any way the photophobic response.

*F. salina* shows two kinds of responses to light stimuli: positive phototaxis (cell displacement towards the light source) and step-down photophobic reaction (upon a sudden decrease in light fluence rate the cells stop, turn and start swimming again in a new direction).[31-32] As previously mentioned, it is not clear at present if these responses are mediated by a rhodopsin-like pigment and/or by a hypericin-like pigment.[25]

Preliminary qualitative observations indicate a UV-induced damage to the photophobic step-down reaction (unpublished data).

*F. salina* phototactic reaction is seriously inhibited by UV-B irradiation.[19] Cell phototaxis decreases with UV-B irradiation time (at the irradiances mentioned above) and completely disappears after one hour (Fig. 14.10). Control samples, as defined in the previous section, show a slight

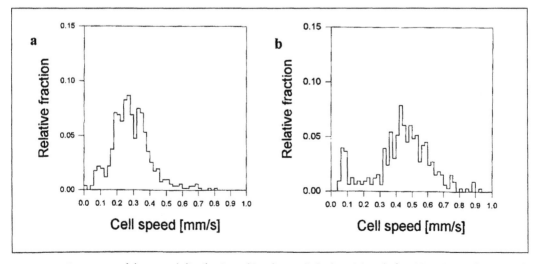

*Fig. 14.8. Histograms of the speed distribution of* F. salina *cells before (a) and after 120 min irradiation with UV-B, UV-A and visible light (b). After irradiation an increase of the fraction of cells swimming at a speed ≤0.1 mm/s (the relative fraction is 0.05 before irradiation and 0.08 after irradiation) is observed.*

decrease in their phototaxis after two hour irradiation, but this decrease is significantly smaller than that of treated samples. The inhibition of phototactic responsiveness of *F. salina* is irreversible, except for the shortest exposure time tested (Fig. 14.11).

From the dose-effect curves of phototactic sensitivity (the phototactic response of a sample as a function of increasing phototaxis stimulating light) for samples irradiated with UV-B, UV-A and visible light for 30 min and for control samples, (Fig. 14.12), it is clear that the cell response of irradiated samples saturates at a much lower level than that of the

controls. It is also clear that a higher fluence rate is required in order to obtain half maximum response, which indicates a reduced light sensitivity in irradiated samples.

## TARGETS OF UV-B RADIATION

A mechanism by which UV radiation can affect living cells is represented by photodynamic responses,[33] in which the excitation energy absorbed by a chromophore molecule is used by either of two different mechanisms, one involving oxygen (type II photodynamic reaction), the other involving other acceptors of the

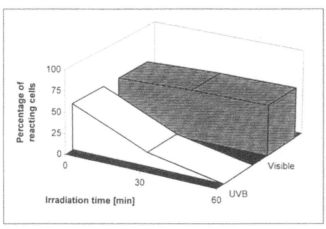

Fig. 14.9. Fraction of motile cells of B. japonicum *showing step-up photophobic response as a function of irradiation time in the "visible" and in the "UV-B". Redrawn from Sgarbossa A et al, J Photochem Photobiol B(Biol) 1995; 27:243-249.*

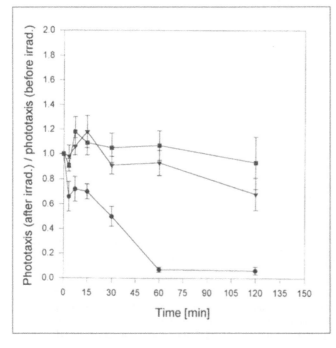

Fig. 14.10. Dose-effect curves of the phototaxis of F. salina *cells in: treated samples, irradiated with UV-B, UV-A and visible radiation (circles); control 1 samples, irradiated with UV-A and visible radiation (squares); control 2 samples, irradiated with visible light only (triangles). The error bars represent the standard error of the mean.*

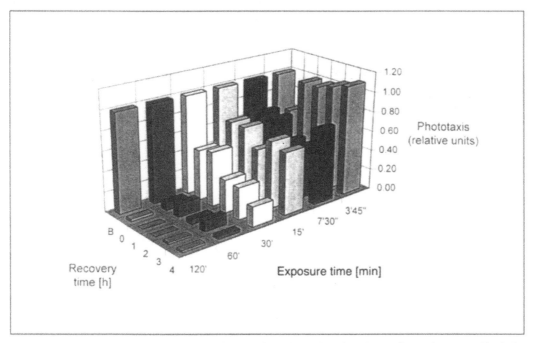

*Fig. 14.11. 3D histogram of the phototaxis of UV-irradiated samples of F. salina cells as a function of both the irradiation time and the time elapsed after the end of irradiation. The x-axis represents the time t (min) of UV irradiation, the y-axis the time T (hours) elapsed after the end of irradiation and the z-axis reports the phototaxis of a sample at (t, T) relative to that of the same sample before irradiation, taken arbitrarily as 1 and shown in the histogram at T = B.*

triplet energy (type I photodynamic reaction).[34-35] Both agents produced in the cell can destroy cellular components.[36]

The UV-induced inhibition of motility observed in *S. coeruleus* seems to be brought about by a photodynamic reaction.[17] We have seen that UV-B radiation is an important contributing factor in the inhibition of motility, but UV-A and visible wavelengths also contribute to the effectiveness of solar radiation. Thus the effect of solar radiation is of the same type of the photodynamic killing caused by artificial white light.[37] On the other hand, the UV-induced retardation of motility in *S. coeruleus* is unlikely due to damage to DNA (an important intracellular target of UV radiation) because the effect can be observed within a few minutes. Experiments carried out with specific quenchers and scavengers of free radicals and singlet oxygen indicate that photodynamic reactions are involved in the reduction of motility. It is still unknown, however, if the mechanism of photodynamic action in *S. coeruleus* is of type I or II.[17]

Photodynamic reactions, on the contrary, are unlikely involved in the UV-B effects in *B. japonicum* and *F. salina*, where target molecules could be intrinsic components of the photoreceptor apparatus and/or the photosensory transduction chain.[18-19]

Also, the lethal effects of UV-B radiation in *B. japonicum* are probably not mediated by a blepharismin-sensitized photodynamic reaction.[18] In fact, preliminary experiments indicate that the effects of UV-B radiation are about the same in both red and blue *B. japonicum* cells—these cells contain an oxidated, non phototoxic form of the pigment not due to undergo photosensitized killing.[38]

Absorption and fluorescence spectra of agar suspension of *B. japonicum* cells, reported in Figs. 14.13 and 14.14, indicate no significant differences between UV-B and visible irradiated cells.[18] The observed apparent increase of fluorescence quantum yield (Fig. 14.14) in visible-irradiated cells can be attributed to an extrusion of the pigment from the cell body, which occurs even in the presence of dim light, probably

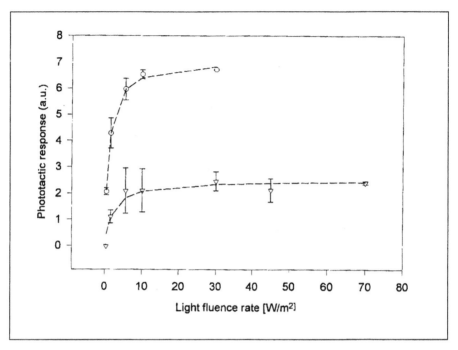

*Fig. 14.12. Dose-effect curves of the phototactic sensitivity of F. salina cells before (circles) and after 30 min irradiation with UV-B, UV-A and visible light (triangles). Phototactic activity in arbitrary units is plotted as a function of light fluence rate at 600 nm. Both curves can be fitted (p>95%) by a hyperbolic function of the form $R = (R_m \cdot I)/(I + I_{R/2})$ (where R is the phototactic response, I is the light fluence rate, $R_m$ is the level of saturation of the response and $I_{R/2}$ is the light fluence rate corresponding to half maximum response) represented in the graph by the dashed lines; the fit shows that the saturation level and the intensity of half-response are significantly different between the two curves.*

to avoid the harmful consequences of blepharismin-sensitized photodynamic reactions;[39] in UV-B-irradiated cells the quantum yield increase could be due to damage of the molecular environment of blepharismin. These spectroscopic measurements thus indicate that UV-B irradiation does not cause a specific molecular transformation of blepharismin.[18] These results have been confirmed recently by time-resolved fluorescence measurements on blepharismin, showing that the fluorescence lifetimes and the relative amplitudes of the different species emitting at 600 nm are not affected by UV-B irradiation.[40]

The UV-B-induced inhibition of photoresponsiveness in *B. japonicum*, therefore, can be explained in terms of damage to some components of the photosensory transduction chain, such as electron/proton transfer systems and/or membrane channels.

The latter hypothesis is supported by the observation that the avoiding reactions to mechanical stimuli are reduced after 30 min and completely suppressed after 60 min of UV-B irradiation. This inhibition, in fact, could be due to damage at the membrane level.[18]

*Fabrea salina* do not undergo photokilling and the effects of UV-B irradiation cannot be mediated by photodynamic reactions. In agreement with the results obtained in *B. japonicum*, fluorescence measurements of the hypericin-like pigment of *F. salina* obtained by cell sonication show that the spectra measured after irradiation with UV-B, UV-A and visible light are qualitatively and quantitatively identical to those of the controls (Fig. 14.15).[19] Thus the UV-B radiation does not vary significantly the fluorescence intensity of the hypericin-like pigment or the patterns of

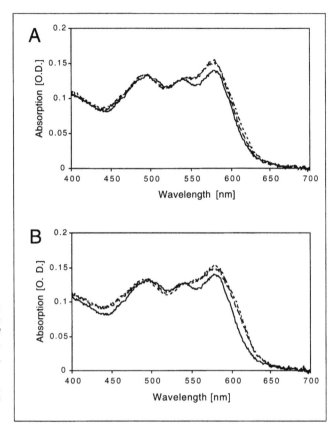

*Fig. 14.13. Optical absorption spectra of suspensions of* B. japonicum *samples dark kept (continuous line), irradiated with visible light (λ ≥ 370 nm, broken line) and with UV-B (λ ≥ 280 nm, dotted line). Irradiation time: A = 30 min, B = 60 min. Redrawn from Sgarbossa A et al, J Photochem Photobiol B(Biol) 1995; 27:243-249.*

excitation and emission of fluorescence, and modifications of the pigment concentration or structure can, therefore, be ruled out.

Interesting data come from the analysis of the dose-effect curves of phototactic sensitivity (reported in Fig. 14.12). The increase, after irradiation, of the fluence rate required in order to obtain half maximum response can indicate a damage to the photoreceptor pigment molecules, with the consequence that a higher number of photons is required in order to obtain the same response. In addition to this reduced light sensitivity *F. salina* irradiated samples show a lower level of saturation of the phototactic response; this suggests damage to some parts of the sensory transduction chain, such that even high fluence rates are not sufficient to elicit the same maximum response as in control samples. These results, therefore, point to an impaired photosensory transduction chain due to possible damage both to the photoreceptor

pigment and to some of the dark reactions following light detection.[19]

The fluorescence measurements mentioned above indicate that the hypericin-like pigment of *F. salina* does not appear to be damaged by UV-B radiation; on the basis of these experimental results, the hypothesis of the rhodopsin-like nature of the photoreceptor pigment is strengthened. Damage to the rhodopsin-like pigment, in addition to damage to some steps of the sensory transduction chain, could, in fact, explain the UV-B-induced reduction of the phototactic response.

Damage at the level of the cell membrane is also possible. The increase of cell speed could, in fact, be ascribed to a long lasting hyperpolarization of the cell membrane potential caused by a UV-B-induced, slowly reversible, alteration of the conductance of some ionic channels, perhaps potassium channels.[19]

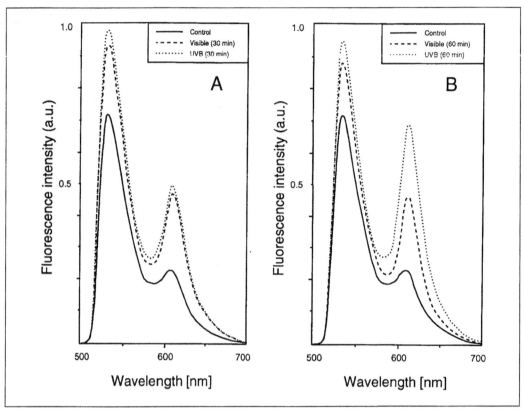

Fig. 14.14. Effect of different irradiation times (A = 30 min, B = 60 min) and wavelengths (continuous line: control, broken line: $\lambda \geq 370$ nm, dotted line: $\lambda \geq 280$ nm) on the fluorescence emission spectra ($\lambda_{exc}$ = 364 nm) of suspensions of B. japonicum. Redrawn from Sgarbossa A et al, J Photochem Photobiol B(Biol) 1995; 27:243-249.

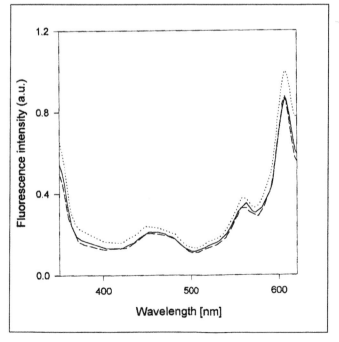

Fig. 14.15. Fluorescence excitation spectra (emission wavelength = 640 nm) of sonicated suspensions of F. salina cells before (solid line) and after 120 min irradiation with UV-A and visible radiation (dashed line) and with UV-B, UV-A and visible radiation (dotted line).

## REFERENCES

1. Dahlback A, Henriksen T, Larsen SHH et al. Biological UV-doses and the effect of an ozone layer depletion. Photochem Photobiol 1989; 49:621-625.
2. Frederick JE. Trends in atmospheric ozone and ultraviolet radiation: mechanisms and observations for the northern hemisphere. Photochem Photobiol 1990; 51:757-763.
3. Frederick JE. Ultraviolet sunlight reaching the earth's surface: a review of recent research. Photochem Photobiol 1993; 57: 175-178.
4. World Meteorological Organization—Global Ozone Research and Monitoring Project—Report No. 37, 1994.
5. Giese AC. Differential susceptibility of a number of protozoans to ultraviolet radiations. J Cell Comp Physiol 1938; 12: 129-138.
6. Giese AC. The ultraviolet action spectrum for retardation of division of *Paramecium*. J Cell Comp Physiol 1945; 26:47-55.
7. Hirshfield H, Giese AC. Ultraviolet radiation effects on growth processes of *Blepharisma undulans*. Exp Cell Res 1953; 4: 283-294.
8. Giese AC. *Blepharisma intermedium*: ultraviolet resistance of pigmented and albino clones. Science 1965; 149:540-541.
9. Giese AC, McCaw B, Cornell R. Retardation of division of three ciliates by intermittent and continuous ultraviolet radiations at different temperatures. J Gen Physiol 1963; 46:1095-1108.
10. Giese AC. Nucleic acid and protein synthesis, as measured by incorporation of tracers, during regeneration in ultraviolet-treated *Blepharisma*. Exp Cell Res 1971; 64: 218-226.
11. Giese AC, Shepard DC, Bennett J et al. Evidence for thermal reactions following exposure of *Didinium* to intermittent ultraviolet radiations. J Gen Physiol 1956; 40:311-325.
12. Lakhanisky T, Hendrickx B, Mouton RF et al. A serological study of removal of UV-induced photoproducts in the DNA of *Thetrahymena pyriformis* GL: influence of caffeine quinacrine and chloroquine. Photochem Photobiol 1979; 29:851-853.

13. Barcelo JA, Calkins J. Positioning of aquatic microorganisms in response to visible light and simulated solar UV-B irradiation. Photochem Photobiol 1979; 29:75-83.
14. Barcelo JA, Calkins J. The relative importance of various environmental factors on the vertical distribution of the aquatic protozoan *Coleps spiralis*. Photochem Photobiol 1980; 31:67-73.
15. Gutierrez JC, Perez-Silva J. Effects of UV-radiation on kinetics of encystment and on morphology of resting cysts of the ciliate *Laurentia acuminata*. J Protozool 1983; 30:715-718.
16. Calkins J, Colley E, Wheeler J. Spectral dependence of some UV-B and UV-C responses of *Tetrahymena pyriformis* irradiated with dye laser generated UV. Photochem Photobiol 1987; 45:389-398.
17. Häder D-P, Häder MA. Effects of solar radiation on motility in *Stentor coeruleus*. Photochem Photobiol 1991; 54:423-428.
18. Sgarbossa A, Lucia S, Lenci F et al. Effects of UV-B irradiation on motility and photoresponsiveness of the coloured ciliate *Blepharisma japonicum*. J Photochem Photobiol B(Biol) 1995; 27:243-249.
19. Martini B, Marangoni R, Gioffré D et al. Effects of UV-B irradiation on motility and photomotility of the marine ciliate *Fabrea salina*. J Photochem Photobiol B(Biol) 1997; in press.
20. Song PS. Protozoans and related photoreceptors: molecular aspects. Ann Rev Biophys Bioeng 1983; 12:35-68.
21. Matsuoka T, Tsuda T, Ishida M et al. Presumed photoreceptor protein and ultrastructure of the photoreceptor organelle in the ciliate protozoan, *Blepharisma*. Photochem Photobiol 1994; 60:598-604.
22. Marangoni R, Gobbi L, Verni F et al. Pigment granules and hypericin-like fluorescence in the marine ciliate *Fabrea salina*. Acta Protozool 1996; 35:177-182.
23. Marangoni R, Cubeddu R, Taroni P et al. Fluorescence microscopy of an endogenous pigment in the marine ciliate *Fabrea salina*. Med Biol Environ 1994; 22:85-89.
24. Marangoni R, Cubeddu R, Taroni P et al. Microspectrofluorometry, fluorescence imaging and confocal microscopy of an

endogenous pigment of the marine ciliate *Fabrea salina*. J Photochem Photobiol B (Biol) 1996; 34:183-189.

25. Marangoni R. Puntoni S. Colombetti G. A model system for photosensory perception in Protozoa: the marine ciliate *Fabrea salina*. In: Taddei-Ferretti C, ed. Biophysics of Photoreception: Molecular and Phototransductive Events. Singapore: World Scientific, (in press).

26. Marangoni R, Puntoni S, Favati L et al. Phototaxis in *Fabrea salina*. I. Action spectrum determination. J Photochem Photobiol B (Biol) 1994; 23:149-154.

27. Podestà A, Marangoni R, Villani C et al. A rhodopsin-like molecule on the plasma membrane of *Fabrea salina*. J Euk Microbiol 1994; 41:565-569.

28. Scevoli P, Bisi F, Colombetti G et al. Photomotile responses of *Blepharisma japonicum*. I. Action spectra determination and time-resolved fluorescence of photoreceptor pigments. J Photochem Photobiol B (Biol) 1987; 1:75-84.

29. Matsuoka T, Matsuoka S, Yamaoka Y et al. Action spectra for step-up photophobic response in *Blepharisma*. J Protozool 1992; 39:498-502.

30. Checcucci G, Damato G, Ghetti F et al. Action spectra of the photophobic response of blue and red forms of *Blepharisma japonicum*. Photochem Photobiol 1993; 57:686-689.

31. Colombetti G, Marangoni R, Machemer H. Phototaxis in *Fabrea salina*. Med Biol Environ 1992; 20:93-100.

32. Colombetti G, Bräucker R, Machemer H. Photobehavior of *Fabrea salina*: responses to directional and diffused gradient-type illumination J Photochem Photobiol B (Biol) 1992; 15:253-257.

33. Ito T. Photodynamic agents as tools for cell biology. In: Smith KC, ed. Photochemical and Photobiological Reviews. Vol. 7. New York: Plenum, 1983:141-188.

34. Maurette MT, Oliveros E, Infelta PP et al. Singlet oxygen and superoxide: experimental differentiation and analysis. Helv Chim Acta, 1983; 66:722-733.

35. Spikes JD. Photosensitization. In: Smith KC, ed. The Science of Photobiology. New York: Plenum, 1977:87-112.

36. Häder D-P, Worrest RC. Effects of enhanced solar ultraviolet radiation on aquatic ecosystems. Photochem Photobiol 1991; 53:717-725.

37. Yang KC, Prusti RK, Walker EB et al. Photodynamic action in *Stentor coeruleus* sensitized by endogenous pigment stentorin. Photochem Photobiol 1986; 43:305-310.

38. Ghetti F, Lenci F, Checcucci G et al. A laser flash photolysis study of the triplet states of the red and the blue forms of *Blepharisma japonicum* pigment. J Photochem Photobiol B (Biol) 1992; 13:315-321.

39. Giese AC. The photobiology of *Blepharisma*. In: Smith KC, ed. Photochemical Photobiological Reviews. New York: Plenum, 1981:139-180.

40. Angelini N, Cubeddu R, Ghetti F et al. In vivo spectroscopic study of photoreceptor pigments of *Blepharisma japonicum* red and blue cells. Biochim Biophys Acta 1995; 1231:247-254.

# CHAPTER 15

# EFFECTS OF ULTRAVIOLET RADIATION ON THE PELAGIC ANTARCTIC ECOSYSTEM

María Vernet and Raymond C. Smith

## ABSTRACT

Ultraviolet radiation (UVR) affects biotic and abiotic factors in marine ecosystems. Effects on organisms are mostly deleterious due to damage to DNA and cellular proteins that are involved in biochemical processes and which ultimately affect growth and reproduction. Differential sensitivity among microalgal species to UVR has been shown to shift community composition. As a result of this shift, the total primary production for the community may be maintained at pre-UVR levels. Similar impacts and mechanisms are expected in Antarctic waters. The overall effect of UVR on the ecosystem needs to include relevant feedback mechanisms which can diminish, and sometimes reverse, deleterious effects on population growth. For example, it has been speculated that UVR can increase iron-limited phytoplankton populations by photoinduced reduction of $Fe^{3+}$ to $Fe^{2+}$, a more soluble form of iron and readily available for algal and bacterial uptake. An equally positive feedback can be attributed to diminished grazing by zooplankton. Thus, energy flow among the trophic levels can decrease as a result of damage to a certain trophic level, but overall biomass and ecosystem production might remain relatively unchanged.

Similar positive and negative feedbacks associated with UVR are related to the dissolved organic matter (DOM) pool, known to be recycled by bacterial activity. Although it could be expected that bacterial production in Antarctic surface waters would decrease when exposed to UVR, this effect can be counteracted by increased substrate nutrient availability. Photolysis of high-molecular weight molecules by UVR produces

*The Effects of Ozone Depletion on Aquatic Ecosystems*, edited by Donat-P. Häder.
© 1997 R.G. Landes Company.

higher availability of low-molecular weight molecules readily taken up by bacteria. This step might be of greater importance in high latitude ecosystems where low bacterial production has been attributed to low substrate availability.

Similarly, increased nutrients for bacterial activity originate from photolysis of high-molecular weight molecules which are known to release $NH_4^+$ and amino acids under UVR. The DOM pool might also increase through phytoplankton excretion of organic matter, a process known to occur under algal stress. On the other hand, a decrease in DOM by diffusion from zooplankton fecal pellets is expected in surface waters due to decreased grazing.

In summary, we argue that the understanding of the effect of UVR on Antarctic ecosystems is more than the sum of the effect of radiation on individual species, given that alteration of interspecific interactions can exacerbate, diminish and sometimes reverse known physiological damage. This, plus complex and nonlinear feedback mechanisms associated with UVR effects make prediction at the ecosystem level uncertain.

## INTRODUCTION

A recent characteristic phenomenon of the Antarctic ecosystem is the well-known springtime decrease in stratospheric ozone, known as the ozone hole. It is confined to the polar vortex over the Antarctic continent, from September to December of each year. However, once the winter/spring vortex breaks down, its effects reach mid latitudes, mostly during the month of December,[1] although it has also been detected in sub-antarctic environments during the spring.[2] There has been significant annual and interannual variability in Antarctic ozone, and, consequently, in changes in ozone-related incident ultraviolet radiation (UVR). During the last two decades major international efforts have focused on the physics and chemistry of the Earth's atmosphere with emphasis on understanding processes that control the ozone layer, while

studies on the effects of UV on the biosphere, in particular at the community and ecosystem level, have been relatively limited.[3]

Interest in UV effects on aquatic ecosystems is increasing because ozone depletion is not restricted to the area over Antarctica and significant reductions have been reported in the Northern Hemisphere.[4-6] Hemispherical trends are superimposed on high interannual variability, as pointed out by Michaels et al,[7] where low ozone during 1992 can be associated with a drop in sunspots, a strong El Niño event and the eruption of Mount Pinatubo, all of which can potentially decrease ozone in the stratosphere. Other populated areas, such as South America, Australia, New Zealand and South Africa are affected, in particular at the time of the vortex disappearance, probably as an effect of dilution.[1,4]

It has been estimated that aquatic ecosystems fix between 30 and 50 Gt of carbon per year, which is roughly half the total global fixation of carbon.[8-10] Consequently, the threat of increased UVR on surface layers of the ocean on marine productivity is of considerable concern. Estimates for the Southern Ocean range from 1-5 Gt C $y^{-1}$.[11] For the Southern Ocean, ice algae are estimated to contribute up to 30% of the total primary production.[12] Traditionally, prediction of UV effects on ecosystems have assumed a linear addition of UV effects on different levels of the food chain where the final effect on higher trophic level predators, such as penguins, whales and seals, have been inferred from the cumulative effect on primary producers and grazers.[13] In other words, the total effect of UV at a given trophic level has been assumed to be the combination of UV effects on the previous trophic level added to the direct effect of UV on the level itself. For example, initial studies on UV effects on marine algal communities reported decreased total primary productivity and shifts between species towards less UV-B-sensitive species as well as a drop in total species diversity, assuming constant grazing.[14-17] In contrast, recent trophic-level assessments suggest that differential UV sensitivity between algae and herbivores may contribute to an increase in algae by exerting a stronger UV influence on the grazers.[18,19] An analogous influence on zooplankton, thus reducing zooplankton grazing, could counteract UV photoinhibition on phytoplankton growth. In addition to biological factors, UVR affects abiotic processes which affect directly or indirectly the food web. These factors are either chemical (e.g. nutrients) or related to the dissolved organic matter (DOM) pool which is intrinsically related to the microbial loop.[20] Such an alteration of the ecosystem functioning would result in a decrease of transfer of energy through the food web.[21]

In this chapter we summarize what is known of the UVR effects on different levels of the Antarctic food web, with emphasis on the relationships between trophic species, and what is known of the UV effects on abiotic processes affecting the food web. Several recent reviews on UVR effects on aquatic and Antarctic ecosystem[13,22] have given excellent summary of the UV photobiology and that information will not be rephrased here. We present evidence to suggest that research required for understanding UV effects on Antarctic ecosystems will necessitate ecosystem studies in addition to detailed determination of UVR on specific processes related to any given trophic level.

## UV RADIATION IN THE SOUTHERN OCEAN

Estimation of quantitative effects of ultraviolet radiation (UVR) on biological systems requires knowledge of the incident spectral irradiance and a biological weighting function (BWF), which provides the wavelength-dependency of biological action. Because BWFs are heavily weighted in the UV-B region of the spectrum, high spectral resolution is required for accurate estimation of effective biological doses. Smith et al[23] have developed a high spec-

tral resolution (1 nm) air and in-water spectroradiometer and Booth et al[24] have developed the U.S. National Science Foundation UV Network which provides high resolution data at three locations in the Antarctic continent. Alternatively, narrow band instruments (e.g. Bio-Spherical Instrument PUV series) can, in conjunction with an adequate full spectral model, be used to estimate incident spectral irradiance with adequate resolution. BWFs, specific to the target unit, have been developed. For Antarctica, stepwise functions for the BWF for photosynthesis have been developed by Helbling et al,[25] Lubin et al,[26] Smith et al[23] and Boucher et al[27] which have yielded results similar to the more detailed determination of Cullen et al.[28] Other BWFs have been developed in temperate areas for plant chloroplasts[29] and DNA.[30] There is a paucity of BWFs for other processes, for other levels of the food chain, not only for Antarctica but everywhere. This is a serious constraint for modeling and predictive purposes.

Actinometry (e.g. refs. 31, 32) has not been used extensively in Antarctic studies. On the other hand, a biological dosimeter, based on the response of an organism to UVR, has been used. This method provides a relative unit to assess potential effects of UV exposure on a specific organism or target molecule. Once the response of the organism to UV is evaluated under standard conditions, i.e. by exposure to natural UV radiation, we can say the organism has been calibrated. A relative estimate of potential UV damage can then be estimated. The potential benefit of the biological dosimeter resides in being a relatively more easy and inexpensive method, once it has been carefully evaluated. The main disadvantage is the exacting dosimetry required for quantitative calibration. It can also be used to compare biological effects on very diverse environments with or without very different UV climatology. Although a biological dosimeter was carefully evaluated for an Antarctic coastal site it has not been used extensively use in the region.[33] Both the actinometry and the biological dosimeter give broad band estimates of UVR unless the incident radiation is differentially screened, usually with filters.[33]

## CLIMATOLOGY OF UV RADIATION

Ultraviolet radiation (UVR) levels are mostly controlled by atmospheric ozone, cloud cover, and solar zenith angle with ozone concentration being relatively specific to the UV-B region.[34] Natural variability in these environmental variables give rise to a very high natural variability in UVR, with ozone primarily affecting the relative ratios of UV-B to UVR, photosynthetic available radiation (PAR), or total irradiance. The dynamic nature of the polar vortex containing the ozone hole has given rise to large changes in these UV-B ratios on time scales of several days or less (Fig. 15.1). The polar vortex, and correspondingly, the ozone hole, is often elongated in shape, giving rise to an uneven distribution of UV-B at locations within the Antarctic continent.[35] The natural-short term variability (hours to days) due to changes in cloud cover and solar zenith angle compounds the difficulty in assessing the influence of increased UV-B levels on natural systems.[23,36] The resultant effect is that natural variability (cloudiness)

can counteract UVR increases. Further, recent work (Gautier et al, University of California Santa Barbara, U.S., personal communication) suggests that the combined influence of cloud cover and surface reflectance influences these UV-B ratios. As not much is known with respect to the effect of this variability on organisms and processes, it is too soon to predict the effect of this variability either to enhance or decrease UV effects on Antarctic ecosystems.

## TRANSMISSION OF UV IN SURFACE WATERS AND ICE

Transmission of UVR within the water column is a key element in assessing UV effects in marine systems. Light transmission is affected by water itself, as well as particulate and dissolved organic matter (POM and DOM, respectively) within the water column. Water is known to be a relatively strong UV absorber[37-39] and spectral attenuation coefficients have been published for clear natural waters.[38] However, in natural waters, particulate and dissolved organic matter strongly absorb UVR and these in-water constituents are highly variable. In blue, more transparent oligotrophic waters, biologically significant UV doses can penetrate several tens of meters. In contrast, more productive coastal waters,

*Fig. 15.1. Daily maximum UV-A irradiances (360-400 nm) from 15 December 1989 to 7 February 1993 at McMurdo Station (77.51°S, 166.40°E) shown as a function of days before and after solstice. Redrawn from Booth et al, 1994.*

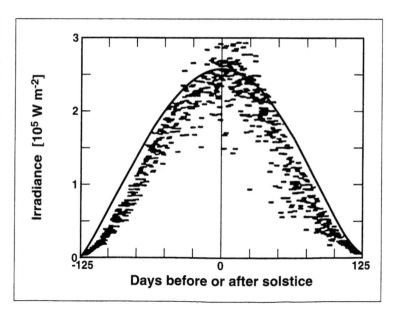

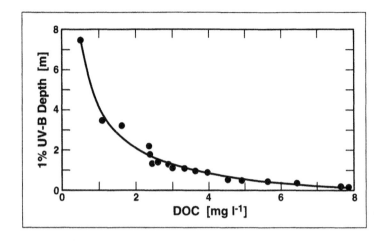

Fig. 15.2. Relationship between the depth of 1% UV incident radiation and dissolved organic carbon (DOC) in lakes. Reprinted with permission from Schindler et al, Nature 1996; 379:706, ©1996 MacMillan Magazines Limited.

with higher particle concentration (e.g. >3 mg chlorophyll *a* m$^{-3}$) can have attenuation coefficients nearly an order of magnitude higher, limiting significant penetration depths to the order of meters.[40] DOM shows an even stronger attenuation in the UVR[40,41] and can effectively limit significant penetration depths to a meter or less. For example, Kramer[42] estimated that the combination of high POM and DOM in Dutch coastal waters would limit UVR transmission in the water column to such an extent that no UV effects or planktonic organisms were expected. High POM absorption in Antarctic waters[43] and probably in ice-edge blooms,[44] would limit UV transmission in late spring and summer due to high production, but not during early spring (e.g. October) where chlorophyll (chl) *a* levels are usually lower than 0.5 mg m$^{-3}$.[45] The paucity of absorption estimates for POM, and in particular for DOM, make it difficult to speculate on their effect in Antarctic waters, although similar levels of DOM as in other parts of the world would support the hypothesis of important UVR absorption by DOM (Fig. 15.2).[46] Estimated UV effects at depths of about 20 m in the vicinity of Palmer might be due in part to the contribution of DOM absorption.[33,47]

The role of DOC in light attenuation is intimately related to other environmental changes. For example, in boreal lakes, the decreased amount of DOC, caused by an increase in average temperature and

acidification in the last 20 years, was related to increased UVR in the water column.[48] In the case of Antarctic waters, a complex mix of competing feedback mechanisms make estimating changes in UVR, due to environmental change, speculative.

There are relatively few direct observations on the optical properties of Antarctic ice and snow. These observations suggest that UV transmission in the ice is maximum in October due to relatively high transparency in spring. Based on these observations, it is expected that ice algae, associated with bottom communities in ice flows, potentially can be exposed to relatively high levels of UV-B. These UV-B levels have increased by as much as an order of magnitude under the ozone hole.[49]

## THE FOOD WEB

### PHYTOPLANKTON

**Photosynthesis**

Deleterious effect of UV-B on photosynthesis has been studied both in cultures and in the field, in particular for Antarctic phytoplankton. The reader is referred to reviews done in the last few years that cover this subject extensively (e.g. refs. 22, 36, 50, 51 and references therein). Overall, UV-B inhibits primary production by 30-50% of shielded samples[52] with a strong depth gradient from surface to about 20-50 m.[23,33,53] All these experiments are based on 6-24 h incubations, either in situ or in incubators exposed to sunlight. On the average for the water column, primary production decreases by 6-12%[23,24] during springtime ozone depletion over Antarctic water resulting in a 2% reduction in the yearly primary production estimates for the marginal ice zone.[23]Helbling et al[54] based on different assumptions and methodology, estimate the decrease in primary production to be 0.15% for the entire ice-free waters south of the Polar Front. A UV inhibition function for photosynthesis has been described by Cullen and Neale.[55] The biological weighting function for Antarctic phytoplankton, necessary to scale UVR to biological effective irradiance, has been determined for natural populations by Lubin et al,[26] Helbling et al,[24] Smith et al,[23] Boucher et al[27] and Neale et al.[56]

## Nutrient uptake

Very little is known of the effect of UVR on nutrient uptake in Antarctic phytoplankton. Studies on temperate species suggest that nitrogenase, the enzyme related to nitrogen assimilation in phytoplankton, is activated by PAR[57] and inactivated by UV-B radiation.[58] In contrast, ammonium uptake seems less affected.[59,60] Overall, amino acid concentration in the cell decreased under UV-B.[61] The effect is also felt on enzymes related to amino acid metabolism. UVR diminishes synthesis and intracellular accumulation of alanine and valine[62] while synthesis and accumulation of glutamic acid increase due to inhibition of glutamate synthase[58] or glutamate dehydrogenase.[60] These results are similar to metabolic changes observed in phytoplankton under nitrogen stress, suggesting that UV-B suppresses nitrogen assimilation into cells.[63] Decreased $NH_4^+$ uptake by *Pavlova* spp. under UV-B and high intensity UV-A was interpreted as reduced supply of ATP and NADPH from direct effects of UV-B on the photosynthetic apparatus and pigment bleaching.[60] Similar effects of UVR on Antarctic species will have to be assumed until experiments are carried out for Antarctic, or at least, polar phytoplankton.

## Exudation

The amount of extracellular carbon produced by phytoplankton has been a controversial subject for several decades.[64-67] Excretion of carbon by photosynthetic organisms is a widespread process associated with photosynthesis.[68] On the average, phytoplankton excretes 5-25% of the carbon incorporated in particulate matter, both in monospecific cultures and in natural populations[65,68] and the amount excreted is a constant proportion of photosynthetic rates. Several studies have pointed out that a large proportion of photosynthetic carbon goes through a DOC phase[69] for at least short periods of time.[70] Under these conditions, between 20-60% of photosynthate must go into the DOC pool to explain the DOC changes observed,[70] mainly during spring bloom events in temperate waters. Additional organic carbon excretion in phytoplankton seems associated with physiological imbalance due to events such as nitrogen limitation,[71-73] in particular under high-light conditions.[73] In the field, the transfer of cells to higher irradiance might produce excess photosynthate.[67,68] Nutrient limitation is observed during late growth stages in batch cultures[74] or at the end of the spring bloom. High DOC concentrations have also been observed after a *Phaeocystis* sp. bloom.[75,76] This excess carbon excreted might be associated with increased intracellular carbohydrate, as in diatoms[74,77] but not observed in dinoflagellates.[72]

Very little is known of exudation by Antarctic phytoplankton and the consequent implication for the DOC pool. Recent results in the Arctic suggest a large amount of extracellular carbon observed seemed to be related to phytoplankton composition (i.e. cells which produce mucilage for colonial formation) and to a lesser extent to in situ nitrate limitation.[78] In Arctic Water, *Chaetoceros socialis* allocated 40% of total carbon incorporated as extracellular under conditions of low silicic acid (<0.2 µM) and measurable nitrate concentrations (0.5-2.5 µM). Similar extracellular carbon production was found in a mixture of *C. socialis* and *P. pouchetii* at the Polar Front and the marginal ice zone with higher nutrient concentrations (5-10 μM nitrate).

These results suggest that species composition and their physiological state may largely control extracellular carbon production in the field.[79] Although low nitrate is known to increase exudation,[74] this effect is not expected in Antarctic open waters; however, this effect might be observed during or after massive coastal blooms.[45,80]

In spite of the obvious importance of phytoplankton exudation on the carbon cycle and as substrate for the microbial loop, no studies have been carried out on the effect of UV-B on exudation, for either temperate or polar phytoplankton. In general, exudation increases when algae are stressed and it can be speculated that UV-B stress would act in a similar way.

## Respiration

Changes in $\delta^{13}C$ in $\Sigma CO_2$ observed in the Bellinghausen Sea in the spring of 1990 combined with changes in cell abundance in the colonial prymnesiophyte *Phaeocystis* sp. suggest that under increased UV-B radiation, as measured under decreased ozone concentration, there is an increase in the ratio of total community respiration to photosynthesis.[81] Heterotrophic respiration increases were attributed to increased bacterial substrate due to cell lysis.

## Growth

The effect of UV-B on marine phytoplankton growth has been shown to be species-specific. For several cultures of temperate species, specific growth rate was affected negatively by UV-B.[82-84] In the diatom *Phaeodactylum tricornutum*, no decrease in UVR sensitivity was observed with time.[82] Similar results were observed on 3D experiments on Antarctic phytoplankton dominated by *Corethron criophylum* where growth rates decreased by 100% on cells exposed to UV-A + UV-B + PAR and by 50% when exposed to UV-A + PAR, as compared to controls exposed to PAR

only.[85] On the other hand, active growth of coastal species was observed for 12 days at Palmer Station where diatom cultures were kept at in situ solar radiation.[86] No difference was found also between treatments (UVR + PAR vs. PAR only) for the colonial prymnesiophyte *Phaeocystis* sp., although these cultures did not grow. This lack of effect was observed in spite of the well-documented inhibition of photosynthesis[23,26-28] for Antarctic phytoplankton in experiments from 2-24 h and points towards different controls of photosynthesis and growth and between short- vs. longterm effects of UV-B. It has been noted for some time that caution must be used when inferring longer term ecological consequences from short-term observations.[87]

Mixing of cells in the upper water column, in particular within the mixed layer, affects the average irradiance in which a cell is exposed during the day.[35,88,89] Several studies have speculated about the possible role of alleviation from UVR in Antarctic waters if cells are mixed deeper in the water column.[50,90,91] Experiments where UVR intensity was manipulated to resemble mixing in the upper water column showed increased production in cloudy days while the effect was opposite on sunny days.[54] Phytoplankton dominated by the diatom *Thalassiosira gravida* showed less photoinhibition when exposed to variable radiation,[92] supporting the hypothesis that mixing might provide UV-B protection.[36]

## Cell size

Coastal waters have, on the average, a higher proportion of larger cells than open waters.[93] For example, more than 80% of the nearshore phytoplankton biomass was associated with cells >10 μm in Terre Adélie during summer while 70 km offshore, cells >10 μm represented only 30% of the total biomass and 59% of the cells were between 1-10 μm.[94] Within coastal waters, high Chl *a* accumulations (i.e. blooms) are dominated by large cells (e.g. >20 μm) while low Chl *a* concentrations are dominated by smaller cells.[80,95] A differential effect of UVR on cell size, as observed for diatom cultures,[96] show higher damage on smaller cells, and we might speculate that oceanic phytoplankton may have a higher sensitivity to UV-B. In addition, UVR increases cell size[82] associated with a concomitant reduction in specific growth rates.

## Species composition

Initial experiments with temperate phytoplankton, showing differential sensitivity to UV-B by different species,[17] suggest a change in species composition in long-term UV-B exposure with more UV-tolerant species ultimately dominating.[16] As mentioned above, there is a wide range of interspecific UV-B sensitivity on growth and survival, with smaller cells being more sensitive, due to a higher surface to volume ratio as a result of cell size and cell shape.[96] In addition to size, an increased UV-B sensitivity in flagellates, as compared with diatoms, was observed in natural populations of Antarctic phytoplankton.[54,97] This difference can be attributed in part to size (flagellates are on the average smaller than Antarctic diatoms) and to increased UV-absorbing properties of diatoms[97] related to the presence of mycosporine-like amino acids which are believed to reduce deleterious effects by UV-B on growth.[84] The predicted shift from less to more resistant species (e.g. from flagellates to diatoms) was observed in a 2-week experiment of natural Antarctic populations exposed to ambient UVR, although similar Chl *a* and particulate carbon accumulation were observed under UVR and UVR + PAR.[52] Under UVR the amount of UV absorbing compounds (e.g. mycosporine-like amino acids) increased as well. As a result of this shift in species composition, a decreased sensitivity of photosynthesis was observed in the phytoplankton exposed to UVR. The higher resistance by diatoms, as compared with flagellates (in particular the colonial prymnesiophyte, *Phaeocystis pouchetii*, ref. 81), seems to be related to a lower effect on photosynthesis as well as nitrate uptake.[59]

Few studies are available on effects of UVR at longer time scales. McMinn et al[98] documented no changes in diatom species composition in laminated sediments in Antarctic anoxic fjords for the last 20 years, coinciding with the decrease of ozone. However, as noted by Bothwell and coworkers[18] the limited data provided by McMinn et al[98] do not substantiate their implied lack of a UV-B effect.

## ZOOPLANKTON

UV effects on zooplankton, under normal and decreased ozone conditions in temperate waters, affect zooplankton survival, reproduction and grazing.[99] It is not clear from these results if decreased grazing would result in a reversal of UV effects on phytoplankton, as observed for a chronomid/diatom interaction in temperate freshwater stream beds (Fig. 15.3). We can expect that a 50% mortality of a grazer would decrease grazing pressure and favor phytoplankton growth. The possibility of grazing reversing deleterious effects of UV on phytoplankton and the relative importance of grazing in controlling phytoplankton population growth in any given community is currently a matter of speculation. Under current UV irradiance, overall decrease in primary production by UV in the Antarctic euphotic zone is estimated at 6-23% of marginal ice zone production.[23,25] The overall result would depend on the effect of UVR on Antarctic grazers, averaged for the euphotic zone, and on time scales representative of phytoplankton accumulation at ambient temperature (days to weeks, if we assume a specific growth rate of 0.1-0.3 d$^{-1}$).[44]

## SEDIMENTATION

Potential changes in grazing pressure will affect sedimentation of particulate matter. In areas where organic matter sedimentation out of the euphotic zone is due to grazer (i.e. krill) fecal pellets,[100] we might expect a shift to cell sedimentation, assuming no change in primary production. Thus, the pulse of organic matter after a bloom could consist mainly of intact cells.

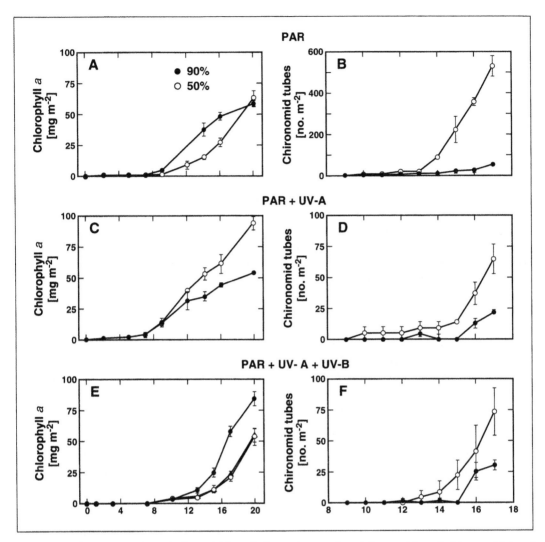

Fig. 15.3. *Changes in phytoplankton (chlorophyll* a *concentration, left panels) and chironomid larval abundance (chironomid tubes, right panels) with time in streams. Experiments carried out at two irradiance levels (filled symbols, 90% of incident irradiance, and open circles, 50% of incident irradiance) at three treatments (PAR: top panels; PAR + UV-A: middle panels; and PAR + UV-A + UV-B: low panels). Reprinted with permission from Bothwell et al, Science 265:97-100. © 1994 American Association for the Advancement of Science.*

This effect will be maximum in coastal areas where larger cells[94] and higher production are found.[45] Secondary effects will include alteration of elemental ratios, heterotrophic substrate and nutrient recycling below the euphotic zone. If, on the other hand, a large proportion of sedimenting matter is due to cell sinking then the quality of organic matter to depth would not be substantially altered.[101] The quantity and timing might be affected if, as dis-

cussed before, UVR would alter species composition and/or species size.

## THE MICROBIAL LOOP

### BACTERIA

Bacterial biomass in Antarctic waters can reach 9% of the net plankton biomass in the top 50 m and increase with depth up to 50%, as measured in Bransfield Strait and Drake Passage in summer.[46] Different from other parts of the ocean, there is no correlation between phytoplankton and bacterial biomass in Antarctic waters[94,102] and the reason for this difference is unclear.[103]

UVR reduces bacterial activity in temperate coastal waters in the top 5 m of the water column, with no indication of higher resistance in surface populations as opposed to those from depth.[104] Inhibition was observed at an irradiance equal to 0.7 W m$^{-2}$. UV-B was also found to photochemically degrade bacterial extracellular enzymes.[104] The combination of decreased bacterial activity and the degradation of extracellular enzymes reduces the flow of energy through the microbial loop. This effect is counteracted, or at least diminished, by the increase in bacterial substrate due to photodegradation of DOM. Increased bacterial activity at low UV-B irradiance with respect to dark uptake (Fig. 15.4) was attributed to this process.

## PHOTO-OXIDATION OF DOM

UV-B interaction with DOM is known to produce oxygen radicals and hydrogen peroxide ($H_2O_2$) which can be considered oxidative agents of biological membranes and have a negative impact on planktonic communities.[105] In addition, multiple studies have documented the photo-oxidation of DOM responsible for degrading high-molecular weight DOM into low-molecu-

lar weight DOM (e.g. Fig. 15.5)[105,106] which is readily available for bacterial consumption.[107,108]

The importance of the size class on bacterial productivity is still a matter of debate, as Amon and Benner[109] found that although bacterial growth efficiencies were higher at low-molecular weight DOM, total bacterial growth and respiration was higher at high-molecular weight DOM (>1000 daltons), resulting in a higher carbon based rate of utilization. It is too early to assess the degree to which UV photo-oxidation of DOM would be of importance in Antarctic surface waters. Given the debate on whether bacterial activity is depressed at low temperature,[110,111] and the potential role of substrate on polar bacterial metabolism,[112] the role of phytoplankton as providers of labile DOC and photo-oxidation of DOM by UVR are both critical to Antarctic ecosystems.

Photochemical production of dissolved amino acids from humic substances have been shown to increase bacterial production in temperate coastal waters.[113] UV-B was found to be the most active portion of the solar spectrum for this process which could be due both to higher energy and higher absorption by the target molecule. Although no or low humic acids are expected in Antarctica, Lara and Thomas[114] have identified recalcitrant DOM production by marine phytoplankton with

*Fig. 15.4. Bacterial secondary production (BSP) as a function of UV-B radiation. Note higher production at low UV-B with respect to dark uptake. Redrawn from Herndl et al. Nature 361:717-719. Copyright, MacMillan Magazines Limited.*

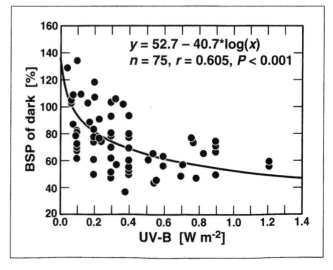

$$y = 52.7 - 40.7*\log(x)$$
$$n = 75, r = 0.605, P < 0.001$$

chemical characteristics previously associated only with humic substances. The source of this pool of DOM seem to be degradation of cellular membranes and can be assumed to be produced anywhere in the ocean.

# NUTRIENTS

## MACRONUTRIENTS

DOM exposed to UV-B releases $NH_4^+$ into the surrounding waters, thus becoming a nutrient source in coastal waters.[113] This larger availability of ammonium, of major importance in areas of nitrogen limitation, can counteract decreased N uptake and metabolism by phytoplankton,[59,63] and potentially bacteria, as a result of UV-B inhibition. In spite of high nitrate concentrations in most Antarctic open waters during the growth season, phytoplankton has shown low specific nitrate uptake rates[115] and differential uptake of $NH_4^+$ when present,[116] suggesting that a potential effect of UV-B in releasing $NH_4^+$ may be of interest in the Southern Ocean.

## MICRONUTRIENTS

The potential interaction of iron (Fe) and UV-B as a source of dissolved iron is important in the Southern Ocean as it has been hypothesized that Fe limitation may be controlling primary production in Antarctic open waters characterized with low chlorophyll accumulation and high macronutrient concentration.[117] For example, the gradient of higher productivity in coastal waters as opposed to open waters observed in the Western Antarctic Peninsula[45,80] is correlated with observed iron concentrations (4.7 nM and 0.16 nM, respectively).[118] A similar approach was taken by de Baar et al[119] to explain high primary productivity at the Polar Front (1200-3000 mg C $m^{-2}$ $d^{-1}$) with high Fe concentration in surface waters (2-4 nM at 60-100 m) as opposed to lower primary production (80-300 mg C $m^{-2}$ $d^{-1}$) at the Antarctic Circumpolar Current with subnanomolar concentrations (0.17 nM at 40 m). On the other hand, de Baar et al[120] and Buma et al[121] did not find rapid Chl *a* accumulation with Fe addition with respect

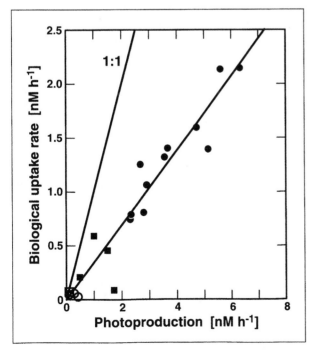

Fig. 15.5. Photochemical production of pyruvate after irradiation of dissolved organic matter (DOM) plotted against the rate of uptake of pyruvate by bacteria in coastal waters (filled circles) and in the Sargasso Sea (open circles). Reprinted with permission from Kieber et al, Nature 1989; 341:637-639, © 1989 MacMillan Magazines Limited.

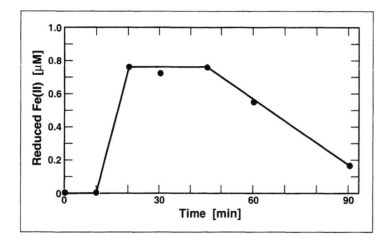

Fig. 15.6. Photoreduction of Fe(III) in seawater (pH 8.0-8.1) in the presence of the diatom Phaeodactylum tricornutum under UVR. Fe(III) concentration of 5 µM; diatom concentration of $10^5$ cells $ml^{-1}$. Redrawn from Kuma et al, *Marine Chemistry* 37:15-27. Copyright 1992, with kind permission from Elsevier NL.

to controls in the Weddell/Scotia Seas (both treatments grew at similar levels). The authors concluded that incubation effects overrode metal, and in particular, Fe addition due in part to the exclusion of large grazers from the experimental vessels. Iron additions shifts phytoplankton composition from flagellates to diatoms, both in Antarctic[121] and in equatorial Pacific waters.[122] Their results were not as dramatic as those observed by Helbling et al[123] who found increased primary productivity and microzooplankton population in surface pelagic waters after addition of Fe. No effect was observed in deep pelagic waters or coastal waters off Seal Island. A shift to larger cells is similar to other experiments of phytoplankton exposed to UVR[19,96] which were attributed to differential cell survival and DNA damage.

In marine oxic waters, $Fe^{3+}$ is the more stable form while $Fe^{2+}$ is more soluble and readily available to phytoplankton and bacterial uptake.[124] The concentration of Fe (III)' (the sum of dissolved inorganic species) is the relevant factor to consider with respect to the uptake of inorganic iron.[125] Its concentration varies from $10^{-8}$ to $10^{-9}$ M. Recent data indicates that 99.9% of the dissolved iron in surface waters is bound within organic complexes, resulting in subpicomolar concentration of dissolved Fe(III). It is believed that the ligands for iron may originate from phytoplankton.[125]

Sunlight increases rates of oxidation and reduction of iron, enhancing labile Fe concentrations and phytoplankton uptake. Although UV-B photoreduces Fe(III) to Fe(II) associated to inorganic ligand complexes, a larger reduction power is expected from organic chromophores.[125] Reduction of organic ligands may occur by the photoproduced superoxide radical ($O^{2-}$). In addition, oxidation of Fe(II) can occur with photoproduced $H_2O_2$.

Photo-reduction of Fe(III) to Fe(II) is also attributed to the action of marine phytoplankton (Fig. 15.6). High concentrations of Fe(II) were observed during phytoplankton spring blooms in Japanese coastal waters.[126] Experiments with filtrate from a diatom culture resulted in photoreduction of Fe(II) after addition of 5 µM Fe(III). This process was attributed to the release of hydrocarboxylic acids by phytoplankton, known to reduce Fe(III) to Fe(II) in the presence of sunlight[124] and is more pronounced at lower temperatures (5° vs. 20°C), important for Antarctic waters (surface water temperature varies from -1.8° to +2.5°C).

## CONCLUSIONS

Two important conclusions can be drawn from this discussion. First, evidence has accumulated to indicate that an assessment of UV effects on Antarctic ecosystems or marine ecosystems in general, will

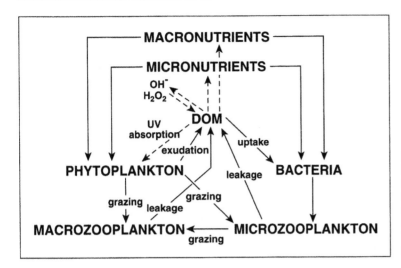

Fig. 15.7. Scheme showing biotic (full line) and abiotic (dashed line) relationships in the upper water column in the ocean, based on interactions discussed in the text. The arrow shows the direction of energy flow.

require experimentation on the ecosystem as a whole, or at least, isolate parts of it which include several interactions (i.e. the microbial loop). The predictive capability of adding effects on individual pools in the system is limited and experiments in temperate areas suggest that this can even be erroneous. Each level or species is not acting in a vacuum and biotic and abiotic interactions will modify its genotypic response to UVR. Second, it is not possible to estimate UV effects on ecosystems without concurrent effort toward understanding environmental and biological forces which drive the system. Thus, UV effects are an added stress upon the system and need to be considered in conjunction with other potential limiting factors, such as nutrients, and other driving forces, such as mixing and ice cover.

In general, we speculate that a more profound and permanent effect of UVR might be the alteration of interaction between singular elements in the ecosystem than the direct effect of UV in inhibition of that same element (Fig. 15.7). For example, changes in species composition might overshadow decrease in total primary production;[16,19] increased substrate for heterotrophic activity might balance UV inhibition of bacterial growth;[105] changes in iron availability[125] could counteract photosynthetic photoinhibition. The consequences are far reaching in that the overall carbon balance might change due to different proportions of carbon burial related to potential changes in cell size, grazing and subsequent sedimentation altering the $CO_2$ interaction between atmosphere and oceans.

## REFERENCES

1. Atkinson RJ, Matthews WA, Newman PA et al. Evidence of the mid-latitude impact of Antarctic ozone depletion. Nature 1989; 340:290-294.
2. Díaz SB, Frederick JE, Lucas T et al. Solar ultraviolet irradiance at Tierra del Fuego: comparisons of measurements and calculations over a full annual cycle. J Geophys Lett 1996; 23:355-358.
3. Frederick JE, Lubin D. Solar ultraviolet irradiance at Palmer Station, Antarctica. In: Weiler CS, Penhale PA, eds. Ultraviolet Radiation in Antarctica: Measurements and Biological Effects. Vol 62. Washington, D.C.: American Geophysical Union: Antarctic Research Series, 1994:43-52.
4. Frederick JE, Soulen PF, Diaz SB et al. Solar ultraviolet irradiance observed from Southern Argentina: September 1990 to March 1991. J Geophys Res 1993; 98:8891-8897.
5. Kerr JB, McElroy CT. Evidence for large upward trends of ultraviolet-B radiation linked to ozone depletion. Science 1993; 262:1032-1034.

6. Kerr JB, McElroy CT. Analyzing ultraviolet-B radiation: Is there a trend? Science 1994; 264:1341-1343.

7. Michaels PJ, Singer SF, Knappenberger PC. Analyzing ultraviolet-B radiation: Is there a trend? Science 1994; 264:1341-1343.

8. Berger WH, Smetacek VS, Wefer G, eds. Productivity of the Ocean: Present and Past. John Wiley and Sons, 1989:85-97.

9. Falkowski PG, Woodhead AD, eds. Primary Productivity and Biogeochemical Cycles in the Sea. New York: Plenum Press, 1992: 213-237.

10. Antoine D, Andre J, Morel A. Oceanic primary production: II. Estimation at global scale from satellite (Coastal Zone Color Scanner) chlorophyll. Glob Biogeochem Cyc 1996; 10:57-69.

11. Smith RC, Baker KS, Byers ML et al. Primary productivity of the Palmer Long-Term Ecological Research Area and the Southern Ocean. J Mar Syst 1996b; (in press).

12. Rivkin RB, Putt M, Alexander SP et al. Biomass and production in polar planktonic and sea ice microbial communities: a comparative study. Mar Biol 1989; 101: 273-283.

13. Häder D-P, Worrest RC, Kumar HD et al. Effects of increased solar ultraviolet radiation on aquatic ecosystems. Ambio 1995; 24:174-180.

14. Lorenzen CJ. Ultraviolet radiation and phytoplankton photosynthesis. Limnol Oceanogr 1979; 24:1117-1120.

15. Worrest RC, Van Dyke H, Thomson BE. Impact of enhanced simulated solar ultraviolet radiation upon a marine community. Photochem Photobiol 1978; 27:471-478.

16. Worrest RC, Wolniakowski KU, Scott JD et al. Sensitivity of marine phytoplankton to UV-B radiation: impact upon a model ecosystem. Photochem Photobiol 1981; 33:223-227.

17. Calkins J, Thordardøttir T. The ecological significance of solar UV radiation on aquatic organisms. Nature 1980; 283:563-566.

18. Bothwell ML, Karentz D, Carpenter EJ. No UV-B effect? Nature 1995; 374:601.

19. Bothwell ML, Sherbot D, Roberge AC et al. Influence of natural ultraviolet radiation on lotic periphytic diatom community

growth, biomass accrual, and species composition: short-term versus long-term effects. J Phycol 1993; 29:24-35.

20. Azam F, Fenchel T, Field JG et al. The ecological role of water-column microbes in the sea. Mar Ecol Prog Ser 1983; 10: 257-263.

21. Voytek MA. Addressing the biological effects of decreased ozone on the Antarctic environment. Ambio 1990; 19:52-61.

22. Weiler CS, Penhale PA, eds. Ultraviolet Radiation in Antarctica: Measurements and Biological Effects. Vol 62. Washington, D.C.: American Geophysical Union, 1994.

23. Smith RC, Prézelin BB, Baker KS et al. Ozone depletion: ultraviolet radiation and phytoplankton biology in Antarctic waters. Science 1992; 255:952-959.

24. Booth CR, Lucas TB, Morrow JH. Ultraviolet radiation in Antarctica: Measurements and biological effects. In: Weiler CS, Penhale PA, eds.Ultraviolet Radiation in Antarctica: Measurements and Biological Effects. Vol 62. Washington, D.C.: American Geophysical Union, Antarctic Research Series 1994:17-37.

25. Helbling EW, Villafañe V, Ferrario M et al. Impact of natural ultraviolet radiation on rates of photosynthesis and on specific marine phytoplankton species. Mar Ecol Prog Ser 1992; 80:89-100.

26. Lubin D, Mitchell BG, Frederick JE et al. contribution toward understanding the biospherical significance of Antarctic ozone depletion. J Geophys Res 1992; 97(D8): 7817-7828.

27. Boucher N, Prézelin BB, Evens T et al. Icecolors '93: biological weighting function for the ultraviolet inhibition of carbon fixation in a natural Antarctic phytoplankton community. Ant J US 1994; XXIX: 272-275.

28. Cullen JJ, Neale PJ, Lesser MP. Biological weighting function for the inhibition of phytoplankton photosynthesis by ultraviolet radiation. Science 1992; 258:646-650.

29. Jones LW, Kok B. Photoinhibition of chloroplast reactions. II. Multiple effects. Plant Physiol 1966; 41:1044-1049.

30. Setlow RB. The wavelengths in sunlight effective in producing skin cancer: A theo-

retical analysis. Proc Nat Acad Sci USA 1974; 71:3363-3366.

31. Wood WF. Photoadaptive responses of the tropical red alga *Eucheuma striatum* Schmitz (Gigartinales) to ultra-violet radiation. Aquatic Botany 1989; 33:41-51.

32. Morales RGE, Jara GP, Cabrera S. Solar ultraviolet radiation measurements by *o*-nitrobenzaldehyde actinometry. Limnol Oceanogr 1993; 38:703-705.

33. Karentz D, Lutze LH. Evaluation of biologically harmful ultraviolet radiation in Antarctica with a biological dosimeter designed for aquatic environments. Limnol Oceanogr 1990; 35:549-561.

34. Roy CR, Gies HP, Tomlinson DW et al. Effects of Ozone Depletion on the Ultraviolet Radiation Environment at the Australian Stations in Antarctica. In: Weiler CS, Penhale PA, eds. Ultraviolet Radiation in Antarctica: Measurements and Biological Effects. Vol 62. Washington, D.C.: American Geophysical Union, 1994:1-15.

35. Smith RC. Ozone, middle ultraviolet radiation and the aquatic environment. Photochem Photobiol 1989; 50:459-468.

36. Smith RC, Cullen JJ. Effects of UV radiation on phytoplankton. Rev Geophys 1995; Supplement:1211-1223.

37. Smith RC, Baker KS. Remote sensing of chlorophyll. In: Godby EA, Otterman J, eds. COSPAR, The Contribution of Space Observations to Global Food Information Systems. Oxford, New York: Pergamon Press, 1978:161-172.

38. Smith RC, Baker KS. Optical properties of the clearest natural waters (200-800 nm). Appl Optics 1981; 20:177-184.

39. Kirk JTO, Hargreaves BR, Morris DP et al. Measurements of UV-B radiation in two freshwater lakes: an instrument intercomparison. Arch Hydrobiol Beih Ergebn Limnol 1994; 43:71-99.

40. Baker KS, Smith RC. Middle ultaviolet irradiance at the ocean surface: measurements and models. In: Calkins J, ed. The Role of Ultraviolet Radiation in Marine Ecosystems. New York: Plenum Publishing Co., 1982:79-91.

41. Bricaud A, Morel A, Prieur L. Absorption by dissolved organic matter of the sea (yellow substance) in the UV and visible domains. Limnol Oceanogr 1981; 26:43-53.

42. Kramer K. Effects of increased solar uv-b radiation on coastal marine ecosystems: An overview. In: Beukema JJ, Wolf WJ, Brouns J, eds. Expected Effects of Climatic Change on Marine Coastal Ecosystems. Boston: Kluwer Academice, 1990:195-210.

43. Mitchell BG, Holm-Hansen O. Observations and modeling of the Antarctic phytoplankton crop in relation to mixing depth. Deep-Sea Res 1991; 38:981-1007.

44. Smith WO, Nelson DM. Phytoplankton bloom produced by a receding ice edge in the Ross Sea: spatial coherence with the density field. Science 1985; 210:163-166.

45. Smith RC, Baker KS, Vernet M. Seasonal and interannual variability of phytplankton biomass west of the Antarctic Peninsula. J Mar Syst 1996a; (in press).

46. Mullins BW, Priddle J. Relationships between bacteria and phytoplankton in the Bransfield Strait and Southern Drake Passage. British Antarctic Survey 1987; 76:51-64.

47. Holm-Hansen O, Mitchell BG, Vernet M. Ultraviolet radiation in Antarctic waters: effect on rates of primary production. Ant J US 1989; 24:177-178.

48. Schindler DW, Curtis PJ, Parker BR et al. Consequences of climate warming and lake acidification for UV-B penetration in North American boreal lakes. Nature 1996; 379:705-708.

49. Trodahl HJ, Buckley RG. Enhanced ultraviolet transmission of Antarctic sea ice during the austral spring. Geophys Res Lett 1990; 17:2177-2179.

50. Holm-Hansen O, Lubin D, Helbling EW. Ultraviolet radiationn and its effects on organisms in aquatic environments. In: Young AR, Björn LO, Moan J, Nultsch W, eds. Environmental UV Photobiology. New York: Plenum Press, 1993:379-425.

51. Cullen JJ, Neale PJ. Ultraviolet radiation, ozone depletion, and marine photosynthesis. Photos Res 1994; 39:303-320.

52. Villafañe VE, Helbling EW, Holm-Hansen O et al. Acclimatization of Antarctic natural phytoplankton assemblages when exposed to solar ultraviolet radiation. J Pl Res 1995; 17:2295-2306.

53. Gieskes W, Kraay GW. Transmission of ultraviolet light in the Weddell Sea: report of the first measurements made in the Antarctic. BIOMASS Newsletter 1990; 12:12-14.

54. Helbling EW, Villafañe V, Holm-Hansen O. Effects of ultraviolet radiation on Antarctic marine phytoplankton photosynthesis with particular attention to the influence of mixing. In: Weiler CS, Penhale P, eds. Ultraviolet Radiation in Antarctica: Measurements and Biological Effects. Vol 62. Washington, D.C.: Antarctic Research Series, 1994:207-227.

55. Cullen JJ, Neale PJ. Quantifying the effects of ultraviolet radiation on aquatic photosynthesis. In: Yamamoto H, Smith CM, eds. Photosynthetic Responses to the Environment. Washington, D.C.: American Society of Plant Physiologists, 1993:45-60.

56. Neale PJ, Lesser MP, Cullen JJ. Effects of ultraviolet radiation on the photosynthesis of phytoplankton in the vicinity of McMurdo Station, Antarctica. In: Weiler CS, Penhale PA, eds. Ultraviolet Radiation in Antarctica: Measurements and Biological Effects. Vol 62. Washington, D.C.: Antarctic Research Series, 1994:125-142.

57. Collos Y, Slawyk G. Nitrogen uptake and assimilation by marine phytoplankton. In: Falkowski PG, ed. Primary Productivity in the Sea. 31. New York: Plenum Press, 1980:195-211.

58. Döhler G. Impact of UV-B radiation on [$^{15}$N]ammonia and [$^{15}$N]nitrate uptake of *Ditylum brightwellii*. Photobiochem Photobiophys 1986; 11:115-121.

59. Döhler G. Impact of UV-B radiation on uptake of $^{15}$N-ammonia and $^{15}$N-nitrate by phytoplankton of the Wadden Sea. Mar Biol 1992; 112:485-489.

60. Döhler G, Buchmann T. Effects of UV-A and UV-B irradiance on pigments and $^{15}$N-ammonium assimilation of the Haptophycean *Pavlova*. J Pl Phys 1995; 146:29-34.

61. Goes JI, Handa N, Taguchi S et al. Impact of UV radiation on the production patterns and composition of dissolved free and combined amino acids in marine phytoplankton. J Plankton Res 1995; 17:1337-1362.

62. Sinha RP, Kumar HD, Kumar A et al. Effects of UV-B irradiation on growth, survival, pigmentation and nitrogen metabolism enzymes in cyanobacteria. Acta Protozool 1995; 34:187-192.

63. Goes JI, Handa N, Taguchi S et al. Changes in the patterns of biosynthesis and composition of amino acids in a marine phytoplankter exposed to ultraviolet-B radiation: nitrogen limitation implicated. Photochem Photobiol 1995; 62:703-710.

64. Sharp JH. Excretion of organic matter by marine phytoplankton: do healthy cells do it? Limnol Oceanogr 1977; 22:381-389.

65. Fogg GE, Nalewajko C, Watt WD. Extracellular products of phytoplankton photosynthesis. Proc R Soc Lond Ser B 1965; 162:517-534.

66. Bjørnsen PK. Phytoplankton exudation of organic matter: why do healthy cells do it? Limnol Oceanogr 1988; 33:151-154.

67. Wood AM, Rai H, Garnier J et al. Practical approaches to algal excretion. Mar Microb Food Webs 1992; 6:21-38.

68. Mague TH, Friberg E, Hughes DJ et al. Extracellular release of carbon by marine phytoplankton; a physiological approach. Limnol Oceanogr 1980; 25:262-279.

69. Williams PJ le B. The importance of losses during microbial growth: commentary on the physiology, measurement and ecology of the release of dissolved organic material. Mar Microb Food Webs 1990; 4:175-206.

70. Kirchman DL, Suzuki Y, Garside C et al. High turnover rates of dissolved organic carbon during a spring phytoplankton bloom. Nature 1991; 352:612-614.

71. Myklestad S, Haug A. Production of carbohydrates by the marine diatom *Chaetoceros affinis* var *willei* (Gran) Husted. I. Effect of the concentration of nutrients in the culture medium. J Exp Mar Biol Ecol 1972; 9:125-136.

72. Sakshaug E, Myklestad S, Krogh T et al. Production of protein and carbohydrate in the Dinoflagellate *Amphidinium carteri*. Some preliminary results. Norw J Bot 1973; 20:211-218.

73. Hellebust JA. Excretion of some organic compounds by marine phytoplankton. Limnol Oceanogr 1965; 10:192-206.

74. Myklestad S. Production of carbohydrates by marine planktonic diatoms. I. Comparison of nine different species in culture. J Exp Mar Biol Ecol 1974; 15:261-274.

75. Veldhuis MJW, Admiraal W. Transfer of photosynthetic products in gelatinous colonies of *Phaeocystis pouchetii* (Haptophyceae) and its effect on the measure. Mar Ecol Prog Ser 1985; 26:301-304.

76. Davidson AT, Marchant HJ. Protist abundance and carbon concentration during a *Phaeocystis*-dominated bloom at an Antarctic site. Polar Biol 1992; 12:387-395.

77. Richardson TL, Cullen JJ. Changes in bouyancy and chemical composition during growth of a coastal marine diatom: ecological and biogeological consequences. Mar Ecol Prog Ser 1995; pp 77-90 and V128, N1-3

78. Vernet M, Matrai PA. Synthesis of particulate and extracellular carbon by phytoplankton in the Barents Sea. J Geophys Res—Oceans 1996; in press.

79. Smith DC, Steward GF, Long RA et al. Bacterial mediation of carbon fluxes during a diatom bloom in a mesocosm. Deep-Sea Res II 1995; 42:75-97.

80. Holm-Hansen O, Mitchell BG. Spatial and temporal distribution of phytoplankton and primary production in the western Bransfield Strait region. Deep-Sea Res 1991; 38,:961-980.

81. Karentz D, Spero HJ. Response of a natural *Phaeocystis* population to ambient fluctuations of UVB radiation caused by Antarctic ozone depletion. J Plankton Res 1995; 17:1771-1789.

82. Behrenfeld MJ, Hardy JT, Lee HI. Chronic effects of ultraviolet-B radiation on growth and cell volume of *Phaeodactylum* (Bacillariophyceae). J Phycol 1992; 28:757-760.

83. Hargraves PE, Zhang J, Wang R et al. Growth characteristics of the diatom *Pseudonitzschia pungens* and *P. fraudulenta* exposed to ultraviolet radiation. Hydrobiologia 1993; 269/270:207-212.

84. Lesser MP. Acclimation of phytoplankton to UV-B radiation: oxidative stress and photoinhibition of photosynthesis are not prevented by UV-absorbing compounds in the dinoflagellate *Prorocentrum micans*. Mar Ecol Prog Ser 1996; 132:287-297.

85. Vernet M. UV radiation in Antarctic waters: response of phytoplankton pigments. In: Mitchell BG, Holm-Hansen O, Sobolev I, eds. Response of marine phytoplankton to natural variations in UV-B flux. Washington, D.C.: Chemical Manufacturers Association, Proceedings of a Workshop, Scripps Institution of Oceanography, La Jolla, CA, April 5., 1990.

86. Karentz D. Ultraviolet tolerance mechanisms in Antarctic marine organisms. In: Weiler CS, Penhale PA, eds. Ultraviolet radiation in Antarctica: Measurements and Biological Effects. Vol 62. Washington, D.C.: American Geophysical Union: Antarctic Research Series, 1994:93-110.

87. Smith RC, Baker KS. Stratospheric ozone, middle ultraviolet radiation and carbon-14 measurements of marine productivity. Science 1980; 208:592-593.

88. Smith RC Baker KS. Assessment of the influence of enhanced UV-B on marine primary productivity. In: Calkins J, ed. The Role of Solar Ultraviolet Radiation in Marine Ecosystems. New York: Plenum Publishing Co., 1982:509-537.

89. Kullenberg G. Note on the role of vertical mixing in relation to effects of UV radiation on the marine environment. In: Calkins J, ed. The Role of Solar Ultraviolet Radiation in Marine Ecosystems. New York: Plenum Press, 1982:283-292.

90. Bidigare RR. Potential effects of UV-B radiation on marine organisms of the southern ocean: distributions of phytoplankton and krill during austral spring. Photochem Photobiol 1989; 50:469-477.

91. Karentz D. Ecological considerations of Antarctic ozone depletion. Antarctic Science 1991; 3:3-11.

92. Ferreyra GA, Schloss IR, Demers S et al. Phytoplankton responses to natural ultraviolet irradiance during early spring in the Weddell-Scotia confluence: an experimental approach. Ant J US 1994; XXIX: 268-270.

93. Malone TC. Size-fractionated primary productivity of marine phytoplankton. In: Falkowski PG, ed. Primary Productivity in the Sea. New York, London: Plenum Press, 1980:301-319.

94. Fiala M, Delille D. Variability and interactions of phytoplankton and bacterioplankton in the Antarctic neritic area. Mar Ecol Prog Ser 1992; 89:135-146.

95. Bidigare RR, Iriarte JL, Kang S-H et al. Phytoplankton: Quantitative and Qualitative Assessments. In: Ross RM, Hofmann EE, Quetin LB, eds. Foundations for Ecological Research West of the Antarctic Peninsula. Washington, D.C.: American Geophysical Union, 1991; 70:173-198.

96. Karentz D, Cleaver JE, Mitchell DL. Cell survival characteristics and molecular responses of Antarctic phytoplankton to ultraviolet-b radiation. J Phycol 1991; 27:326-341.

97. Vernet M, Brody EA, Holm-Hansen O et al. The response of Antarctic phytoplankton to ultraviolet light: absorption, photosynthesis, and taxonomic composition. In: Weiler CS, Penhale PA, eds. Ultra Violet Radiation in Antarctica: Measurements and Biological Effects. Vol 62. Washington, D.C.: American Geophysical Union, 1994: 143-158.

98. McMinn AD, Heijnis H, Hodgson D. Minimal effects of UVB radiation Antarctic diatoms over the past 20 years. Nature 1994; 370:547-549.

99. Hunter JR, Kaupp SE, Taylor JH. Assessment of effects of UV radiation on marine fish larvae. In: Calkins J, ed. The Role of Solar Ultraviolet Radiation in Marine Ecosystems. New York: Plenum Publishing Co, 1982:459-497.

100. von Bodungen B, Smetacek V, Tilzer B et al. Primary production and sedimentation during spring in the Antarctic Peninsula region. Deep-Sea Res 1986; 33:177-194.

101. Wassmann P, Vernet M, Mitchell BG et al. Mass sedimentation of *Phaeocystis pouchetii* in the Barents Sea. Mar Ecol Prog Ser 1990; 66:183-195.

102. Bird DF, Karl DM. Bacterial growth, abundance and loss due to protozoan grazing during the 1989 spring bloom. Ant J US 1990; 25:156-157.

103. Karl DM. Microbial processes in the Southern Ocean. In: Friedmann EI, ed. Antarctic Microbiology. New York: John Wiley and Sons, Inc., 1992.

104. Herndl GJ, Muller-Niklas G, Frick J. Major role of ultraviolet-B in controlling bacterioplankton growth in the surface layer of the ocean. Nature 1993; 361:717-719.

105. Mopper K, Zhou X, Kieber R J et al. Photochemical degradation of dissolved organic carbon and its impact on the oceanic carbon cycle. Nature 1991; 353:60-62.

106. Kieber DJ, McDaniel J, Mopper K. Photochemical source of biological substrates in sea water: implications for carbon cycling. Nature 1989; 341:637-639.

107. Moran MA, Hodson RE. Support of bacterioplankton production by dissolved humic substances from three marine environments. Mar Ecol Prog Ser 1994; 110:241-247.

108. Morris DP, Zagarese H, Williams CE et al. The attenuation of solar UV radiation in lakes and the role of dissolved organic carbon. Limnol Oceanogr 1995; 40:1381-1391.

109. Amon RMW, Benner R. Bacterial utilization of different size classes of dissolved organic matter. Limnol Oceanogr 1996; 41:41-51.

110. Pomeroy LR, Diebel D. Temperature regulation of bacterial activity during the spring bloom in Newfoundland coastal waters. Science 1986; 233:359-361.

111. Thingstad F, Billen G. Microbial degradation of *Phaeocystis* material in the water column. J Mar Syst 1994; 5:55-65.

112. Pomeroy LR, Weibe WG. Energy sources for microbial food web. Mar Microb Food Webs 1993; 7:101-118.

113. Bushaw KL, Zepp RG, Tarr MA et al. Photochemical release of biologically available nitrogen from aquatic dissolved organic matter. Nature 1996; 381:404-407.

114. Lara RJ, Thomas DN. Formation of recalcitrant organic matter: humification dynamics of algal derived dissolved organic carbon and its hydrophobic fractions. Mar Chem 1995; 51:193-199.

115. Dugdale RC, Wilkerson FP. Low specific nitrate uptake rate: a common feature of high-nutrient, low-chlorophyll marine ecosystems. Limnol Oceanogr 1991; 36: 1678-1688.

116. Tupas LM, Koike I, Karl DM et al. Nitrogen metabolism by heterotrophic bacterial

assemblages in Antarctic coastal waters. Polar Biol 1994; 14:195-204.

117. Martin JH, Gordon RM, Fitzwater SE. The case for iron. Limnol Oceanogr 1991; 36:1793-1802.

118. Martin JH, Gordon RM, Fitzwater SE. Iron in Antarctic waters. Nature 1990; 345: 156-158.

119. De Baar HJW, Buma AGJ, Nölting RF et al. On iron limitation in the Southern Ocean: experimental observations in the Weddell and Scotia Seas. Mar Ecol Prog Ser 1990; 65:105-122.

120. De Baar HJW, de Jong JTM, Baaker DCE et al. Importance of iron for plankton blooms and carbon dioxide drawdown in the Southern Ocean. Nature 1995; 373:412-415.

121. Buma AGJ, De Baar HJW, Nöltig RF et al. Metal enrichment experiments in the Weddell-Scotia Seas: Effects of iron and manganese on various plankton communities. Limnol Oceanogr 1991; 36:1865-1878.

122. Chavez FP, Buck KR, Coale KH et al. Growth rates, grazing, sinking, and iron limitation of equatorial Pacific phytoplankton. Limnol Oceanogr 1991; 36:1816-1833.

123. Helbling EW, Villafañe V, Holm-Hansen O Effect of iron on productivity and size distribution of Antarctic phytoplankton. Limnol Oceanogr 1991; 36:1879-1885.

124. Kuma K, Nakabayashi S, Suzuki Y et al. Photo-reduction of Fe(III) by dissolved organic substances and existence of Fe(II) in seawater during spring blooms. Mar Chem 1992; 37:15-27.

125. Wells ML, Price NM, Bruland KW. Iron chemistry in seawater and its relationship to phytoplankton: a workshop report. Mar Chem 1995; 48:157-182.

126. Nakabayashi S, Kudo I, Kuma K et al. Existence of dissolved $Fe^{2+}$ in a spring bloom at Funka Bay. Jpn Soc Fish Oceanogr 1989; 53:649-680.

# INDEX

Printed and bound by CPI Group (UK) Ltd, Croydon, CR0 4YY

03/10/2024

01040317-0014